PRACTICAL
THERMOFORMING

PLASTICS ENGINEERING

Founding Editor

Donald E. Hudgin

Professor
Clemson University
Clemson, South Carolina

PRACTICAL THERMOFORMING
PRINCIPLES AND APPLICATIONS
SECOND EDITION, REVISED AND EXPANDED

JOHN FLORIAN
Consulting Design Engineer
Bakersfield, California

CRC Press
Taylor & Francis Group
Boca Raton London New York

CRC Press is an imprint of the
Taylor & Francis Group, an **informa** business

Preface to the Second Edition

The rapid and accelerated growth of the thermoforming industry has not slowed despite the economic recession of the late 1980s and early 1990s. The economic downturn and environmental issues caused havoc, and unfortunately both entered into the business picture at the same time, which had a devastating effect on some packaging manufacturers and suppliers. Among the hardest hit were thermoform product suppliers catering to high-volume product lines for fast-food packaging. Investment for new equipment for the first time in many years experienced some weakness as well. Despite the setbacks, the thermoforming industry has remained one of the fastest growing due to its low initial investment requirements, superior cycle repeatability, and adaptability to various product configurations.

The temporary setback in product losses in the fast-food areas were quickly replaced with other equally high-volume products. Such abrupt changes in the packaging field may have caused some temporary oversupply and marketing territory takeovers; however, soon competitive forces leveled off. The heavy-gauge thermoforming processors were not affected equally by the environmental concerns or the general recession, which were regional, touching some areas of the country more than others.

The current outlook is bright; there is plenty of opportunity for growth for all thermoformers. In my opinion, thermoforming manufacturing methods have not yet reached 50% of their growth potential. What was thought impossible to make in the past can easily be made today. We can only wonder what we will be able to produce in the future.

The first edition of this book has been most successful. It is found on many thermoformers' desks and is the foundation of my seminar programs. The entire field of thermoforming, including supportive and peripheral areas, is laid out for the reader. To date, there is no book on this subject comparable in quality or detail. The few books that have been written on this subject either concentrate on a specific topic or give overwhelmingly complicated explanations of the technical concepts and include theoretical formulas that often do not produce real-life results.

This second edition has been revised to update information that has changed since the first edition was published and expanded to incorporate subject areas that have come about since and those that will be of greatest interest in the future. For example, I have covered the issue of plastics recycling, which in recent years has come into the spotlight. Recycling concerns must be addressed as they will remain with us due to environmental and legislative influences. They will affect all thermoformers as well as the end users of their products.

The goal of the second edition remains the same: to provide the best and most detailed information, references, and guidance for the thermoformer. The uniqueness of my book is that it is written in a simple, easily comprehensible style and meets the needs of the widest range of people involved in the thermoforming industry. It is especially recommended for owners, managers, and design engineers, as they are the decision makers, as well as for newcomers to the industry. It could be used as a reference guide by practicing engineers, technicians, quality control and customer service personnel, equipment designers and builders, thermoplastic suppliers, sales and purchasing agents, and even thermoforming machine operating personnel.

The illustrations and tables in this book are prepared to emphasize only the concepts discussed. To achieve that, the figures of equipment and molds exclude all nonessential components not directly involved with the specific explanation. Including all the necessary supporting, safety, and actuating mechanisms actually present would have made the illustrations difficult to understand. The tables are also prepared with easy reading and referencing in mind.

Over thirty-five years of involvement with the thermoforming process and its products has provided me with valuable knowledge and understanding of this manufacturing business. Working with many large and small firms throughout North America, from Puerto Rico to Hawaii and Canada to Mexico, I have encountered many different manufacturing conditions, equipment variations, and technical challenges. In the last sixteen years I have been involved with education—teaching the thermoforming process in various private classes and seminars and at the University of Wisconsin—Madison. This wide range of experience has given me the opportunity to gain insight into the most pertinent and practical information on the thermoforming molding process. This book covers it all, from the basic concepts to the latest innovations. The process methods described constitute fundamental techniques that suggest numerous variations and combinations for the practical application of the thermoformer. The book also contains comprehensive explanations and possible causes of problems occurring during thermoforming.

In this book, I left untouched the information that has not changed as well as the basic rules of the thermoforming process. I chose not to include brand and trade names to avoid any commercialization. I mentioned a trade name only when a specific item is known solely by that name. I am confident that most makers and users of the materials and equipment will recognize their specific products, materials, equipment, and machinery, by either concept or function.

One more time, I would like to take this opportunity to thank my wife, Judy, for her many years of understanding and support throughout my career. Patiently giving up her personal time, she not only typed the manuscript but also helped shape the text to provide the clearest possible explanations that would be understandable to everyone, even those unfamiliar with the subject area. With God's gracious help, we have accomplished this once again for the second edition.

John Florian

Preface to the First Edition

The thermoforming process has become one of the fastest growing plastics manufacturing methods since the commercialization of thermoplastic materials due to its superior adaptability, low initial investment, and excellent repeating qualities. In just four decades, thermoformed goods have become part of many existing product lines. Thermoforming has completely replaced other processing techniques for some product lines and has even created new plastic product lines. Every year rapid developments and clever implementations of this manufacturing method allow it to compete with more established methods and to penetrate many new product categories. The possibilities for this industry are extensive and the business opportunities within it are enormous.

The rapid growth of the thermoforming industry and the keen competitiveness among thermoforming processors has kept many of the technical details and developments and the practical know-how in secrecy. In recent years, with the crossover of personnel between companies, the flow of information has increased. The literature on thermoforming, however, still lags behind the industry's projected boom, which has created the need for more education, training, and documentation.

The thermoforming process can be used for material from the thinnest (thin-gauge) to the thickest (heavy-gauge) material to manufacture products from the smallest to the largest possible size and to produce anywhere from just a few pieces to an enormous quantity. Covering all the possible process principles within a single work is therefore not an easy task. But, of course, the similarities and overlapping process behaviors within these variations form an essential body of knowledge that will be of interest to all working in this field.

The goal of this book is to provide information, references, and guidance on the thermoforming of thermoplastics in a simple and easily understood format. The book is intended to meet the needs of a wide range of people involved in thermoforming: business owners and managers, new and experienced engineers, product designers, quality control engineers and customer service personnel, equipment designers and builders, material and machinery suppliers, sales and purchasing agents, and the actual operating technicians.

The illustrations and tables for this book have been prepared to emphasize the basics. The illustrations generally exclude all components that are not directly involved with the specific explanation in order to emphasize the particular thermoforming or equipment functions. Under actual working conditions, with all supportive, safety and actuating mechanisms, these principles would not be so clearly illustrated. The tables are also prepared with easy reading and referencing in mind.

Through my many years of involvement with the thermoforming process, working closely with many highly knowledgeable and skilled colleagues, I have gathered the knowledge and understanding of this process that have enabled me to complete this book. I would like to thank the many equipment and material suppliers who contributed information on their own products, machinery, and instrumentation. Although specific brand and trade names have been avoided throughout this book in order to minimize commercialization, their product concepts and functions are thoroughly represented and explained here. I would like to thank Gerald Sitser for his help in editing and for filtering out my native language influences from this writing.

Last but certainly not least, I thank my devoted wife, Judy, for her long, full understanding and total support throughout my career and for her tireless hours typing, proofreading, and reviewing this entire project, which took up all of our extra time but, with God's help, we were finally able to complete. I also would like to thank the editors of Marcel Dekker, Inc., for their interest, encouragement, and patience in working with me.

John Florian

Contents

1
Introduction

I. THE BASIC CONCEPT OF THERMOFORMING

The thermoforming process is only one of many manufacturing methods that converts plastic resin material into numerous products. Yet in our modern life-style, we are coming to rely more and more on the benefits of thermoforming and make extensive use of plastic products produced by this process.

Thermoforming is the amalgamated description of the various thermoplastic sheet-forming techniques, such as vacuum forming, pressure forming, matched mold forming, and their combinations. All of these forming techniques require a premanufactured thermoplastic sheet, which is clamped, heated, and shaped into or over a mold. Products made by this process are generally finished after the trimming operation and are ready to be used. However, some thermoformed products are designed as components of larger items, such as boats, aircraft, automobiles, or even smaller items such as product display trays. These parts may require additional work after trimming, such as painting, printing, heat sealing, and gluing.

The thermoforming process offers fast and uniform forming and therefore lends itself to automation and long-term production runs. With its relatively fast molding cycles and comparatively inexpensive mold costs, the thermoforming process is often chosen as the most cost-effective manufacturing method over all the other processes. The scrap created by the normal edge trim and spacing allowances can be reprocessed and recycled along with the rejected articles.

The basic principle of the thermoforming process bears some similarity to metal stamping techniques, but thermoforming has far fewer limitations in stretching, depth-of-draw ratio, and material distributions. As the process becomes more and more sophisticated and the equipment more controllable, the limitations of thermoforming will diminish.

These are the basic components of the thermoforming process:

1. Thermoplastic sheet
2. Clamping mechanism
3. Heating systems
4. Molds
5. Forming force
6. Trim apparatus

These components are equally important, yet all can affect one another directly or indirectly. Their individual aspects and interrelating ties and dependencies hold the key to the success of thermoforming. The first three components on this list are absolutely mandatory to the process. However, rarely can any of the last three be left out of the process without disqualifying it from being a genuine thermoforming method. These thermoforming process components are discussed in greater detail in Chapter 2.

II. HISTORY OF THE THERMOFORMING INDUSTRY

Some sheet-forming techniques have been known since the turn of the century. However, those techniques involved metal, glass, and natural fiber materials. The real roots of the thermoforming process began with the development of thermoplastics, which came about during World War II, although formulations of some synthetic material can be credited to earlier years. The development of the synthetic material process grew out of the initial development of synthetic rubber. The postwar years offered volume commercialization of these newly developed plastic materials and witnessed the rapid development of equipment capable of adapting to modern manufacturing methods to produce useful and salable products. By the 1950s, the volume production of thermoplastics and products made from them reached impressively high levels. Among the various industries at the time, sheet manufacturing was the most productive. Larger plants for casting, calendaring, and extrusion were organized to handle the greater demand for converting thermoplastic resin into film and sheet forms. In a similar growth pattern, the conversion of the sheets into finished products became its own highly specialized industry responsible for such varied commodities as tarps, raincoats, tents, and upholstery materials. Simultaneously, the packaging industry gobbled up large quantities of film and sheet stock as wrapping and protective materials. Heat-sealing and heat-shaping techniques quickly replaced the older methods of gluing and sewing. Using the forces of vacuum and heat to shape plastic film or sheet into needed forms was the natural progression of these developments. Numerous manufacturers rapidly learned and adopted these forming techniques. The cycle repeatability and reasonable product uniformity of the thermoforming process rapidly evolved into improved production speeds and larger product output volumes, resulting in reduced product pricing. By the 1960s, both blister packaging and individual-portion food packaging were developed into high-volume markets. At the same time, sign manufacturers and the appliance industry adopted vacuum forming into their manufacturing processes.

The 1960s was an era that set the thermoforming industry into an ambitious pattern and defined its future trends. Roughly at this time, the specialization and separation of "thin"- and "thick"-gauge industries was established. Any newly organized thermoforming firm received full technical support and advice from large resin manufacturing companies.

By the 1970s, both volume usage and competition had increased to the point that extremely high volume output and high-speed machines were in need. The equipment manufacturers met such demands with thermoforming equipment capable of producing nearly 100,000 individual thermoformed containers per hour. Such equipment was placed into multiline operations and a plant so equipped became capable of supplying large portions of the country. The increased production rates allowed the thermoforming industry to go head to head with and even take over markets from established industries such as paper and pulp. At the same time, such high-volume production output rates demanded much tightening of production controls and strong adherence to quality control procedures. Most of these controls and procedures were brought over from other industries and custom-fitted into thermoforming. Since the thermoforming process is an extreme scrap producer, scrap handling and some type of reprocessing technique had to be developed and implemented, as plant size and location permitted. Thermoformers quickly learned not only how to handle their scrap but also developed ways of reducing it. Some went as far as complete recycling or simply selling the scrap back to the sheet producer. In some cases, new products were introduced just to use up the scrap materials.

By the late 1970s and early 1980s, thermoformers had gained such confidence in their process that they freely went after new product ideas and territories. Many customary and established product lines have since been converted to thermoforming. Bold and nontraditional equipment lineups began to be offered from a single source. Such equipment is capable of producing finished products starting with a supply of plastic pellets, and it would recycle its own scrap, all with little required attendance. The machinery, by its electronics and computer controls, performs self-monitoring and diagnostic functions. Usually, only one person need be in attendance to oversee the manufacturing process. Large-volume users are the best candidates for this type of machinery purchase. These include large dairy companies, centralized supermarket chains, fruit juice packagers, and so on. If and when the thermoformed containers can be made in proximity to the product filling lines, even the shipping, storage, and inventory costs can be reduced. For the production of different container capacities, the manufacturer has only to reduce the container height by placement of inserts into the deepest mold cavities while retaining the same-size opening. For example, a dairy can use the same containers for several of their products just by changing the labels. For a capacity change, the removal or addition of mold inserts can produce $1/2$-, 1- and $1^1/2$-lb-capacity containers with the same tooling. With this capability the costs of tooling and mold changes are also reduced.

In the early 1980s, when the demand for this type of equipment really started to increase, the equipment manufacturers were ready. However, equipment costs and high interest rates made many prospective buyers shy away from purchasing pellet-to-product equipment. In the meantime, both the equipment makers and the thermoformers were making major advances in equipment and process controls. The progress not only provided improved product quality and reduced cost for thermoforming but allowed thermoformers to enter into new product territories that could not previously be attempted.

As we entered the 1990s persistent sluggish economies worldwide slowed down the chance of rapid advancement for most manufacturing including the thermoforming industry. Firms losing business automatically halted purchasing new equipment. Even businesses not hit by a slowdown have been frightened to expand, not knowing if such a loss of business would hit them as well. On top of this, general recessional effects, environmental issues, and related consumer concerns come into play. No matter which side of the issue was taken up, it rapidly affected everyone. For example, some of the largest fast food companies succumbed to public and environmental demands to discontinue serving take-out food in certain packaging materials. Such a decision has been regional, by locality, or by specific countries, but not implemented worldwide and has been limited to mass produced specific food types of products. When such changes are implemented it unintentionally but directly causes a major product loss for some of the regional packaging product makers. They had to scramble and fill idling production equipment to remain in business, while long range product development was unprepared for the extra work load and in the given short time could not meet the demand. To fill the gap in the supply of new product ideas, many manufacturers started to look at existing product lines of other thermoformers to grab a share of someone else's business. As soon as this happens it usually causes a chain-reaction effecting those product manufacturers whose business has been targeted by the takeover intruders. The duplication of products in the market place not only creates extra competition but an oversupply which usually destroys the weaker business. The aftershocks of the changes are still felt into the mid-1990s and hopefully will dissipate soon. Only people who managed to justify and purchase new equipment can expect positive business and product growth patterns. Equipment and new mold sales were back in track by 1995.

As we approach the end of the 90s and continue to move into the next century there are all indications for more improvements, product demands, and innovations. One area of improvement I see as a coming choice will be more computerization offering more precise control over the entire process, limiting the dependence on human personnel and decision making. Remote control of production lines will be made, perhaps by via telephone, by hook-ups between the equipment, and perhaps by laptop computer. Such control will be managed some distances apart and will allow highly trained personnel to be off site. At the same time fewer trained personnel will physically man the actual machinery. Equipment, diagnostic, and process changes will be accomplished by assigned personnel, who may not be present at the site, but have the capability via the computer to key into the equipment's controls. Of course such futuristic development currently is in the experimental stage and needs to be perfected. In the meantime the most recently developed and available improvements in both equipment automation and marketing territories will affect the traditional plastic business.

As I predicted in the late 1980s, some old and established companies will disappear while new enterprises are organized to take over their business. The takeover and outright purchasing of thermoforming operations will be commonplace. It is the easiest and most cost effective way to eliminate competition and at the same time gain market share or even acquire new business opportunities, or badly needed technical know-how or brand new process techniques. All one has to do is to scan the pages of the trade journals and find acquisitions after acquisition taking place. Even the largest companies can be purchased. This just happened to the number one, top ranked thermoformer, Mobil Chemical Company's Plastic Division, that recently announced its sale.

With this ongoing reformation, the job market will experience a shortage of technically trained and experienced personnel. Increasing job transfers and personnel movement within the industry will end the traditional plastic company's secretive hold on knowledge. Training and education, conferences, and meetings will all serve to broaden the general knowledge of thermoforming and turn it into a more mature and wider-based industry.

III. PRODUCTS MADE BY THERMOFORMING

Many thermoformed products in use today have been made to replace earlier forms of the same products. This substitution is coming about so quickly that many of us have forgotten what these products used to look like. For example, it will take some memory search to recall how carryout foods were packaged before the advent of one-piece hinged foam containers, or what material was used to finish the inside of a refrigerator cabinet.

A list of the various industries that rely on the resulting products of thermoforming can be compiled in a variety of ways using a number of statistical factors. Thermoformed products can be classified by the thickness of their materials. If the list of thermoformed products were to be made on the basis of consumed plastic weight, the highest poundage rating would probably belong to the automotive industry. However, for a more meaningful classification, product volume should be considered. The list should start with the highest-volume producers and work down to the small job-shop type of thermoforming operation. In this way, not only are volume categories established, but machinery types and even marketing areas can be pinpointed.

It is difficult to collect accurate production data for a particular industry because often, conflicting numbers are obtained. Some product lines can easily be miscategorized or even overlooked. The list of products should be followed only as a guideline, one that is constantly changing and requiring revision. The following list has been compiled with this in mind. Readers should continually update the list for their own market research. The list begins with the estimated highest-thermoforming-volume producer and works progressively downward toward the smallest-volume producer: packaging industry, fast-food (take-out-food) industry, retail food industry, transportation industry, sign manufacturers, appliance industry, institutional food service industry, medical industry, nursery (horticulture) industry, recreation industry, housing and construction industry, luggage industry, photographic equipment industry, food processors, assembling production lines, and funeral industry.

A. Packaging Industry

Since the beginning of thermoforming, the packaging industry has remained the highest-volume user of thermoformed items. The production of blister packs alone numbers in the millions of units for the packaging of retail goods. The thermoformed blister provides protection, unitization, and even good merchandising qualities to most package products. It displays the product while reducing the chances for pilferage. Most important, these benefits are realized in the most cost-efficient way.

The thermoformed blister units are usually attached to a cardboard backing by heat sealing, gluing, or stapling. This captures the packaged good in the cavity of the blister bubble. The thermoformed blister can be designed in such a way as to cover the entire packaged product, or it can be made to leave certain portions of the product exposed for customer inspection. Today, most blister packaging is filled using high-speed automatic equipment, and the blisters are therefore also thermoformed on high-speed thermoformers. Equipment is also available that combines all aspects of the blister thermoforming, filling, and sealing process. Such equipment is called "form, fill, and seal." This multiduty equipment has gained interest because of its ability to minimize pre- and post-inventory conditions. Such equipment can rather easily be switched from one type of package production to another. Form, fill and seal lines are usually slower than traditional lines and their production outputs are smaller than those of single-duty high-speed machines. However, their speed and output are often acceptable in many smaller-volume operations. Today, all types of products are handled with simple but effective blister packaging, from common hardware items and cosmetic products to food products such as presliced salami, ham, and other cold cuts.

Besides containing products in thermoformed blister packs, the packaging industry uses thermoformed goods for displaying these products. There are many different types of display stands, store display units, and components of display racks that are produced by thermoforming. Such displays may have tiered steps formed in them to allow stacking and displaying products in optimal fashion. Cosmetic and candy products can be displayed in the same manner. Soft drink and credit card advertisements are among the items thermoformed in this manner.

B. Fast-Food (Take-Out-Food) Industry

In this fast-growing industry enormous quantities of thermoformed products are utilized, from tumblers and their lids to single-item and full-meal serving containers. Soft-drink tumblers alone represent a huge quantity because of the variable sizes and styles in use. There are sandwich and hamburger packages as well as hinged-lid or carry-out food trays, which are produced with stringent specifications. Firms in the fast-food industry purchase their supplies in such high volume that buyers need several manufacturing sources to fill their needs. All suppliers must meet identical specifications in order to produce identical product quality.

The fast-food industry prefers printed thermoformed containers. The printing can be done prior to or after thermoforming. If printing is done prior to thermoforming, it can utilize a random pattern on the web so that some printing will fall on the desirable area. The printing can also be made to fall on the registered area on the formed article. If the registered area is not flat, a preestimated distortion has to be built into the artwork to compensate for the material distortion at forming. Small production runs are generally post-printed for reasons of economy.

C. Retail Food Industry

The supermarket is a large-volume user of thermoformed containers. Presently, fresh meat, eggs, and large quantities of fresh fruit and produce are prepackaged in thermoformed plastic trays and containers (tubs). The trays are usually made of polystyrene foam but are also manufactured of high-impact polystyrenes or clear or colored polystyrene. The trays are usually overwrapped with clear film to capture the product and create the complete package. Egg cartons and tubs with lids are usually printed after thermoforming. For best economy, the mouths and lids of round containers are made to a size uniform to accommodate containers of different depths. To further improve packaging cost, it is best to produce the thermoformed containers close to or perhaps at the same plant facility where the food is prepared and the containers filled. Such a combination of product filling and thermoforming operations can eliminate shipping, temporary packaging, and premanufacturing inventory costs. However, it is feasible only for large-quantity users.

D. Transportation Industry

Both public and private transportation vehicles are equipped with numerous thermoformed plastic components. Most of these components are used for interior finishing. However, more and more exterior parts are finding acceptance but not in structural components. Presently, bus and train seats, aircraft seat backing, armrests, and fold-down tray tables are made by thermoforming. Most of the interior lining of an aircraft is produced this way, as are contoured plastic windshields.

In recent years, more automobile companies have switched to thermoformed articles for the finishing touches to the cab interior. Automobile head liners (roof liners), door panels, dashboards, trunk liners, and even floor coverings are produced and shaped by thermoforming. Heavy-walled single-piece pickup-truck bed liners, instrument cluster protectors, spare tire covers, bumper caps, air dams, splash guards, spoilers, and hatchback window louvers are also made using thermoforming techniques.

E. Sign Manufacturers

Sign makers are also a large-volume producer of thermoformed products. The volume of signs may be as impressive as the poundage of raw material used in the sign making. Today's modern sign construction is capable of producing large one-piece signs formed, with raised lettering, out of a single piece of acrylic sheet stock. The signs are usually made of clear acrylic, then color painted from the inside with acrylic-base paint. Using acrylic on the outside makes the signs weather resistant and virtually maintenance free, capable of withstanding the onslaughts of sunshine and of extremely hot and cold weather.

F. Appliance Industry

Appliance makers manufacture, purchase, and install high-volume thermoformed parts. Today, all refrigerator and freezer door liners and most cabinet liners are thermoformed. When the cabinet liners are made of one-piece construction, the details of the adjustable shelf supports are formed-in simultaneously. Some dishwashers, clothes dryers, window air conditioners, humidifiers, home computers, radio and television cabinets, and their components are also made by thermoforming.

G. Institutional Food Service Industry

The food service industry is not only one of the oldest but also one of the largest users of thermoformed products. Food serving trays and disposable dishes are used in enormous quantities. However, present consumption is far from the potential volume that could be reached. Hospitals, nursing homes, school feeding programs, fairs, and conventions are heavy users of such products. Picnicware and catering supplies also belong in this category. Military, forest firefighting, and disaster relief organizations could also become potential users of thermoformed products.

H. Medical Industry

The medical industry is a large consumer of thermoformed products. Various types of presterilized surgical equipment, tools, syringes, needles, tubing, and vials are packaged for hospitals, clinics, and doctors offices. Most of these packages are disposable, intended for one-time use. These packages are usually produced in a "clean-room" environment and sterilized afterward. The integrity of such packages must be perfect, with no chance of failure. Some of this packaging is also capable of alerting users to sterility loss by means of color changes. The packaging materials used by this industry must meet stringent raw material and other specification standards. The use of recycled and reprocessed materials is unacceptable.

The industry also uses other thermoformed products, such as contoured surgical tables, dental chairs, and exercise platforms. Numerous prostheses and their components are produced by the thermoforming shaping process.

I. Nursery (Horticulture) Industry

Almost all seedling plants are grown in trays made by thermoforming. Commercial growers generally choose to use reusable trays that last several growing seasons. Plants are made available to gardeners in disposable trays or containers. Such trays, often referred to as "bedding plant trays" or "pony-packs," may have six to eight cavities, each housing a single seedling per cavity. These trays, produced largely from recycled material, are very inexpensive. Product integrity is not a major issue; minor cracks and fractures or weakened bottom structures are not a major concern. The other portion of this industry is that of flowerpot production, more and more of which has been switched over from injection molding to thermoforming. The higher production rate thus made possible and the reduced mold cost prompted the change.

J. Recreation Industry

The recreation industry utilizes many products made by thermoforming: small fishing boat hulls, canoes, sailboats, and their components. In addition, contoured windshields for boats, snowmobiles, and motorcycles are produced by thermoforming, and small swimming pools, spas, and kiddie pools are popular items that are easily thermoformed out of thicker thermoplastic sheets.

K. Housing and Construction Industry

The construction industry has been using thermoformed products for many years, but recently, the popularity of such products has accelerated rapidly. There are many products that can easily be substituted for thermoformed items, but others cannot be produced any other way. Such products as the one-piece skylight dome and contoured windows lend themselves ideally to thermoforming. Bathtubs, bathtub liners, and shower enclosures made from plastic are good examples of traditional materials that are being replaced by plastic. Others among the numerous products used are carved wood door panels, decorative building structures, and full-sized thermoformed garage-door face panels made up to 16 ft wide and glued to a wooden door structure.

L. Luggage Industry

Several nation-wide companies as well as numerous smaller luggage-making firms have opted to use the thermoforming method for suitcase and briefcase bodies. Thermoformed luggage is usually lighter in weight and more rugged than luggage made by other methods. Although thermoforming methods cannot produce the luggage body and their feet out of the same sheet, as can be done with injection molding, they provide more benefits in other respects, such as stress-free molding (less chance for corner fractures) and impact resistance.

M. Photographic Equipment Industry

One of the oldest products made for the photographic equipment industry is the thermoformed developing tray. Such trays are formed out of a medium-thickness sheet, the thickness gradually increasing for larger trays. Many other thermoformed products are also in use, such as the components of flash cubes and flash bars (the metallized reflector and the clear plastic housing). The latest adaptation of thermoforming is for the film cartridges of disk cameras. Their production requires a finely detailed high-tolerance thermoforming technique.

N. Food Processors

When foods are processed, seasoned, pickled, or stored for packaging, large vats and tubs are used. Most such containers are designed so, when empty, they can be stored tightly nested into each other; while containing the product, they can be stacked. In this way, food can easily be segregated by seasoning or flavor until final portioning or packaging is accomplished.

O. Assembling Production Lines

Numerous large-scale manufacturers use part-carrying trays on their assembly lines. Usually, the trays are custom made to fit specific requirements. They could have a single cavity, multiple cavities, or specifically shaped cavities that match the particular parts configurations. The components or parts are placed in the trays at the warehouse and travel on a belt system to the assemblers. When all components are installed, the trays are returned to the warehouse for refilling. The use of such trays can provide an excellent inventory control or guide for proper sequencing for assembly. Locks, clocks, electrical devices, and instruments are good candidates for such trays.

P. Funeral Industry

Many coffins are now produced by thermoforming rather than by the traditional wood construction. One-piece thermoformed coffins may even have imitation wood graining. Casket liners are also made by this process, with surface finishes resembling wood, silk, or velvet.

IV. THE THERMOFORMING INDUSTRY

In the last three decades thermoforming has grown from a small job-shop operation to a sophisticated high-volume, high-quality manufacturing process. Today, we have companies with regional multiplant locations that supply products for the entire country. These companies manufacture products around the clock. Many equipment suppliers make specialized machines for specific types of thermoforming techniques. They not only compete among themselves but also among foreign suppliers that have entered our domestic markets. Since the packaging industry is the largest user of thermoformed products, it has dominated the thermoforming industry to the extent that many thermoformers have adopted a number of packaging industry practices. Such domination has not only had positive side effects but has some shortcomings as well—such as tight secrecy between practitioners—which have hindered progress.

It is very difficult to pinpoint the exact number of thermoforming practitioners in a given area. Telephone books, catalogs, professional listings, and trade publications provide fewer than the actual number of thermoforming processors. It is estimated that for every company listed, at least two additional companies are engaged in product making in a given area. Companies may be listed under different product-making categories, obscuring their possible activity in other fields. Some captive operations are not listed under specific product identifications. This occurs frequently with paper and packaging houses, chemical companies, and so on, which may not be listed under any of the thermoforming categories although heavily involved with the process. Finally, almost all thermoplastic resin manufacturers are involved to some degree with product research, manufacture, and testing, either for themselves or for their customers. Any knowledge gained can lead to improvements in both resin compounding and future business opportunities. Some of the larger chemical companies that are heavily involved with plastic resin manufacturing and supply maintain large research staffs and allocate large budget outlays for product development and testing. Most of these companies accumulate new product files over the years. From these files, interested manufacturers may find product ideas that would be helpful when resin purchases are made.

It should be kept in mind that the ultimate goal of all manufacturers, including thermoformers, is to produce useful and desirable products that are salable. The chemical companies, thermoforming equipment manufacturers, and all the supportive equipment and additive suppliers cooperate with the thermoforming processors to achieve this goal.

2
Components of the Thermoforming Process

I. THERMOPLASTIC SHEETS

A. Thermoplastic Resins

The thermoplastic sheet is the most important element of the thermoforming process. Without the thermoplastic sheet, thermoforming would not exist. Thermoplastics are also called polymers or resins; they are unique man-made compounds. If exposed to elevated temperatures, they become soft and eventually even liquefy. If cooled down, they harden and set up firm. The temperature at which this occurs is the setup or crystallinity point. Changes in softening and hardening can be achieved repeatedly with temperature fluctuation. Such a physical alteration is characteristic of all thermoplastics.

The plastic polymers are manufactured from low-molecular-weight compounds called monomers, which are derived from crude oil, coal, and natural gas. Crude oil and coal must undergo a process of distillation, cracking, or solvent extraction. In natural gas production some of the trapped liquids can be moved directly into the process. Monomers produced in this way include ethylene (C_2H_4), propylene (C_3H_6), and benzene (C_6H_6), the raw materials for many common plastics (Figure 1).

In the polymerization process, monomers, with the help of catalysts, high heat, and high pressure, react together chemically to create high-molecular-weight chainlike molecules (Figure 2). Thermoplastics have strong covalent

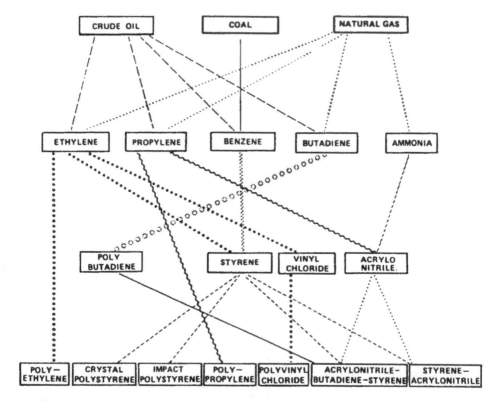

Figure 1 Major raw materials.

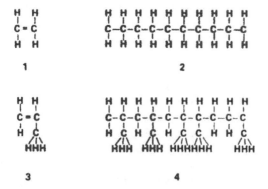

Figure 2 Chemical formulas of polyethylenes and polypropylenes: (1) ethylene monomer; (2) polyethylene polymer; (3) propylene monomer; (4) polypropylene polymer.

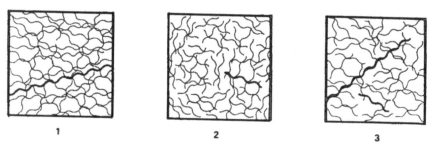

Figure 3 Molecular lengths and entanglement: (1) long-molecular-chain resin (high molecular weight); (2) short-molecular-chain resin (low molecular weight); (3) mixed long- and short-molecular-chain resin.

chemical bonds within the independent long-chain molecules, and they also have strong tangled or woven intermolecular bonds (Figure 3). The length of these chains and their entanglement are determining factors in the strength and resiliency of the thermoplastic produced.

A polymer's molecular chain length and entanglements are key factors in the process of thermoforming and will be mentioned repeatedly throughout the book. Depending on the manufacturing technique, different degrees of side branching, different chain lengths (i.e., molecular weight), and various ratios of blends can be achieved. The longer the molecular chains, the higher the molecular weight, and thus the stronger or tougher the plastic produced. Short molecular chains (low in molecular weight) tend to be brittle and easily fracturable.

Virgin polymers generally show the highest molecular weights (longest molecular chains). However, different polymerization processes do achieve different qualities of thermoplastic resin. In the process, it is likely that all monomers will not link up to a chain of polymers. Such "free monomers" could be capsulized between the entanglement of polymer chains (Figure 4). The processing of thermoplastics frees most if not all of the free monomers, which are distinguishable by their strong "plastic-like" odor. Such free monomers are most evident in the styrenic family of thermoplastics.

Free monomer levels are controllable by the resin manufacturers, and minimum standards are set by the Food and Drug Administration. The basic regulation, described in the FDA's Title 21, Table 1 of 177.1640(c), reads: "Polystyrene basic polymers shall contain not more than 0.5 weight percent of total residual styrene monomer."

The free monomer levels set by the FDA ruling are higher than those some food-container users accept; they set lower standards for their own use. "Low" free monomer levels contain 0.1 weight percent of total residual styrene

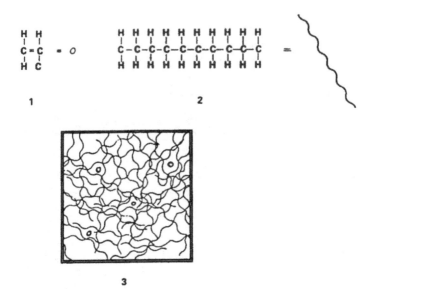

Figure 4   (1) Free monomer; (2) polymer; (3) polymer resin with free monomers.

monomer. "Super-low" free monomer levels contain 0.05 weight percent total residual styrene monomer. Containers used with foods containing a high level of animal fat, such as butter and lard, have a tendency to solventize and to pick up the free monomers or synthetic rubber molecules. The result of this chemical reaction is that such foods (i.e., cookies containing butter) pick up plastic-like off-flavors and odors during warehousing and storage.

Nevertheless, occasionally, higher-free-monomer-level resins end up on market shelves as food-packaging products. A strong styrenic smell immediately indicates a high level of free monomer content. Conscientious manufacturers have little problem keeping and maintaining the free monomer level well below the FDA standard. High free monomer levels should concern only thermoformed container manufacturers whose products will be used in food preparation and packaging. The sheet-making and thermoforming operations in the heat cycles alleviate large portions of free monomers. Such manufacturing areas often have to be equipped with additional ventilation to remove the strong odors.

The key demand in thermoplastics when used in the thermoforming process is that the resin be of high molecular weight with the best intermolecular entanglements available. Since the thermoforming process uses a preformed plastic sheet clamped and held only by its edges, the actual material stretching and material distribution (thinning) are highly dependent on that molecular relation. Initially, the thermoplastic sheet should contain molecules with maximum molecular weight (long chains) to permit optimum stretching without rupturing or splitting. Again, the long molecular chains have the strongest intermolecular entanglements. On the other hand, if the chains are considerably smaller, the resulting poor intermolecular characteristics could affect the quality of the plastic sheet, sheet formability, and even the quality of the finished product.

Blends of high and low molecular weights can be made which aim to improve the intermolecular characteristics or to permit use of resins of lower quality. A proper blend ratio is required to achieve uniformity from one production run to another. Disappointing results can be encountered when materials are supplied by a variety of sources. To secure identical results, it is necessary to obtain all material from one supplier; otherwise, thorough testing is required before making a commitment to purchase material from another supplier.

A fact that is often ignored is that degradation and scission of molecules take place throughout the thermoforming process. Actual mechanical shearing of the molecules happens during pelletization, at the extruder screw, when material is melted and pushed through at the die, and during scrap-making and recycling process. This degradation is normally negligible, depending on the type of end product involved. Some products do not accept any recycled material without damage to their integrity. Generally, manufacturers accept limited quantities of scrap blends. There are many products that can be manufactured completely from reclaimed material.

It is necessary to keep in mind that the degradation and scission processes are cumulative. The recycled polymers are recycled repeatedly, and at some point can no longer be processed but must be discarded. Fortunately, there are early indicators which forewarn that the thermoplastic resin is about spent.

B. Plastic Sheet Making

The thermoplastic sheet is essential to the thermoforming process. All the known thermoplastic resins have to be converted into sheet form, by various methods, resulting in precut panels or continuous sheets which are then wound onto coiled rolls. Thermoplastic sheets are manufactured in a wide variety of widths and lengths and come in a variety of predetermined thickness.

Three basic techniques are used in plastic sheet making, each achieving the same result: converting the customary thermoplastic pellets into sheet form. Each technique has unique features which make it more adaptable to specific thermoplastic resins. In many instances, a particular resin would not lend itself to any other technique; therefore, only that method can be used to make the thermoplastic sheet.

The three basic sheet-making methods are calendaring, casting, and extruding. In all three techniques, there is a basic similarity. They are all based on the introduction of heat into the sheet-making process, in order to soften and melt the thermoplastic resins. When the sheet is formed, the heat must be extracted, allowing the thermoplastics to "set up" in final form.

The fundamental characteristics of the original thermoplastic resin are retained in the sheet form. However, the additional, unique characteristics imprinted into the plastic sheet are affected by the sheet-making method. The combined characteristics will remain and be retained by the plastic sheet and are easily identifiable both prior to the thermoforming and throughout the thermoforming process. This built-in characteristic could either enhance or create difficulties in the thermoforming process. Further, it could affect the finished product's functionality.

When particular sheet-making characteristics are judged as advantages for the thermoforming process, the thermoplastic sheet manufacturer limits its sheet making to the method that best produces them. For example, vinyls or polyurethanes are most often calendared, and ABS, polyethylene, polypropylene, and rubber-modified styrenes can also be calendared. In recent years most sheets used for thermoforming have been manufactured (estimated by weight or volume output) by the basic extrusion process or one of its numerous variations, each aimed at improved sheet characteristics, higher productivity, or additional economic benefits.

1. Calendaring

Calendaring is a very simple process that is easily adaptable to thermoplastics. This process, which replicates the well-established sheet-rolling and rubber-sheet-making process, is the same technique as that used in mechanized pizza-dough-making apparatus.

The calendars consist of a series of rollers which are rather large in diameter (approximately 12 to 48 in.) and heated (325 to 450°F). They are counterrotating and mechanically driven. A fixed gap or roller distance is set between the initial rollers and is then continuously reduced between subsequent rollers. The function of the initial two rollers is to provide for mixing, blending, heating, and metering of the softened plastics. The remaining rollers are used to size or "gauge down" the thickness of the sheet to the final gauge required. The final rollers are basically cooling rollers which allow for the removal of heat and the setting or firming up of the sheet. The final rollers are also used to apply various surface finishes to the sheet. A polished roller will give a smooth, shiny finish; an etched roller will transfer a dull finish; and an engraved roller will make a very detailed textured surface.

The calendar rollers are temperature controlled by oil or other types of fluids, which circulate constantly throughout the calendars. Plastics are introduced to the calendaring equipment in a preheated and presoftened pellet or chunk form. This feeding procedure can be a very simple manual operation or can be achieved through a more sophisticated automated system, depending on the size of the manufacturing facility. Whichever way the feeding is accomplished, the plastic supply must be maintained adequately to avoid a shortage of supply. If such a diminishing of supply occurs, large voids or holes will result in the sheet which ultimately can interrupt sheet continuity. In fact, one of the most frequent problems with these types of calendared sheets is that small pinholes are often manufactured in the sheet. These constitute escape holes for the vacuum or air pressure used in the thermoforming process and do not allow some areas of the sheeting to be formed. Such faulty areas must be rejected. Today's highly skilled sheet manufacturers minimize pinholing effects to the point where such problems are rarely encountered. However, this predicament does occur, and it is important to recognize its cause.

During the calendaring process, the edges of the continuous sheet may become uneven and wavy. Such edges are trimmed off and refed to the initial calendar rolls (Figure 5). Generally, plastic sheets made by the calendaring process are wound onto rolls. However, the manufactured sheeting can also be cut and stacked in individual panels, according to need.

The arrangement and array of calendar rollers vary from manufacturer to manufacturer. Calendar rollers are also selected according to resin type, sheet size, feed method, and other individual requirements. The calendar roller

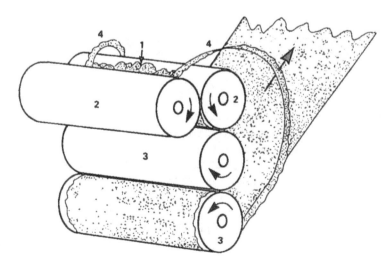

Figure 5 Calendaring. This is a reversed-L-shaped roller stack. (1) Thermoplastic material; (2) feed rollers; (3) sizing and surfacing rollers; (4) edge trim.

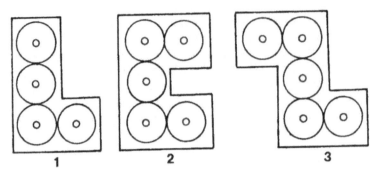

Figure 6 Calender roller arrangements: (1) straight L-shaped roller stack; (2) C-shaped roller stack; (3) Z-shaped roller stack.

stacks may form the most popular inverted L shape, straight L shape, or C or Z shapes (Figure 6).

Plastic sheets made by the calendaring method have no significant molecular orientation. As soon as they are exposed to the heating cycle in the thermoforming process, the sheets have a tendency to sag immediately. Sagging is generally undesirable in the thermoforming process. The problem of sagging is discussed in Section I.A of Chapter 3.

2. Casting

The casting of plastic sheets and films is also an accepted method of manufacturing thermoplastic sheets for the thermoforming industry. For particular resin materials, the casting method may be the only method available for the production of sheet forms.

The basic casting technique has two variations, each used for specific purposes. The first method is almost always used with acrylic or acrylic-type materials. The casting process is accomplished by pouring the base monomers, or in many instances, partially polymerized syrups or their blends into suitable sheet molds or onto moving stainless steel belts. When individual clampable sheet molds are used, the process is called a "batch process." When using moving stainless steel belts, the process is called a "continuous process." In both cases, the liquids are used to fill a void between two surfaces. When heat is applied, final polymerization will take place. In the heating cycle the amount of heat has to be carefully controlled and monitored. Polymerization itself generates heat that can easily lead to rapid acceleration and cause a runaway process, resulting in the breakdown of the polymer. The formation of bubbles in the castings is an indication of polymer breakdown and gasification. By rapid cooling of the castings, the acrylic sheet is salvaged. However, the bubbles remain trapped inside the sheet and may affect further processing of the sheet. Bubbles definitely interfere with the optical clarity of the finished product.

Special surface treatments of the casting mold surfaces can be adopted which will result in a mirror, satin, or matte finish on the plastic sheet. Batch castings can usually produce sheets 0.03 to 4.25 in. thick with sizes up to 120 by 144 in. The continuous process is capable of producing thicknesses of 0.08 to 0.375 in. and widths up to 9 ft. Heavier thicknesses are avoided because the sheets cannot then be wound onto rolls.

When the sheets are made, tissue interleaf is applied to avoid scratching the surfaces. The main application of acrylic sheets made by this casting process is in the forming and fabrication of signs, the building of displays, and for skylight windows and tub enclosures.

The second type of casting method is called "solvent casting." This type of casting could well be the oldest method of producing sheet forms. The solvent casting method is used mostly with polyvinyl chloride (PVC) materials to produce films. The basic resin material and the necessary additives are solventized: dissolved to form a pourable homogeneous liquid. This liquid, in heated form, is pumped through a specially designed flat die opening, whose size can be controlled, thus controlling material flow. The curtainlike liquid plastic will flow onto a moving stainless steel belt. As with most liquids, the liquid plastic will spread and establish an even level, creating a uniform thickness. The gauge of these films produces thicknesses of a fraction of a mil up to 20 mils, and can be controlled by the material flow, die opening, liquid back pressure, viscosity, and travel speed of the stainless steel belt. With this many changeable factors, not only good gauge control but also quick gauge changes can be achieved. The stainless steel belt with the predetermined layer of plastic on top travels through an oven, and in this manner is heated. The oven heat, plus fanned air, vaporizes the solvent completely, leaving only the plastic residue as a film. The vaporized solvent is pumped through a solvent recovery system and almost all solvent is trapped, recovered, and reused (Figure 7). The plastic films made by the solvent casting method have good clarity and fairly good overall gauge control. Slight dimple effects on the upper surfaces can result, due to the self-leveling effect in the plastic's liquid form.

All sheets made by both casting methods are typically free of pinholes, have no hidden strains, and have equal strength in both directions of the sheet. In addition to those characteristics, the sheet or film made of clear resin displays excellent optical properties. The casting method as a production technique is often the only possible way that certain types of thermoplastic sheets or films can be manufactured.

Just like calendaring, the casting method fails to create a molecular orientation in plastic sheets. The lack of orientation does cause the sheet, when exposed to heat in the thermoforming process, to sag almost immediately.

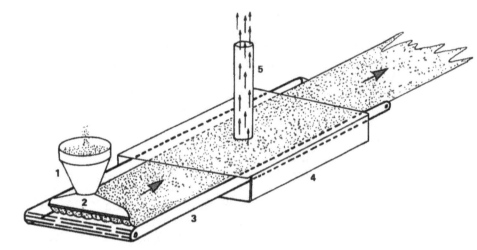

Figure 7 Sheet casting lines: (1) liquid feed hopper; (2) flow die; (3) moving stainless steel belt; (4) solvent drying oven; (5) solvent vapor removal.

Generally, this is a disadvantage. In the thermoforming process sagging alters the distance between the sheet and the heaters, which results in uneven heat distribution over the entire sheet surface. In one instance sagging is beneficial in thermoforming—when the sag can be used as a means of prestretching the sheet, making it form better and more uniformly into an openface female mold. Such forming techniques are discussed in Chapter 3.

3. Extruding

By far the most common and most versatile sheet manufacturing technique for the thermoforming industry is the extrusion process. More pounds of thermoplastic resins are converted into sheet form by this method than by all other methods combined.

The extrusion process is designed and implemented in such a way that it is capable of converting the raw thermoplastic materials, which come in pellet, powder, or granular form, into a continuous melt flow. When made by the extruder, this melted plastic flow is forced through a flat or round die system that forms it into a continuous sheet. The sheet is cooled, sized to proper gauge, and cut into individual panels or wound onto a continuous coil (Figure 8).

The extrusion process always starts with the basic resin supply, which is delivered in bags for the small-quantity user and in Gaylord boxes or in bulk (such as rail cars) for the large-quantity user. The thermoplastic resin materials are introduced to the feed hoppers by mechanical pumps which lift and carry

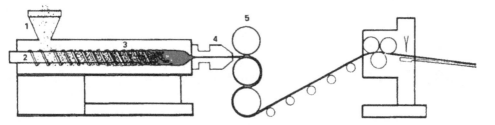

Figure 8   Sheet extrusion line: (1) feed hopper; (2) extruder screw; (3) extruder barrel; (4) die; (5) roller stack.

the material from a smaller box reservoir or central storage silo system, or directly from a rail car. On very rare occasions, individual bags are emptied into the feed hoppers by hand.

The actual extrusion process could use "virgin resins," which usually means materials supplied from the resin manufacturer, or could use reprocessed materials, or a blend of the two. Both resin materials generally come in the customary pellet form but occasionally come in powder or granule form. When blends are made using different forms or different particle sizes, it is a constant challenge to maintain fixed blend ratios. With standard feed hoppers it is easy to lose predetermined blend ratios. The feed hopper acts as a built-in mechanical separator. Because of the different flow characteristics of different particle sizes and shapes, the most perfect blends will segregate out, changing the originally planned blend ratio. This blending has to be watched and controlled closely if material standards and uniformity are critical.

By its nature the thermoforming process will always be a high scrap producer (trim scrap). In most cases, the scrap materials are reprocessed and blended back into the sheet. When predetermined ratios are set for maintaining the quality standards of the thermoformed products, any segregation of the blends would automatically alter the intended ratio, resulting in an inferior product, or would cause production problems. The reprocessed material comes not only from the trim scrap of thermoforming, but could come from startup or production errors. The extrusion process can also generate sizable quantities of reprocessible material. The extrusion process itself produces reclaimable material. Until the extruder is tuned to the desired specifications and the process is set, the materials that have been produced are reclaimed. The uneven edges that are trimmed away are also continuously reclaimed. In addition to this, when rolled sheets are slit into specific widths, considerable unwanted, leftover material results. This material is reclaimed and reprocessed together with the scrap produced by the thermoformer. The art of scrap handling and its limitations are discussed in Section II.C of Chapter 6.

There are many, varied uses for the extrusion process, but only those important to the thermoforming process will be presented. We begin by describing the components of the process, of which four are especially important. The first is the extruder, the second is the die, the third is the roller stack, and the fourth is the "takeoff" system, comprising the sheeter/stacker and the winder/slitter.

The extruder is the main component, the heart of the entire system. Besides being the most costly component, this piece of machinery uses the most energy. The extruder receives the solid thermoplastic pellets, granules, or powders, then melts, blends, and pressurizes the resins for introduction into the die. All extruders consist of an extruder barrel, whose first section is externally heated, the remaining sections being externally cooled. Inside the extruder barrel is a close-fitting, rotating helical screw. The simplest screw design, which is widely used for extruding the common thermoplastics, is a single-stage design having three distinct sections. The first section of the screw is called, appropriately, the "feed" section. The thermoplastic resin is fed into this section, usually by gravity, from an overhead feed hopper. This is the critical point, at which the different particle sizes or particle shapes have a tendency to segregate, resulting in an unwanted blend ratio.

The second section is called the "transition" section. This section is where the solid thermoplastic particles are compressed, heated, and further compressed and liquefied. The liquefication is caused by internal friction between the particles and compression between the screw and the barrel surfaces, combined with the externally applied heat.

The third section is the "metering" section, which is relatively the longest and has the minimum screw flight depth. This section completes the melting procedure, creating a well-mixed, homogeneous thermoplastic flow. The final portion of this section acts as a metering pump, giving uniform pressure and flow characteristics to the molten plastic. Selection of the most ideal screw designs is important to achieve the best results with different types of thermoplastics. Many different and far more complex screw designs have been developed, not only to improve extruder performance but to enhance the thermoforming characteristics of specific thermoplastic resins. Besides the complexity of various screw designs in a single-screw extruder, there are twin-screw and multiscrew extruders, all of which have been developed for specialized purposes. In general, the multiscrew extruders are costlier and require more maintenance but offer superior performance with certain plastics. However, the single-screw extruder is still the "workhorse" of the industry.

Extruders are characterized by two key dimensions: the bore diameter, D, and the length of the barrel, L. This is called the "L/D ratio." For example, an extruder that has a $4^{1}/_{2}$-in. bore size and a length of 153 in. has a 34:1 ratio. The common commercial extruder sizes range in bore size from 2 to 8 in. However, small laboratory-size extruders are available from $^{3}/_{4}$ in., and very large commercial extruders have been built up to a 24-in. bore size.

Extruder speed, heat input, and heat removal in multiple zones can be well controlled for optimum results. In the most advanced systems, control is computerized. Such systems can respond and adjust to the slightest changes and demands (e.g., as close as ±1°F). These systems also interact with the downstream equipment, thus controlling them. Warning alarms and data recorders can be incorporated in these systems.

Some thermoplastic resins contain a small amount of free monomers, moisture, or entrapped gases or air, which are usually discharged through vents built into the extruder barrel. As the extruder pushes the molten thermoplastic out at the end of the barrel, the flow is directed through a built-in filter screen before entering the die. The purpose of the filter is to remove any contaminants that have been incorporated into the resin by previous processing or handling.

As the thermoplastic resin travels through its various stages, the extrusion process slightly alters the molecular chain lengths, reducing the original molecular weights. The alteration is so small that the difference between virgin and first-run plastics cannot be detected. However, it should be remembered that most thermoplastics run through an extruder multiple times, with continuously diminishing quality.

The second component of the extrusion system is the die (Figure 9). At this point, the final shape of the melt flow is determined. Generally, for sheet forms, used mostly for thermoforming, the molten resin is forced through a flat die. The die has a "coat hanger"-shaped manifold and a restrictor bar to maintain uniform material flow throughout. The die lips are adjustable and control the die gap (opening), thereby regulating the thickness of the material as it is extruded from the die.

The die lip is adjusted with a sequence of Allen-head screws which are built into the die lips at approximately 1-in. intervals. To close down the die gap for the purpose of reducing the sheet thickness, the screws must be tightened. The screws force the die lips to close. To open the die gap, the screws are backed out, allowing the die lips to open and the material to flow in greater quantity. Adjusting the die is tedious, and great care must be taken to achieve ultimate uniformity in the die opening. If the screws are not adjusted correctly, gauge variations will be obtained in the extruded sheet and will be produced continuously. Gauge variation can produce a "tapered sheet," which is the result of having an extruded sheet with greater thickness on one side than on the other. Also, a "gauge band" could result, causing a band of greater or lesser extrusion thickness in the center portions of the extruded sheet. The nipping effects of the roller stack will reduce these errors but may not eliminate them

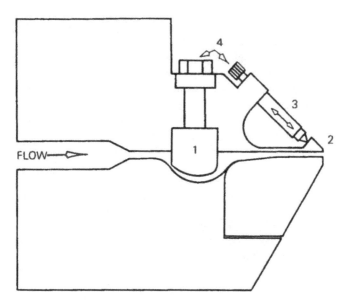

Figure 9   Extrusion die: (1) restrictor bar; (2) adjustable lip; (3) adjustment pusher rod; (4) adjustment screw heads.

entirely. The dies have several built-in, individually controllable heating zones across their widths, thus allowing a uniform material flow characteristic throughout the melt.

The third component of the extrusion process is the roller stack. This equipment is similar to that used in the calendaring process. The melt flow coming out of the flat die is nipped between two polished rollers and then threaded through an S shape on the two consecutive rollers. This combination of rollers is called the "roller stack" (or "roll stack") or "S wrap." The purpose of this vertical stack of rollers is first, to nip and set the desired sheet thickness, and second, to cool it to a crystalline form. The rollers are temperature controlled by circulating fluids and maintained at a constant temperature. The surfaces of the first two rollers are highly polished. The remaining roller surfaces determine the finished surface patterns of the thermoplastic sheet. When all the rollers have a highly polished (chrome-plated) surface, the sheet will have a glossy finish; an etched roller surface will produce a satin-sheet finish; and an engraved roller will transfer its design and will display a reversed pattern in the thermoplastic sheet surface. Rollers can transfer surface designs for only one side of the thermoplastic sheet or for both sides. Actual roller diameters range from 8 to 24 in. The larger rollers have definite advantages in the cooling procedure because of larger surface contacts. The roller-stack speed is adjustable in a wide range to match the output rate of the extruder. Speed is often controlled automatically, although it can be adjusted manually. Such speed variation, especially when the roller stack is running slightly faster than the extruder, is of great concern. The thermoplastic sheet gauge can be reduced easily, not only by reduction in the die gap, but by stretching with the roller stack. When the roller stack is run at a higher speed than that of the extruder, the sheet will be pulled mechanically. This procedure will not only reduce the gauge but will stretch the molecular entanglement and structures, producing an extra mouth of orientation to the plastic sheet in the machine direction. Such orientation imbalance is not visible until the sheet is exposed to heat. In most cases, unbalanced orientation in the thermoplastic sheet presents difficulties in the thermoforming process. Evidence of this problem will surface in the thermoforming process when complicated shapes or multidepth or extremely deep parts are thermoformed.

When the thermoplastic sheets are extruded and nipped, then cooled by the roller stack, the continuous edges are not even, having a slight waviness. Such an edge is almost always slit away in a continuous strip and either granulated right there or threaded through a feed system which carries it to a postgranulator. In both cases the groundup edge trim is reextruded with the virgin material.

The fourth component of the extrusion process is the sheet-takeoff system, made to handle the manufactured sheets. The sheet coming out of the S wrap must be accumulated and stored. There are two ways to handle the extruded sheets. One is to feed the continuous sheet into a guillotine cutter, which will cut the sheet to a predetermined length. The instrument is preset to given lengths and when such lengths have traveled past the instrument, it will actuate the cutting blade, which will shear the entire sheet width. Such panels can then be stacked on top of each other, either manually or by use of automatic equipment. Extruded sheets are usually cut into panels in the higher gauges and when thermoforming is planned on sheet-fed machines. Sheets from 30 mils to $^3/_4$ in. (the upper commercial limit of extruded sheets) are usually cut into panels. The only time that lesser gauges are cut into panels is when the material does not lend itself to coiling or when the equipment can handle only individual sheets. Material of less than 30 mils is generally wound onto rolls. The rolls usually start out with a paper core on which the plastics are wound. Cores of 4-, 6-, and 8-in. diameter are used, and the cores remain with the plastic rolls until the thermoplastic sheets are used up.

Thermoplastic sheets wound into roll form right after the extrusion process retain some of their residual heat. This heat, combined with the newly processed plasticity of the resin, makes the sheet soft and yielding, a condition often called the "green" plastic stage. When green plastic sheets are wound onto tight coils, the curvature of the coiling will be adopted by the plastic sheet, and during storage this curvature will be set in. This phenomenon, called "roll set," becomes more severe close to the center of the rolls, where the radius is smallest. This condition may affect the unwinding of the rolls when continuous feeding of the thermoforming equipment takes place, or during slitting operations. In some instances the problem could be so severe that the thermoplastic sheet would snap and break across the sheet width. Of course, if the plastic sheets can be fed into the heating area, the heat of the sheet will eliminate the roll set and remove any previous "memory" from the sheet.

Besides producing straight thermoplastic sheets, extruders perform many other important tasks. Some of these are in the range of sheet making and are used in the thermoforming process. Individually produced thermoplastic sheets are often laminated together. The thermoplastic sheets could be manufactured on the same extruder at different times or be produced on separate extrusion equipment. Two or more sheets laminated together, usually on secondary process equipment, are called laminating lines. Laminating generally entails applying heat to the sheets and may or may not require additional bonding agents. The purpose of lamination is to gain needed benefits, which may be physical, chemical, or even economical. For example, in a laminate one layer may display excellent impact resistance, while the other layer may have outstanding scratch resistance. On another laminate, the individual layers may have different chemical resistance, and whereas individually, the layers provide only a partial chemical barrier to certain chemicals, in lamination they are completely protected against chemical migration. As an example of economic factors, one less costly layer combined with a higher-priced layer can reduce the overall cost of the thermoplastic sheet while providing high quality or weather resistance on one side of the sheet.

In a particular laminating process, when the individual layers are combined with an additional material, an extruder is often used to force a thermoplastic melt in between the individual webs. This process is called "extrusion lamination."

The process of laminating materials together can only be achieved with the thicker materials which display sufficient strength and self-support. Thinner materials do not have such self-supporting strength and cannot be fed into laminating equipment. Often, a thermoplastic sheet requires only a very thin · surface coating to achieve the needed improvements. Extremely thin layers cannot be laminated or made in any other manner except by a process called "extrusion coating." In this process an earlier manufactured base thermoplastic sheet is usually fed through a sheet transport system. A separate extruder is used with a very narrow die gap to lay a thin coating on the top of the base sheet. Should coating be desired on both sides, the base sheet is turned upside down and passed through the equipment a second time. In higher-volume production the second coating would be applied in line by flip-turning the base sheet over and employing a second extruder coater. Extrusion coating is not an easy task. Having a cold base sheet cools the very thin coating layer rather rapidly, and this temperature difference often limits the adhesion of the layers, resulting in delayed delamination.

The cost of coextrusion versus lamination is heavily in favor of coextrusion. Scrap recovery and interference with the different thermoplastic skins or layers and surface layer colorants due to their lesser thickness also favor the coextrusion process. When different thermoplastics are chosen for coextrusion, various basic factors must match in order to achieve compatibility in a multilayered sheet. The thermoplastics used for coextrusion and thermoforming must have the same or compatible rheology, melt index, softening temperatures, crystallinity, shear, and surface tension. Thermoforming coextruded sheets is discussed in Chapter 3. As food packaging gets more and more sophisticated and longer shelf life is demanded, only coextruded materials can meet the criteria. Several different barrier characteristic materials combined into a multilayered sheet prevent various gases and moisture from entering or exiting a package. Barrier material formulation and modification are often proprietary information. When the multilayers in the coextrusion are combined, the individual layers and their chemical compositions are difficult to trace back to their origin. With the food industry's demand for longer and better shelf life, the development of new coextrudable barrier materials will be stepped up radically. Work is under way to substitute coextruded thermoformed containers for many of the metal cans used at present. For the time being, lids will continue to be made of aluminum, using the customary crimping methods and lift top openers.

a. **Biaxial orientation:**   In the extrusion of thermoplastic sheets for the thermoforming process, flexible thermoplastics are used most often. Occasionally, synthetic rubber additives are used with styrenic materials, such as styrene-butadiene rubbers to enhance their resiliency. Such additives improve the strength and impact resistance of the basic resins or add needed improvement to reclaimed, reprocessed plastic materials. Flexibility and impact resistance are naturally common with some thermoplastics, while others are improved only with additives. The inclusion of additives does, however, result in the loss of perfect clarity, making the sheets hazy.

Let's use crystal polystyrene as an example. It has all the qualities needed to be ideal for packaging and food serving. It is readily available at reasonable cost, easily moldable, crystal clear, and has more than satisfactory food-holding and chemical resistance qualities. Its only drawback is that it is much too fragile to thermoform. At the slightest flexing the thin walls of the sheet or formed parts fracture and break into pieces. These negative effects make this ideal material useless. However, when such sheets are subjected to a biaxial orientation procedure after extrusion, they are made ideally flexible and somewhat fracture resistant. To implement such biaxial orientation, the crystal polystyrene is first extruded into sheet form (Figure 10). The sheet is maintained continuously warm by a series of heated rollers, then introduced into a temperature-controlled enclosed oven.

In this oven the temperature of the sheet is held slightly below the melting point of the crystal polystyrene, so that the plastic will not flow. Chain grippers grip and clamp the very edges of the plastic sheet and biaxially (in a wishbone pattern) stretch the crystal polystyrene sheet in both the machine direction (MD) and the transverse direction (TD). Then the sheet is cooled, which sets or locks this physical orientation into the sheet. In the orientation procedure the long and "curly" individual molecules exit from the extruder die with numerous curls and entanglements being physically stretched out. This phenomenon is a change of molecular shape which makes a very rigid thermoplastic such as crystal polystyrene flexible enough to be acceptable for use in the thermoforming process, and to make useful products out of it. Biaxially oriented polystyrene trays for food packaging and food serving are ideal. They have FDA approval (Title 21, 177.1640) and are available in large quantities at reasonably low cost. Oriented polystyrene (OPS) has good structural strength both at room temperature and at refrigerator and freezer temperatures. It has excellent clarity, which is ideal for showing the food content of the package. In addition to those advantages, it demonstrates fairly good moisture and abrasion

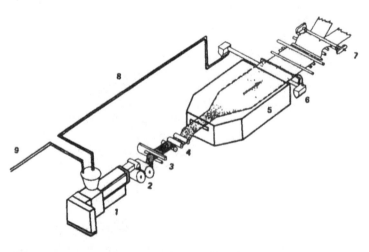

Figure 10 Biaxially oriented sheet-making line: (1) extruder; (2) casting rolls; (3) thickness gauge; (4) MD stretcher rollers; (5) TD stretcher oven; (6) edge trim; (7) slitter; (8) edge-trim scrap return; (9) virgin resin supply.

resistance. It has many benefits, but without the biaxial orientation technique, crystal polystyrene as a sheet material would prove to be useless in the thermoforming process.

There are other thermoplastic resin materials that can be biaxially oriented: polypropylenes, polycarbonates, and acrylics. These materials can be biaxially oriented in both a continuous manner and a batch orientation, panel by panel (tender frame process). The purpose of biaxial orientation of these materials is to increase their impact resistance and gain additional strength. For example, cell-cast acrylics have the highest density and thus provide the best scratch resistance; at the same time, they provide poor impact resistance. Extruded and biaxially oriented acrylic sheets prove to have completely the opposite features and when laminated together in a multilayered structure, all the best features can be expected in the combined sheet. This type of laminated structure thermoformed into a contoured shape is currently used in many aircraft windshields. Since orientation is a physical change in which molecular elongation has been frozen into the sheet, exposure to the same heat levels can easily cause the molecules to return to their original curly stage. This process, called "deorientation," immediately wipes out the benefits of the original orientation. The temperature of deorientation and that of the softening of the thermoplastic sheet are often very close. Having the sheet clamped on all four sides does retard deorientation, and only the pressure-forming technique can be implemented, vacuum forming being to slow and generally not being of adequate force to overcome the orientation elasticity of the biaxially oriented thermoplastic sheet. Because sheet softening and deorientation temperatures are very close, it is mandatory to have very close control in the thermoforming process. Fluctuations may result in a poor-quality thermoformed product; out-of-control conditions result in a ruined product. Products that undergo deorientation usually display high brittleness and/or excessive and irregular material thickness variations. Since the orientation process is made with very close heat control and the biaxial stretching is done with only the edges of the sheet gripped, constant uniformity of orientation may not result. A visual check of the oriented sheet may not reveal flaws in the orientation unless the flaws are very severe. Only when the sheet is heated can problems be detected. Generally, a biaxially oriented sheet should have balanced orientation in both the TD and MD directions. The greater the spread between those directional orientations, the more difficulties they cause in the thermoforming process, and the difficulties increase with sophisticated mold shapes. One basic rule to remember: The ideal properties of a sheet can easily be ruined in a heat cycle, but a poor or badly made sheet can never be improved. This holds true particularly for biaxially oriented thermoplastic sheets.

b. Foam extruding: Foaming of thermoplastics has become very popular to reduce material costs and to gain desirable mechanical properties. Foaming can be accomplished rather routinely and most uniformly using an extruder. Extruding thermoplastic foams, particularly for the purpose of thermoforming, allowed manufacturers to produce products that are light in weight, with excellent cushioning characteristics and outstanding insulating abilities. Products made from such thermoplastic foams may not have to satisfy all three characteristics, but one or two of those is enough to justify the use of foaming.

The fact that there is a reduction in thermoplastic use in foamed material as opposed to solid material has been a welcome relief for product manufacturers. As the rising cost of raw materials dictates, more and more products will be made of materials that are entirely foamed or contain a layer or two which are laminated or coextruded, with a foam thermoplastic base layer.

When protective cushioning is desired, as in egg cartons, foaming of the thermoplastic resin is highly desirable. Foamed vinyl is used in the production of soft, cushiony surface materials such as seating, inner surface paneling, and automobile dashboards.

If heat retention is desired when hot foods are packaged, foamed polystyrenes are ideal, examples being thermoformed hamburger containers and carry-out hinged-lid trays. All of the thermoplastics lend themselves to foaming. However, to date, only polystyrenes, ABS, polyethylenes, polyvinylchlorides, and polypropylenes have gained popularity. By far the largest-volume application is that of polystyrene foam for use in meat and produce trays, egg cartons, snack food and sandwich containers, and carry-out trays. There are many other useful applications of foamed sheets, but only the ones that pertain to thermoforming will be covered here.

Extruding foamed thermoplastic materials in sheet form for the purpose of thermoforming is not a simple task, especially when continuous uniformity and flawless specifications are desired. Thermoplastics are foamed in a complex extrusion system wherein various blowing-agent materials are introduced into the melt (melted thermoplastic). Direct injection of blowing agents into the extruder is the most widespread and economical method of producing low-density extruded thermoplastic foams. These foams usually have a density range of 3 to 10 lb/ft^3 (0.05 to 0.15 specific gravity). In the implementation of direct injection of blowing agents into the extruder, a liquid blowing agent is injected that will boil at the extrusion temperature, and the blowing agent is thoroughly mixed with the melt. At the point of injection of the blowing agent into the extruder barrel a check-valve system is installed which only permits the premetered blowing agent to enter into the melt, restricting the backflow of melt into the metering system. To achieve uniform, fine-cell structures in the foam, nucleating agents are mixed with the thermoplastic resins. Talcum powder or sodium bicarbonates with citric acids are popular nucleating agents. Pentane or fluorocarbons [e.g., Freon (a tradename of Du Pont)] are commonly used blowing agents for polystyrene and polyethylene foam making.

The extrusion rates in many polystyrene and polyethylene thermoplastic foam sheets are in the range 300 to 1000 lb/hr. Output rates depend primarily on the type of foam sheet and its quality, in addition to the size and type of extrusion equipment.

To extrude foamed thermoplastic sheets, an extra-long single-screw extruder or twin-screw foam extruders can be used. However, such extruders are not used much in the United States, due to their lower output rates and the limited availability of such equipment. The most widely used system is the tandem foam extrusion system, consisting of a primary extruder connected to a secondary larger extruder. The two extruders can be aligned in parallel (Figure 11), or to save space, could form a right angle or even be front to back. The duties of the primary extruder are to liquefy the thermoplastic resins, mixing and blending them with the nucleating and blowing agents. The proper blends of these chemicals, in a melt form, extruded through a screen filter system (which removes all contaminants), are forced into the secondary extruder. The secondary extruder receives the highly liquefied melt at feed pres-

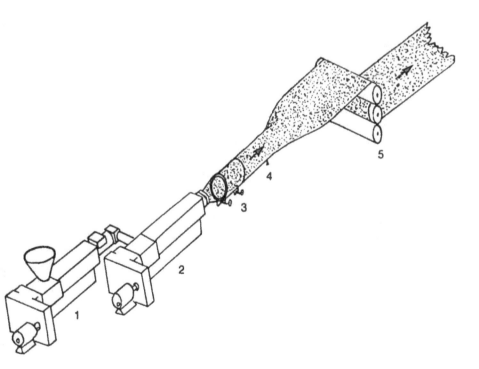

Figure 11   Foam extruding line: (1) primary extruder; (2) secondary extruder; (3) sizing drum; (4) slitter; (5) S-shaped wrap drive.

sures usually in the range 3000 to 4000 psi. The secondary extruder's main duty is to cool the melt and reduce its pressure. This secondary cooling extruder is usually larger than the primary plasticating extruder. The most widely used foam extrusion system consists of a $4^1/_2$-in. primary extruder with a 6-in. secondary extruder barrel size. The secondary extruder has a longer barrel and is equipped with multiple-zoned cooling jackets which provide the absorbing cooling. The extruder barrel provides precise temperature control to the melt. When the melt exits the die to the atmosphere is very important in achieving uniformity of the foam rise and cell formation. Foam sheets are almost always extruded through a circular die, and such dies are designed with unique features. Some design features may be proprietary. Normally, the exiting foam achieves a 1:4 blowup ratio from the die-lip opening to the extruded foam sheet. The extruded tubular foam is then stretched over a cooled sizing "drum" (often called a "can"), which in addition to producing further cooling and sizing of the foam sheet, performs an orientation procedure. When proper can and die ratios (4:1) are used, combined with ideal sheet pull and close temperature control, an outstanding biaxial orientation can be set into the foam sheet. In addition to ideal conditions when the melt is exiting, temperatures are closely controlled at the lowest possible temperature. Actual cell walls can be made with exceptional molecular orientation. When sheet orientation and cell wall orientation are achieved, the foam sheet can display outstanding structural strength and flexibility without the usual foam fracture and collapse. When internal or external air-cooling rings are employed very close to the die exit, providing chilling to the sheet surfaces, a smooth skin can be produced on the foam surfaces. Such a skin increases the thermoformability, adds to the appearance, and provides a better printing surface for the foam sheet. The tubular-shaped foam is usually slit at one or two locations after the sizing drum, flattened into sheet form, then wound onto rolls. Such a direct-injection foam extrusion system is shown in Figure 11. The direct-injection foam manufacturing process is ideal only for continuous manufacturing, due to the production line's lengthy startup period and the stabilization of production conditions. As the extruder is running and specifications are altered, unwanted foam sheet materials are produced. Startup time and tuning of a foam extrusion line may take from 1 to 6 hr during which time foam sheet material is being produced, which may add up to several thousand pounds. Such unwanted, "out of spec" materials are reprocessed and recycled into the products, which must be able to absorb this reclaimed material.

The extruded foam sheets in roll form have to be stored at least 24 hr after extrusion and subsequently aged. This procedure is needed to allow the blowing agents to migrate and evaporate. Unless sufficient aging time is provided, problems will develop when the foam sheets are thermoformed. The sheets usually do not perform well, display a slightly "gummy" (very soft) texture, and the products show lack of detail. Full aging of the foam sheets takes longer when temperatures are lower (e.g., in winter) or when outdoor aging (i.e., in rain) is used. A rainwater blanket and cold temperature minimize gas evaporation.

In addition to direct injection of blowing agents into an extrusion system, there are other ways to produce thermoplastic foam sheets for thermoforming. However, none of these are as popular as the direct injection process because they may have less favorable economics. The use of chemical blowing agents produces a medium- to higher-density foam, 20 to 50 ft^3/lb (0.3 to 0.8 specific gravity). In this technique a dry powder chemical blowing agent which is blended with the thermoplastic resins is generally used. The blowing agents decompose and gasify at the resin melting temperature, thereby producing the intended foaming structures in the plastic.

The other method of producing foamed thermoplastic sheets involves using pentane-impregnated polystyrene beads (such materials are used for expendable polystyrene moldings). This material can be extruded into a foamed sheet form using conventional extruding equipment. However, the economics are not very favorable for this type of foam sheet manufacturing, due to the slow output rate plus the shorter-storage-condition limitations of pentane impregnations.

It should be pointed out, that most foamed thermoplastic sheet manufacturers are no longer allowed to use, and therefore have eliminated CFC (chlorofluorocarbon) formulations as their blowing agent (for example, blowing agent freon 12 dichlorodifluoromethane, CCl_2F_2). This was based on the 1987 International Agreement of the Montreal Protocol, at which time 47 nations agreed that CFC substances deplete the atmospheric ozone layer; therefore its manufacture and use should be curtailed. This agreement established a timetable to reduce production levels and applications of this materials. These limits were further tightened again after the 1990 meeting and it has been decided that they will be totally phased out by the year of 2000. The latest reports coming out of the National Oceanic and Atmospheric Administration in Boulder, Colorado, are that the Montreal Protocol was working well and if all plans are followed we can expect the hole in the ozone layer to disappear by 2050.

For the time being, and due to the long timetable for the development of a viable substitute, and until other blowing agents are developed or invented, HCFC (hydrochlorofluorocarbon) formulations are allowed to be used. The HCFC formulations have been found chemically far less stable than CFCs as they tend to breakdown before reaching the ozone layer and cause further depletion. Other blowing agent choices are still available. However, they may not offer the same properties as the above discussed blowing agents offer. Such materials are liquified pentane and butane gases. The drawback of these materials are, number one, that they are extremely flammable. The storage, use, and application of these materials require special precautions. Plant environment, equipment, and even the surrounding area must be made sparkproof, including the finished foam material aging and storage area.

Secondly, in addition to this fire hazard problem these chemicals are hydrocarbon formulations, that can and will contribute to air pollution. When emitted to the atmosphere it will produce smog, when it is converted by sunlight. Emission levels are regulated by the EPA (Environmental Protection Agency) and by many local Air Quality Resources Boards.

In the long run we still need to search for other and hopefully better substitutes for blowing agents.

C. Types of Thermoplastic Sheets Used in Thermoforming

The thermoplastic sheets produced for the thermoforming process and made by all the ways discussed previously, can be produced with variations in addition to those involved in the basic manufacturing techniques. Some sheet raw material can come directly from polymerization plants, or various additives can be blended into the raw materials. Additives can change the color, enhance material characteristics, reduce raw material need, or increase product life expectancy. The use of additives or surface treatments or even a mere change in color could affect the thermoformability of a specific sheet stock. We discuss next the interaction of the following thermoplastic sheets with the thermoforming process: natural, oriented, tinted, pigmented, filled, foamed, textured, laminated, coextruded, preprinted, flocked, metallized, and specialty thermoplastic sheets. These materials are available from scores of sheet manufacturers in a multitude of variations and combinations. The materials should be looked at individually, and their advantages and disadvantages should be known prior to attempting thermoforming.

1. Natural Thermoplastic Sheets

This type of thermoplastic sheet is made of resins without additives or fillers following the polymerization process. As to appearance in both resin and sheet form, the plastics are crystal clear or somewhat translucent. The sheets are easy to thermoform, and both types have outstanding flow characteristics. The only slight drawback is that when they are exposed to infrared heat systems, some of the heat energy will travel through the material. However, the use of heat reflectors or dual sandwich heaters totally eliminates this, their only disadvantage.

2. Oriented Thermoplastic Sheets

Oriented thermoplastic sheets are one of the most sensitive sheet materials to handle in the heating and forming cycles of the thermoforming process. As discussed earlier, loss of orientation is very close to the softening temperature of the thermoplastic sheets, so precise and fast heating and forming cycles must be maintained. When tinting or pigmentation is added, actual cycle times are altered, but preciseness of the cycle time controls must still be maintained very accurately. Evenly balanced biaxial orientation is the ideal condition. However, fluctuations of unevenness are often encountered, which means that the orientation stretch in the machine direction does not match that in the transverse direction. When such an unevenly balanced biaxially oriented thermoplastic sheet is exposed to heat, the deorientation forces will distort the sheet. The levels of unbalanced orientation and its limits of usefulness should be judged differently with each product type. Some products are less vulnerable than others with regard to using out-of-balance thermoplastic sheets. Usually, the deeper drawn, more intricately designed parts demand a better-balanced orientation than do shallow, simple thermoformed parts. There are various ways to check for imbalanced orientation and to establish imbalanced biaxial orientation limits for particular products. Imbalance testing procedures are discussed in Chapter 6.

3. Tinted Thermoplastic Sheets

To obtain colored transparent or translucent thermoplastic sheets for thermoforming, tints or dyes are usually added to the natural clear or translucent resin materials. This method of coloring has minimal effects on the plastic and provides a slight improvement in the heating cycle of thermoforming. With the addition of higher quantities of dye material, there is an increase in the intensity of the thermoplastic sheet color. However, this color addition only can be achieved up to a point: when dye saturations are at maximum intensity and any additional quantities of dye loading would be a waste of dye and money. When tinting has reached its maximum limits, the thermoplastic sheet would appear to be almost black.

4. Pigmented Thermoplastic Sheets

When colored products are desired, pigmentation or coloring concentrates are usually added in the sheet manufacturing process. The pigmentation is thoroughly blended with the base thermoplastics so that coloring is uniform throughout the mass. Pigments usually do not change the physical characteristics of the thermoplastic, except for its responsiveness to heating with radiant heat energy. It is easy to recognize that black plastics are outstandingly absorbent to radiant heat energy and that white thermoplastic sheets, due to their high reflectiveness, are less absorbent to radiant heat energy. For example, when pigments are chosen for a whitish color, one would find a talc (which gives a slightly off-white color) more heat absorbent than the white pigments of titanium dioxide (which give a snow-white color). There is a definite difference in heat absorbancy between green and blue. Greens will generally warm up faster than will blue thermoplastic. The sheet's heat reflectiveness or absorptiveness directly affects the time needed to heat a colored thermoplastic sheet. This time increase or decrease is part of the actual cycle time. The overall results are that in a continuous high-speed operation a color change in a product could increase or decrease the actual cycle times of the production line. Changes that occur when colors are changed could affect initial production-cost estimations and cause profitability changes. Naturally, with smaller-volume production the color changes and alteration of production cycle times are insignificant.

5. Filled Thermoplastic Sheets

Aside from additions for coloring purposes, materials are added to thermoplastic resins chiefly to alter their physical abilities or to load the resin with filler materials to reduce costs. The sole purpose of such filler materials, often called "resin extenders," is to make smaller quantities of pure, higher-priced thermoplastic resins, thus producing more units of product for less. Some coloring pigments have this double duty: to add their own color to the plastic and to provide body filling to the resin. Additional duties of fillers are to enhance the plastic's physical qualities, add strength, improve rigidity, offer more impact resistance, reduce heat distortion, provide extra fracture resistance, and many others.

The use of filler materials is beneficial for most of the foregoing reasons; however, the same fillers could also affect the behavior of the thermoplastic sheet in the thermoforming process. Some filler materials, such as talc, wood, and walnut shell powders, act as heat-insulation or heat-absorption-resistant materials. Such heat resistants could significantly slow the heating cycle of the thermoforming process. However, the poor heat conductivity will also affect the cooling cycle: for example, when the formed parts resist giving up their acquired heat and set up for part removal from the mold. This poor heat conductivity overall demands slower thermoforming cycles. On the other hand, some fillers which may be used for other purposes improve heat conductivity and heat absorption. Materials such as glass fibers or graphite powders of fibers act as heat-conducting channels within the plastic. Naturally, the surface colors of those filled plastic sheets would further affect the heat cycle of the thermoplastic sheet. In the scope of the thermoforming process, whenever fillers are used for physical improvement, actual alterations in cycle times can be expected. Such a change may be slight, depending on the filler materials' makeup.

6. Foamed Thermoplastic Sheets

Foaming of thermoplastic materials is a very common practice, and many of the popular thermoplastics are produced in a foamed sheet form which has been adapted to the thermoforming process. Foaming is done for various reasons: to produce a soft, cushiony material, to gain heat-insulating qualities, or simply for resin-extending purposes. When thermoplastics are foamed, depending on the foam density, a reduced quantity of thermoplastic resin is needed to create sheet material for thermoforming.

Thermoplastic foam sheets come in pure white which does not require pigmentation to achieve the white coloring. The foaming structure itself provides the color. When colored foams are desired, only light coloring can be achieved. When a darkly colored thermoplastic resin is foamed, the coloring always tends to be reduced to a pastel. Adding more coloring to the resin would not darken the foam further, just add unnecessary cost.

There are flexible foam sheets and rather rigid foam sheets made of different thermoplastics resins for different purposes. The cell structures and sizes can be predetermined in the foam extrusion process. Coarse-cell foam versus fine-cell foam or anything in between the two extremes displays unique thermoforming characteristics which have to be identified and respected throughout the thermoforming process. The cell structure of a foam sheet for thermoforming must be one of closed cell structures and should not contain more than 10% open cells. In open-cell foam or excessive open-cell-count foam, the cells have a tendency to collapse when exposed to thermoforming heat cycles. Usually, the foams that have an open-cell count of higher than 10% do not have proper heat-expansion ratios, which results in loss of detail at the thermoforming operation or a product of reduced strength. To overcome their natural insulating qualities, foam sheets are usually heated from both sides and require a greater degree of precision in handling.

In addition to being used as single sheets, foam sheets are often laminated with other materials. These laminations may have solid thermoplastic layers made of the same resins as those of which the foam is made, or could be produced of different thermoplastic resin. Other types of materials have also found popularity in laminations with foam: for example, various types of woven cloth materials. When such materials are combined with a foam layer and subjected to the thermoforming process, extra care must be taken and limitations on the depth of draw must be tied in with the limits of stretchability of the woven materials.

Foam sheets often display a density difference within the same sheet, especially in the direction of extrusion (MD). This difference could be very slight or more severe, often resembling fine wood grain. When this graininess is minimal it should not affect the thermoforming process. However, when the graininess is more pronounced, it may affect the appearance and weaken the overall strength of the thermoformed products. When grain formation is apparent, one would find maximum strength across the grain and weaker strength parallel to the grain. To obtain maximum strength in rectangular products, it is best to locate the longer sides of the product parallel to the grain. Products can be made across the grain direction (TD) but will display less strength and have a tendency to distort and bow.

7. Textured Thermoplastic Sheets

Most of the solid thermoplastic sheet materials discussed above—natural, colored, pigmented, and filled—can be produced with a textured or predetermined patterned surface. The textured finish is usually added to the thermoplastic sheet when the sheet is produced. In the various sheet-producing techniques, when the thermoplastic sheet is made, the sheet form is directed through a series of cooling and sizing rollers. When these have a patterned surface, they transfer a reversed impression to the hot thermoplastic sheet surface. The surface patterns are either engraved or electrolytically etched into the rollers. The pattern designs can resemble leather grain finishes or various types of repeated geometric designs. The textured finish is usually applied to one side of the thermoplastic sheet. However, if desired, both sides can be textured and identical or different texture patterns can be applied. Texturing sheet surfaces is a cost-effective way to produce textured finished products, completely eliminating costly textured tooling. Texturing does not affect the thermoforming process, but the thermoforming process slightly affects the texturing of the sheet. When using a textured thermoplastic sheet, the only precaution is that a small, tight radius must be avoided. When a textured plastic sheet is formed over a small radius, the forming forces tend to stretch out the textured pattern, giving it a slightly worn or polished appearance. When a textured sheet is formed into a cavity of small radius, it tends to gather or close the pattern and give a different appearance to the radiused portion. If small radii are avoided, excellent products can be produced (e.g., automobile dashboards, luggage, briefcases, etc.).

8. Laminated Thermoplastic Sheets

There are many types of laminated thermoplastic sheets in use. Most of the laminations are made for the purpose of reducing the cost of sheet materials. Naturally, when using a laminated material that contains a highly desirable surface laminate together with a less costly subbase, the product would cost less overall than if the entire sheet were made of the higher-priced material. Surface-laminated thermoplastic materials generally offer better and longer terms of weather resistance or may have improved chemical resistance or better physical properties. Laminated materials are judged mostly by their degree of continuous adherence to each other. If delamination takes effect prior to or after thermoforming, the products are useless and must be discarded. The lamination adhesion must withstand the heat cycles and forming forces in the thermoforming process. Early delamination could present a problem when small volumes of air become trapped between the layers and cause further delamination, or in extreme cases, bulging between layers. Laminated materials will be used more and more, and laminates combined with thermoplastic foam sheets are the materials of the future.

9. Coextruded Thermoplastic Sheets

The principal benefits of this type of sheet material are its fast manufacture and material economics. In the coextrusion process, an individual sheet a fraction of a mil in thickness can be laid on top of a heavier base layer. This very thin coextruded layer provides the necessary barrier qualities without causing heavy interference in the scrap-recycling portions of the thermoforming process. Coextruded thermoplastic sheet may come in the simple form of two layers: a base layer with one side having a thinner coextruded coating. The most popular coextruded thermoplastic materials are used to make weather-resistant (ultraviolet-ray protected) products or a mar- and scratch-resistant surface coating for products subjected to repeated handling. Heavier thicknesses of coextrusions are also made, and their popularity is strongest in the foam coextrusions. There are multilayer coextruded sheet materials in which three or more layers are coextruded into single-sheet form. Coextrusion can make a sandwich-like sheet where a base sheet is captured between the two coextruded layers. Coextrusions can also be made in which each individual layer is a different thermoplastic material and each has a specific purpose and barrier quality. There are also multilayer coextruded sheets that contain up to five or more layers of coextrusions, each layer providing a specialized barrier quality or bulk strength or surface coating. Such multilayered coextruded materials are needed for food packaging and for longer-shelf-life preservation of foods. A multitude of coextruded materials and sheet combinations are used and "secretized." When multilayered coextruded materials are thermoformed in the desired shapes, the original coextruded sheet material is stretched into a shape, causing the coextruded layers to be stretched and thus the layers to thin. The barrier qualities of these layers are as good as the thinnest portion of the particular layer can provide. When during thermoforming, the thinning of a particular layer drops below the needed barrier qualities and no longer provides the requisite protection, heavier-layered coextrusion is demanded. When a coextruded sheet is produced, it is not easy to establish layer thicknesses after sheet production, and it is an even greater task to establish layer thicknesses after thermoforming.

It is safe to state that coextruded materials presently constitute the most highly demanded sheet materials for thermoforming, and the demand for these materials and subsequent progressive research will multiply future demand. Coextruded materials are the materials of the future.

10. Printed Thermoplastic Sheets

All the thermoplastic sheet materials discussed previously can be printed. The printing in the thermoforming process should be judged as a surface coating. The primary purpose of printing sheet stock is decorative. However, printing has also been utilized for product identification, labeling, or simply product finishing. Printing can be done using a continuous printing method which applies a printed surface onto a roll sheet material. In this case rolls are unwound, printed, dried, and rewound back into roll form. The printing design is either a continuous pattern, solidly covers the surface, or appears as a random pattern in which the images are placed randomly on the thermoplastic sheet surface and are eventually repeated. For more precise jobs, printing is done in an exact repeated pattern, the location of each image being in a fixed, predetermined location. The same printing processes can also be applied with thermoplastics in sheet form, and such a batch printing process is usually favored for smaller production runs.

When used for thermoforming, preprinted thermoplastic sheets usually affect only two areas of the thermoforming process. First, the color of the printing inks will change the heat cycle and may increase or decrease the heat absorbancy of the thermoplastic sheet material. Second, the color of the printing ink definitely affects recycling of the scrap materials. To achieve printing legibility, printing ink colors are usually intensified, or entirely different colors are used as the base color of the sheet material. When such different and intense color prints are applied and the scrap is recycled, a change in color can be expected. To avoid excessive color alterations or undesired scrap, a preplanned thermoplastic sheet color and printing-ink color must be used. Often, the color of the base sheet material is unimportant, because it is not visible in the finished product; only the printed surface is exposed. Therefore, when recycling, color differences between printing ink and base sheet are ignored. For example, when inner wall panels for aircraft or buses are produced, the top surfaces are printed with protective vinyl paint which has a continuous print pattern (resembling that of fine-quality wallpaper). The base sheet color is not exposed for viewing and therefore can be made out of thermoplastic material of any color. On the other hand, when wood-grain imitations are made, the printer starts with a beige or light brown base material and continuously prints it with a dark brown or black wood-grain pattern. To finish the illusion of grain, a woodlike texturing is put on the surface. Using this type of a preprinted thermoplastic sheet, products can be thermoformed that resemble fine wood products, such as picture frames, plaques, displays, or simple wooden trays.

Preprinted sheets are also produced for labeling purposes, where the preprinted sheets are located exactly on specific thermoforming areas. Special indexing marks are printed on the sheet and an indexing mechanism provides the exact location of the print to the matching location on the thermoforming tool. Often, the artwork of the printing plate is distorted purposely to accommodate the distortion caused by the thermoforming process. This application usually happens when the printing is made in a flat form but is subsequently located on a curved, stretched, or reshaped part of a thermoformed product. The precise location and its repeatability provide a constant challenge to the thermoformer, as high quality and continuous uniformity of the thermoplastic sheet are always needed. Quality control and constant rechecking of sheets, plus the maintenance of reduced scrap factors, keeps manufacturers of this type of sheet permanently on their toes.

11. Flocked Thermoplastic Sheets

Quite often when higher-priced gift items such as writing instruments, watches, and jewelry are packaged, the boxes have well-contoured fitted box liners. These are made with a velvetlike finish and are produced by thermoforming. Such gift box inserts are made of solid thermoplastic sheets with a flock coating. The flock coating is applied to the sheet in a continuous process, the web being unwound and rewound after the flocking. Flock coating can also be applied to individual sheets, but this process is slow and uneconomical.

Flock coating is usually applied to thermoplastic sheets with a thin layer of plastisol adhesive. The flock itself can be made out of various fiber materials, including acrylic, nylon, polyester, polypropylene, and natural fibers. The flock materials come in rich colors and have either a sheen or a dull finish, whichever is desired. The flock fibers are cut into equal lengths and supplied in bagged packaging. Flocking of the thermoplastic sheet is a somewhat delicate process. First the adhesive is applied to the sheet in a precise thin layer; then the flock hopper spreads a thin layer of flock material directly on top of the adhesive. Excess flock distribution is controlled by the applicator, also by rubberized wiper blades. The thermoplastic sheet can also be threaded through a mechanical vibrating surface, which further enhances flock distribution and flock orientation. The next step is the application of an electrostatic charge, which forces the flock material to orient and stand upright. After the mechanical and electrostatical process, the flocking shows a uniform velvety appearance throughout the thermoplastic sheet surface. The final step in the flocking process is to dry and cure the adhesive that holds the flock fibers. The excess fibers are wiped and vacuumed off and recycled back to the feed hopper.

It is strongly advisable to choose the same color for the thermoplastic sheet and the flock because this will help to hide slight discrepancies in the flock coating or in the flocked material when it is stretched, after thermoforming. Flocked surfaces can be applied to both sides of a thermoplastic sheet; however, single-sided flocking is most popular. The flocked surfaces create a velvety finish, and with a high-sheen flock material actually create a very elegant finished appearance.

In the thermoforming process, the flocked material has to be heated with care. Melting and burning of the flock can easily be encountered when too-rapid heat cycles are applied. The scrap materials produced when flocked thermoplastics are thermoformed should be discarded, and such material disposal should be recognized when the cost factors of this type of sheet manufacture are calculated.

In the thermoforming process excessively deep-drawn parts and small-radius curves should be avoided because such elaborate designs will stretch the thermoplastic base sheet above the limits of the flock coating.

12. Metallized Thermoplastic Sheets

Metallized (shiny mirrorlike finish) thermoplastic sheets can also be thermoformed; however, their popularity is limited. Trays formed of metallized thermoplastic sheets are found primarily in Europe to package high-priced candy. They are also used in the photographic field in the flash reflector backing in flash cubes and flash bars. These products can be produced of premetallized material or made by thermoforming followed by metallization. The second procedure ensures higher-quality metallization.

The thermoplastic sheet is usually metallized by a vacuum process in a batch or continuous system. Traditionally in vacuum metallization, a metal vapor, which is primarily aluminum, is applied to the surface of a thermoplastic sheet. The process may require an intermediate coating of some type, which permits better adhesion of metal particles to the base sheet. It is also possible that a second cover coating may be used to eliminate metal particle flaking or chemical deterioration and discoloration. When metallized thermoplastic sheets are produced for thermoforming, a clear thermoplastic resin material, such as polystyrene, polypropylene, or polyester is generally chosen. When metallized thermoplastics are used in the thermoforming process, a slightly longer heating cycle is usually needed, due to the sheet's reflectiveness to radiant heat energy. These types of sheets are also vulnerable to thinning of the metallized coating or even cracking of the coating when subjected to excessive stretching in the forming process. Deep-drawn parts should be avoided, as this will minimize such unwanted effects. When thermoforming a metallized sheet, thinning of the metallized layers can be expected, just as material thickness is reduced at the deepest parts of the product. Thinning of the metallized surface after thermoforming would provide a see-through characteristic as the metal particles are pulled apart. The lessening of this metallized effect can generally be ignored, as the effect shows most in the bottom area of each cavity, and the packaged product usually covers these areas. Only after the product has been removed can one notice the lessening of metallization.

Continuous printing with metallized printing ink can simulate a shiny metal-like surface. The printed surface produces the same reduction in quality after thermoforming as that of the metallized sheet, and never duplicates the mirrorlike appearance of the metallized sheet. It is possible to introduce electrical conductivity to the metallized sheet after thermoforming, but this process is not presently employed. However, it could become important in the future.

13. Specialty Thermoplastic Sheets

All the special and unique thermoplastic sheet materials are lumped together in this category. Some of the materials have been in existence for many years and their production is intended for purposes other than thermoforming. Some of the other thermoplastic sheet materials are presently under investigation for thermoforming purposes, but they have not been commercialized. Some of the thermoplastic sheet materials that hold future promise are still in the developmental stages. These materials, which require either specialized thermoforming equipment or specialized thermoforming techniques, offer unique advantages which cannot be obtained with any other materials. Thermoplastic sheet materials that have a double-walled construction are made on a sheet extruder with a profile extrusion-like die. The sheet has a multitude of flutelike ties over the length of the sheet. These tie-patterned sheets are available in 2-mm, 3-mm, 4-mm, and up to 1-in. or larger formats. Such sheet constructions are thermoformable, but to perform the thermoforming process and avoid collapse of the double-wall construction, an airtight clamping mechanism must be implemented. This special airtight feature will retain the inner air space and not permit the double-walled construction to be destroyed by the heat and softening of the thermoplastic resin.

Another unique material currently under investigation is an aluminum foil laminated together with a thermoplastic sheet which is thermoformable. The foil somewhat limits the depth of draw on this type of material, to the limits of the aluminum foil. However, this material combines the benefits of the foil and the adhering thermoplastic.

There are other thermoplastic sheet materials which are easily thermoformable, such as crystallizable polyethylene teraphthalate (PET). With such materials, just after the forming process, instead of entering the cooling cycle, the formed parts are heated further. This extra heating process changes the molecular structure and crystallizes the plastic and the thermoplastic becomes heat resistant (350 to 400°F). This specialized PET type of thermoplastic is adaptable for use in heatable and reheatable food containers, replacing metal or glass.

There are other thermoplastic sheet materials which are made to resist heat distortion, such as styrene-maleic anhydride and the sulfone polymers. Thermoplastic elastomers allow the production of articles simulating vulcanized rubber, but thermoformable.

Many others types of materials are under investigation or development and will become available in the near future. Variations and combinations, and newly developed materials, will be introduced at a quickening pace. As we process and handle sheet materials in the thermoforming process (such materials did not exist five years ago), we can be sure that we will soon work with materials that are presently unknown.

We have now discussed all the presently used and known thermoformable thermoplastic sheets, including the particular sheet manufacturing methods that have been adopted to produce sheet forms. Most of the information and data have been presented from the specific viewpoint of a thermoformer, providing information as to how the sheet material has been made, what to look for, and what types of sensitivities can be expected. In addition to their distinctive differences, the various thermoplastic sheet materials may have unexpected or unknown reactions. Such reactions occur in the thermoforming process or in the fabrication end of the process, in product containment, or in the final end use. A thorough investigation and testing procedure must be implemented to ensure safety, proper handling, and proper use of the intended product.

All of the thermoplastic sheet materials produced and used in the thermoforming process can be categorized into two major groups: thermoforming on "sheet-fed" machines, in which the sheet material is produced and used in precut panels; and thermoforming on continuous "roll-fed" machines, in which the sheet material is produced and used from continuous rolls.

II. CLAMPING MECHANISMS

To carry out a successful thermoforming process, all thermoplastic sheet materials must be held securely on all four sides. This four-sided grip must be maintained throughout all phases of the thermoforming process. The securing of the sheet materials is equally necessary for thin and flexible sheets and for heavy, more rigid materials. All thermoplastic sheet materials have a tendency to soften and lose their self-supportiveness. They also distort and may sag when exposed to the heat cycle. When heated, the thermoplastic sheet will react and yield to natural thermal expansion, loss of crystallinity, and relief of molecular orientation, all of which tend to cause sheet movement and sheet distortion. In the heat cycle, when the thermoplastic sheet is not fully supported by its four sides, the unheld and unsupported side will usually distort and will most often tend to pull away from the clamping area. This causes a shift in material thickness and material gathering, with thickening and gathering that concentrates toward the center of the sheet. In the most severe cases, the sheet material shift could be so extensive that the sheet will no longer cover the entire mold area, and will therefore produce an incomplete product.

In the remainder of the thermoforming cycles, the clamping of the sheet will provide the necessary sheet transport, which will then provide movement of the sheet through the various process cycles. This movement consists of (1) transporting the heated sheet to the molding area, (2) aiding in the forming process by helping in the prestretch or stretching of the clamped sheet over the mold, (3) firmly holding the sheet in the cooling cycle, (4) implementing product removal (stripping) from the mold, and (5) in the last steps of the thermoforming process transferring (and in some cases helping to locate) the formed panel into the trimming operation. Proper clamping which is positive on all four sides is a "must" in the thermoforming operation and its repeatability ensures a successful process. There are various types of clamping apparatus, and all of them are judged to be satisfactory as long as their sheet-holding capabilities are maintained throughout the process. There are different types of clamping or sheet-holding and sheet-transporting mechanisms used. Such variations depend on the type of thermoforming operation and the type of machinery used. In the thermoforming process using sheet-fed machines, the clamping mechanisms used must be capable of clamping and handling individual panels. On the other hand, continuous roll-fed machines require continuous chain-type sheet transportation and sheet-holding mechanisms.

A. Clamp-Frame Mechanisms

There are many variations of the basic clamp-frame mechanism. All the clamp frames follow the basic frame appearance and have a matching counterframe. Usually, one frame is made stationary while the other tends to open with hinges on the back side (Figure 12). The thermoplastic sheet is clamped between the two frames. In the clamping action the two frames are closed (with the sheet in between) together and capture the precut sheet on all four sides. The clamp frames may hold the sheet just by using the pressures of the two opposing frames, which will provide enough surface friction. This method usually works well with the majority of thermoplastic sheets. However, some of the sheet could have very smooth surfaces, or be made of elastic materials, or too much pulling force is obtained and the frictional clamping is not enough. If that is the case, stronger gripping methods and clamp-frame designs have to be implemented. One improvement in clamp-frame construction involves riveted pin placement into one or both clamp frames with matching relief holes. Some

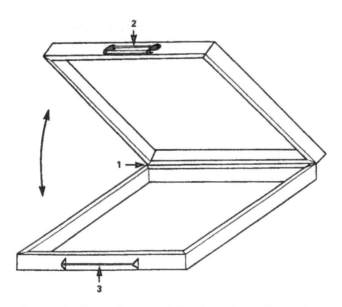

Figure 12   Clamp frame (with flat clamp-frame lip): (1) hinge; (2) clamp latch; (3) clamp lock bar.

clamp-frames have machined-in grooves on one clamp frame while the other clamp frame has matching located protrusions. This type of clamp frame tends to pinch the thermoplastic sheet between the two interacting designed frames, permitting no slip and no movement of the sheet. Some unique and cleverly designed sheet-capturing clamp frames have patented features that tend to be a "sales tool" for the particular machinery manufacturer. Clamp-frame sizes range from 12 in. by 12 in. to as large as 20 ft by 30 ft. Sizes are dependent on the products the thermoformer intends to make and the size of equipment being used, although a larger machine with a larger clamp frame may have specially designed clamp-frame reducers. These clamp-frame reducers may operate as independent frame unit inserts or could be made of snap-in adapters. The purpose of these clamp-frame size reducers is to allow a smaller sheet format to be used with the larger thermoforming equipment, therefore not demanding the thermoformer to produce excessive scrap around the product.

Thermoplastic sheet clamping can be implemented with a manually operated latch, or in more automated operations, with pneumatically or hydraulically actuated clamping mechanisms. In the simplest clamp-frame construction, the following steps are made to secure the thermoplastic sheet between the clamps:

1. The clamp frame is opened manually.
2. The precut thermoplastic is placed between the two hinged counter-members or clamp frame.
3. The clamp frame is closed.
4. A center latch or several latches are closed by hand to secure the sheet clamping.
5. The thermoforming procedures are executed.
6. The clamp frame is opened manually and the thermoformed sheet with the formed part is removed.
7. The cycle is repeated.

In the more automated systems a push of a button would actuate the sheet-clamping mechanisms, which are powered either by pressurized air or hydraulic forces. When larger clamp frames are used, a series of clamp-latching mechanisms are placed around the sides of the clamp frames. This automated clamp-frame latching device has to be designed and implemented such that the mechanisms are not made bulky and cumbersome and so interfering with sheet placement and the removal of formed parts. The first of these two criteria is most important because locating flexible, somewhat flimsy sheet stock between the clamp frame with all the latching mechanisms is not an easy task. In the case of extremely large panels, or in the more sophisticated automated thermoforming operations, sheet placement transports and product removal automation are attached to the thermoforming machinery. The supportive equipment is designed with total interaction with the clamp-frame mechanism, and only a push of the button is required to actuate the equipment. The sheet-loading and thermoformed-part-removal tasks are usually implemented with motorized moving-belt systems or with a series of suction cups. This mechanization can be made so elaborate that the thermoforming equipment becomes fully automated. In sheet-fed thermoforming, a complete function, from sheet loading to thermoformed-part removal, consists of a cycle, and when such cycles are made on a recurrent basis, a production run is established.

B. Transport-Chain Mechanisms

When a sheet-fed machine is used in a continuous manner, a transport-chain system performs the duties of a clamp frame. The roll-stock material is unwound and carried through a partial or complete thermoforming program by this transport chain. The exact task of this transport-chain system entails capturing the two sides of the thermoplastic sheet and advancing the sheet through the various cycles of thermoforming. However, the sheet advance is a continuously forward movement, although sheet advancement must match the reciprocating action of the molding cycle. With this transport-chain system, the unwound sheet material is carried into the oven and stopped and moved two or more times before advancing into the mold area. After molding, the transport-chain system carries the formed sheet to the rest of the operation. On some types of machinery, the sheet will travel out of the thermoformer; it then relocates into any of the popular secondary trimming operations.

The transport-chain system in the thermoforming equipment functions as the two parallel sides of the clamp frame. For front and back ends clamping, the continuity of the web will act as the clamp; otherwise, individual clamp bars have to be installed. The use of clamp bars is most necessary when intricate and deep-drawn parts are formed. The installation of clamp bars will probably prevent distortion and pull-in of the sheet. The pull-in or pull-back caused by the stretch forces (stretch forces caused by the molding procedure when the plastic sheet is formed into the desired shape). The pull-in or pull-back effect occurs throughout the continuity of the sheet. The stretching forces are extended far into the previously formed area, and also into the area that is heated and made ready for the next "forming shot." For positive elimination of this pull-in and pull-back sheet distortion, mechanical clamp bars have to be added, which are located close to the front and back of the mold. In the most ideal clamp-bar designs, the counter-poising clamp-bar edge surfaces should not simply be made smooth and flat but should be made with interrelated wedge-like grooves which would end in pinching and securely gripping the thermoplastic sheet between the two members.

In many instances the stretch forces are so extensive that the thermoplastic sheet could also be pulled or even ripped out of the transport chains. To alleviate the problem of pulling the sheet off the chain, the first step the operator can take is to increase the thermoplastic web width. The increased material strip between the chain rail and the mold will absorb the stretch forces and eliminate sheet pull-out. When the stretching forces are so great that the extra web width does not remedy the sheet pull-out, or other factors make the sheet width increase less desirable, a complete clamping mechanism has to be built around the mold. These clamping mechanisms would capture a specified portion of the web around a given mold. The capture of the sheet is made so precise that there is no movement of the sheet through these clamps. Actuation of the clamp mechanism must be implemented prior to the mold movement or just before the stretch forces come into effect. The clamping mechanisms can be actuated by springs or by pneumatic or hydraulic forces, and the only thing that has to be watched is that the actuation timing does not get delayed or "lazy." Spring-type actuators in continuous use have a tendency to become relaxed and somewhat compressed, therefore lacking crisp, positive clamping. Pneumatic and hydraulic clamp-frame actuators in continuous operation last much longer but still tend to get "lazy" and slow down. Some of the individual cylinders may move moments later or more slowly. Uneven speed will cause an unparallel movement and binding to the clamp frame or clamp bars. Instead of wearing itself in, the clamp-frame actuation get worse and less dependable in time. Those clamp-frame mechanisms and their functions must be maintained and monitored throughout production.

In deep-drawn multicavity molds, the stretch forces within the inside of the clamp frame can also create an irregular stretch with the sheet material. The stretch forces can shift material thickness somewhat uncontrollably from one area to the other area, causing material thickness variations. Such variations are usually not the same from cycle to cycle. To remedy such uncontrollable shifts in material thickness, the only option the thermoformer has is to create a grid-patterned clamp-frame mechanism which would clamp the individual areas of each to-be-formed part. With this grid clamping method, the individual areas of which the product will be formed is securely trapped, ensuring uniformity to all the individual forming and to the end product. When round or semiround parts are thermoformed, it is highly advisable that the clamping grid design follow the contours of the mold cavities. For example, a square clamp pattern (Figure 13) that provides equal material entrapment for

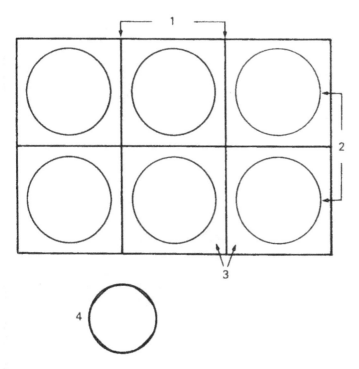

Figure 13 (1) Square clamping grid; (2) rounded mold cavities; (3) excess corner material to draw; (4) cross-sectional view of wall thickness.

individual round products does not provide even material access for round-product thermoforming. In round-shape thermoforming, more material will be drawn from the corner areas than from the side areas. This will result in a slightly thicker "riblike" access material on the four sides next to the corner areas. The only way to eliminate such a pattern on the sidewalls of the parts is to build a clamping grid that matches the contours of the thermoformed part at equal distances from the cavity edge (Figure 14). Naturally, such intricate clamping-grid designs are often difficult, if not impossible or very costly, to build.

The actual transport-chain system is available in two basic types. (1) The pressure transport-chain system has two opposing chain loops facing each other, creating a pinching pressure between the two chain loops. The pressure chain system is found mostly on older equipment and is chiefly adapted for use with foam-type materials. (2) The most popular transport-chain systems at present are the "pin-chain" sheet transports. These systems have pinlike

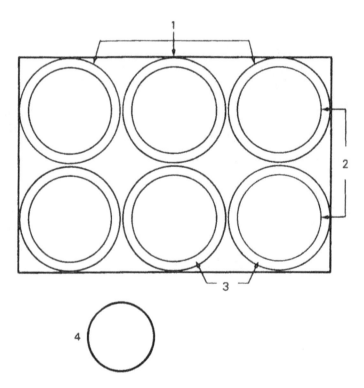

Figure 14   (1) Rounded clamping grid; (2) rounded mold cavities; (3) equal material space to draw; (4) cross-sectional view of even wall thickness.

protrusions which at continuous intervals penetrate the thermoplastic sheet. The pins are located in every other link or at several linkage intervals and could be so made that the pin is an actual part of the chain linkage, or a separate pin is riveted to a small flange of the chain linkage. The pins penetrate the thermoplastic sheet at the beginning of the process and with a series of repeated pins and penetration in equally spaced intervals will create a sheet-capturing clamp-frame-like system. The most popular pin-chain sheet transport system today is made of single-piece chain links and has the pin grounded out of the same chain linkage (Figure 15). This type of pin provides the least problem; it does not work loose at all. The pin points may lose their sharpness; however, they can be reground. In time, a chain-transport system will wear out and can be subjected to abuse to the point that the chain linkage will be stretched. Lubrication and proper tension adjustments will extend the life of these transport chains. When chains are worn and stretched, they should be replaced, together with new sprockets on both sides of the machine. A minute difference between sides can cause out-of-time sequencing and indexing of the transported sheet.

The pins of the transport-chain system may encounter difficulty in penetrating the thicker, more rigid thermoplastic materials. It is also often observed that after penetration, the pins have a tendency to push out a small piece of plastic. Such tiny chips coming loose will vibrate and dance on the top of the sheet surface and eventually will be carried into the molding process. The fine chips of plastic do not interfere with the thermoforming process; however, they will imprint themselves into the formed product. At first observation most thermoformers will suspect a previously damaged mold to be causing this dimpling on the thermoformed product surface. Close examination of the mold quickly shows that this is not so and that the culprit is obviously the pin pene-

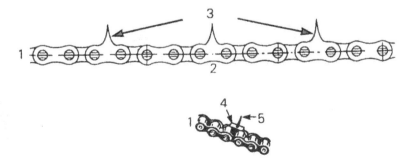

Figure 15 Sheet transport chains: (1) roller chain; (2) standard chain link; (3) one-piece pin link; (4) flanged link; (5) riveted pin.

tration. To remedy this situation, especially when the sheet temperatures or plant ambient temperatures are very low, it is necessary to install a strip heater close to the edge of the sheet. When the heaters are turned on, the problem is eliminated almost immediately.

In a continuous thermoforming operation, both clamp frame and transport chain will travel repeatedly through the heating ovens. Most of the heat absorbed by the sheet-holding systems is well dissipated into the frame and the rest of the machinery. If the cycles are rapidly repeated or the thermoplastic sheet requires longer exposures to the heat, the same heat levels that make the sheet soft and formable make the clamp frames or chain sheet transports too hot. The excessive heat buildup must be eliminated or removed because eventually this heat will melt the thermoplastic sheet or cause it to stick to the transport. Cooling rails or cooling channels with coolant fluids pumped through them are sometimes added to thermoforming machines to counteract this. Cooling rails are usually not attached to the sheet transport-chain mechanisms, but placed close to them, creating a heat-absorbing shield. The cooling channels are built into or attached right on the clamp frames, and liquid coolants are pumped through them, which will absorb and carry the excess heat away. Maintaining a relatively cold sheet clamping and sheet transport system is a necessary aspect of thermoforming. Maintaining cold sheet edges is needed for good sheet-holding qualities, and such a precaution will provide uniformity to the thermoforming cycles and product duplications. Firm, positive edge holding benefits all downstream operations, such as reindexing or relocating in the trimming or fabricating phases of thermoforming.

Sheet clamping and sheet transport are two of the most important elements of thermoforming operations; they cannot be neglected or ignored. Often, success slips away from the thermoforming processor when such systems are not maintained at their highest peak of operation, and a failing system rapidly weakens the entire thermoforming process.

III. HEATING SYSTEMS

The very name "thermoforming" indicates that this forming method is based on the use of heat. The materials used, called "thermoplastics," when exposed to heat at their various softening temperatures, yield to forming and shaping. The heat used for softening and forming thermoplastic sheet materials gives them considerably more material stretching ability than that of cold forming techniques.

The heat applied to this process could come from a number of heat sources. The exposure of the thermoplastic sheet to heat must be made precisely and uniformly to provide identical conditions from cycle to cycle. There are several types of heat sources, and various methods are used to apply heat to the thermoforming process. Some heating methods are general-purpose heating systems; others may only be adaptable for specific purposes and techniques.

All of the heating methods, using any of the existing heat sources, demand energy. It is well accepted in the thermoforming industry that the heating phase of the thermoforming process can consume up to 80% of total energy demand. With the high cost of energy the thermoformer processor must realize that the bulk of the cost is in the heat cycle of the process. At the same time, most energy waste is made in the same cycle. Thus the most money can be wasted or the greatest cost savings realized in the heat cycle of the thermoforming process. Cost-saving methods are discussed in Section V of Chapter 6.

The two basic energy sources used for heating thermoplastic sheets are gas and electricity. Both can be adapted to all the thermoforming methods. However, their varying qualifications and adaptability to specific thermoforming processes distinguish them from each other, and discrimination must be shown in selection.

When gas is used as the heat energy source, the gas is usually burned in a controlled manner. Gas as a heat energy source is relatively inexpensive compared with the cost of electric power. However, temperature control is much more difficult and far less precise than in electric heating systems. It is also necessary to provide proper ventilation, exhaust removal, and discharge systems because burning gas produces carbon monoxide. In the case of very large instillations, burning and exhaust discharge approvals and authorizations have to be obtained from regulating agencies. Although an electric power source is more costly, it is the most widely used heating energy source. The availability of electric power and its unlimited controllability, together with the multitudes of instrumentation and heating element types, make the electrical heating technique the most versatile and adaptable system for thermoforming. More than 90% of thermoforming is made using electric power to heat the thermoplastic sheet for the thermoforming process. The electrical energy is converted into heat when electricity is run through a highly resistant wire. The heat is generated by the resistance and conducted to an externally built jacket. The shape and form of the heater elements, plus the size of the external jacket, determine the type of heater unit recommended for a particular use.

Despite the wide variation of heating methods and heater types, there are three basic heating methods used in the thermoforming process: gas-fired convection ovens, contact heating, and radiant heating. We discuss next these three basic methods together with all the available variations of heater types.

A. Gas-Fired Convection Ovens

The high cost of electrical energy and the high volume of heat demand for very large area heating for large thermoformed products (8 by 10 ft and larger) heavily favors the use of gas as an energy source. Gas-fired convection ovens have a definite place in thermoforming. They heat extra-large panel and extra-heavy sheet thicknesses in the most effective way. The heat energy of the gas is localized in a smaller burning chamber which heats the air in an outer chamber. The heated air is blown and forced by blowers and fans into a large oven which contains the clamped panel of the thermoplastic sheet. The hot air heats the plastic sheet by convection. Proper circulation of the heated air and its recirculation back to the heating chamber ensures continuous convectional heating of the thermoplastic sheet. The large area and bulk of the thick thermoplastic sheet requires a longer resident time in the heating oven, and that is why gas energy is the most cost-effective heating method. In this type of heating process gas will provide the most British thermal units (Btu) for the money. (One Btu is the quantity of heat required to raise the temperature of 1 pound of water 1°F at 39.2°F.) For very large panels or very thick plastic sheets a high Btu value is required to produce proper forming temperatures. For example, the 4-in.-thick acrylic windshields mentioned in Section III of Chapter 1 require a 24-hr heating cycle. It is obvious that this type of heating would be tremendously cost-effective when gas is used as the energy source. It is also understandable that although gas-fired heating is somewhat less controllable than electrical heating, the size and bulk of these large thermoplastic sheets do not warrant such close control.

B. Contact Heating

Contact heating of thermoplastic sheets for thermoforming has been around since the origination of the concept of thermoforming. By contact, when materials are placed against or on a heated surface, the heat will travel into the thermoplastic sheet and in time raise its temperature to match that of the heated surface. For efficient contact heating, the heated surface usually consists of a rather large, thick metal plate. The plate can be heated from its back side with gas flames. However, the plate is usually heated with cartridge heaters which are an equal distance apart and embedded into the plates. The plates can also be heated with a series of strip heaters attached to their back sides. electric heater units heat the entire plate, which in turn will heat the thermoplastic sheet when it comes into contact with the plate. The temperature of the heated plate is usually regulated by an embedded thermostat, which controls the power to the heater elements. Temperature fluctuation should not be a problem because the large mass of the metal plate has enough heat sink and retention qualities to level any temperature oscillation. When the heated plate is brought to the desired temperature, the thermoplastic sheet is held against it. The actual heating is done with surface contact between the sheet and the heated plate. Contact heating is generally done with only one-sided contact.

However, two-sided contact heating can also be used. In this type of heating the heated plates from both sides maintain some pressure on the thermoplastic sheet. It is of the utmost importance that the position of the two plates and their pressure be exactly parallel to the sheet and that the pressure not be overintensive. Too much pressure will drag the sheet between the plates when it is advanced into the molding area and could cause surface damage such as scratching, abrasion, and markings. On the other hand, if the surface contact with the heated plate is poor, as when there are trapped air bubbles under the sheet, imperfections on the sheet, gauge variation, texturing, or imperfections on the heated plate surface, poor heat transfer will occur. All of this can cause interruptions in the heat transfer between the heated plate and the thermoplastic sheet. If such poor heat transfer conditions are encountered, a longer resident time has to be implemented, which will increase the overall cycle time. In extreme cases, the cold spots even affect the quality of the thermoformed product, which would then have to be rejected. The completeness of surface contact between the heated plate and the thermoplastic sheet is the key to the success of using contact heating for the thermoforming process. With contact heating it usually takes longer to heat the thermoplastic sheet for forming; therefore, its use is limited to sheet thicknesses of less than 30 mils. Contact heating is also limited to thermoforming techniques where uniformly heated sheets are desired. With this method of heating, no temperature variation within the sheet (preprogrammed heating) can be obtained. Contact heating can be used satisfactorily when textured or foamed thermoplastic sheets are heating for thermoforming. To increase the heating efficiency of contact heating methods, the thermoplastic sheet can be preheated with radiant heaters just before advancing it into the contact heaters.

C. Radiant Heating

Heating with radiant energy essentially uses infrared-light-spectrum wavelengths. When these waves are radiated toward an object they will heat its surface and then through conduction will continually heat the rest of the body. Radiant energy heaters do not need to be in physical contact with the subject. The radiation will penetrate a given air space. Infrared radiant heat energy is put out by the sun or even from a small fire. Infrared radiant energy is provided by a wide spectrum of wavelengths, from 1.2 micron (μ), which is the closest wavelength to visible light, to a very long wavelength of 8 μ, which is at the point of inefficiency for this heating method. The best conversion efficiency from electrical power to radiant energy peaks at wavelengths of 3.00 to 3.50 μ. Knowledge of actual heater-coil temperatures is not detrimental to the thermoforming process; however, their colors may be good indicators of the most efficient outputs. Heater units can radiate from a black heat-emitting stage to glowing white. The best efficiency (at wavelengths of 3.00 to 3.50 μ) comes when the heater glows "cherry red." The rate of infrared energy emission subsides when heater element darkening is observed. In controlling infrared radiant energy we have to time the exposure of the thermoplastic sheet or reduce the power going to the heater elements, which changes the infrared heat output of the heater elements. Infrared radiation is a hotter energy source than convectional or contact heat sources. Because of its controllability and higher energy efficiency, infrared radiant heating is the most popular heating method in the field of thermoforming.

There are two basic energy sources for radiant heat energy: gas and electric power. Electric heating systems have the most variety and are the most popular. Electricity is readily available, and controls and instrumentation and the wide selection of heating elements available make it very convenient.

1. Gas Radiant Heaters

As a heat energy source, in addition to the convectional heating method, gas is available as an infrared radiant heating system. In this system the gas Btu energy is converted into infrared radiant energy by burning the gas in a controlled manner (inside fine-cellular-structured ceramic panels). Individual panels consist of a fine-cellular structure which resembles a honeycomb pattern, having openings to the surface. The gas and air supplies are mixed through the open face. Inside the cellular structure the burning takes place. The heat generated makes the ceramic structure turn glowing red; at this point the radiating infrared heat energy is emitted. Ovens for thermoforming are equipped with a bank of this type of gas-fired radiant infrared heating panel; panels provide a less expensive energy source. Some of the limitations of gas-fired radiant heaters easily offset the cost savings. Keeping these types of heating units absolutely clean and in good working order is important in obtaining uniform heat control and is essential for safety purposes. The panels cannot provide programmed heating, only uniform heat throughout their surfaces. If melted plastic or any residue covers or plugs up portions of the panels, the other portions will receive excessive amounts of gas, which could create a torchlike flame. This flame could become very dangerous because it can easily ignite the thermoplastic sheet, and such a fire could easily get out of control. The burning area has to be well ventilated because the burning process could produce an overabundance of carbon monoxide.

The newest heater apparatus adopted for the thermoforming process when converting the gas energy into a radiant heat energy source is made by the catalytic infrared heater elements. In this type of heater the gas supply is channeled into an enclosed chamber together with sufficient air supply. The chamber is preheated by electric cartridge heater elements to an elevated operating temperature. The mixture of gas and air enter into the chamber of the heat panels via an intricate plumbing system. The chambers contain a catalyst which is usually made of platinum shavings or in a "steel wool"–like form. The catalyst will trigger the oxidation of the gas without burning, flamelessly. As this oxidation takes effect it will generate infrared heat without producing any of the conventional flames. The very same method is used and installed under most modern automobiles where it is used as an emission control device, removing or oxidizing any unburned fuel contained in the exhaust gas. (It is the Catalytic Converter, which is coupled into the exhaust tail pipe system.)

Since these heater units are using a capsulated chamber to oxidize the gas, only a limited amount of exhaust and byproduct is generated by the process. The radiant heat output of these heater units are only found in the longer wavelength spectrum of the infrared range. These heater units not made to glow red hot, like the electric heater units; instead they will remain dark and will emit the longer wave length spectrum radiant heat waves (wavelengths of $5.60–10.00\mu$). Such longer wavelength output rates can be found most effective especially when uniform heating levels are desired. Also some specific types of plastic materials do accept the longer wavelength heat levels better and will not induce a rapid scorching of the sensitive surface of the plastic sheet. On the other hand this type of heater unit can never induce such forceful heating levels as can be produced by a glowing red hot electric heater element. The moderate heat output of this type of heater unit can be overcome by either extending the exposure time of the sheet or extending the oven length. Increasing the distance of the oven's heat tunnel increases the sheet travel time, thus ensuring the necessary heat exposure time.

Catalytic infrared heater units are best suited for providing uniform heating of the thermoplastic sheet. On the other hand, such catalytic radiant heaters should not be considered where programmed heating is desired for different heat levels to be produced at various locations of the same sheet. Also, when rapid or high levels of focused heating are needed for achieving critical heating conditions for certain types of plastic materials, this type of heater unit is not found beneficial.

The real benefit of this type of catalytic heater unit is found in the definite energy cost savings. A wide variety of studies showed that using gas as an energy source can cut the energy cost by about half or better, that of course may depend on your locality and the local pricing sources of electricity versus the cost of gas supply. Thermoformers must investigate all aspects, before commitments are made, as to which types of heaters will fit better into their particularly situation.

2. Electrical Radiant Heaters

Infrared radiant heat can be produced by a multitude of types and specially built electrically powered heating elements. The radiant energy in all heater types comes from a core of high-resistance wire (nickel-chrome alloy wire). When electric power is channeled through, the resistance makes the wire glow red hot. The heat generated by the resistance is transferred to the surrounding area or surrounding jacket materials and in conjunction, the jacketing will turn red hot. Several radiant heater elements are available which have been built with a specific purpose or application in mind. The infrared radiant heat energy in all cases provides a hotter energy source than that of convectional heat. Infrared radiant heat is always applied with direct exposure of the thermoplastic sheet to the heat source. Overexposure, underexposure, shading of the heater elements, or change in distance of the exposed sheet from the heat source can result in different heating levels in the thermoplastic sheet. Heater intensities can also be altered by the power input, which would change the infrared wavelength spectrum and directly affect the time needed for heating the thermoplastic sheet. There are many types of radiant heater element variations, each with unique differences and special characteristics. Some heater units can be classified as general-purpose units, while others are made for specialized service. The following group of radiant heaters represents the heater elements commonly used in the thermoforming process: open resistor type, tubular or rod type, flat strip, ceramic, Pyrex glass, quartz, heat-emitting panel, infrared flat quartz heaters, and heat lamps.

a. **Open resistor heaters:** Open resistor heating elements have just about been completely phased out. Only a few obsolete machines and some home-built equipment still utilize these elements. The open resistor electrical wire or ribbon-type heaters proved to be dangerous to operate due to the hazard of electric shock, plus the constant threat to short out and emit sparks. The heating efficiency offers only poor quality and adjustment opportunities.

b. **Tubular or rod heaters:** The tubular heating element, often called Calrod (a trade name of General Electric Co.), has been around for many years and is still the "workhorse" of the thermoforming industry. The tubes are produced in many lengths, wattages, and in all the popular voltages. These tubes are usually produced as a straight length of tube which can be installed straight or bent into any configuration. The tubes consist of a fine thin-gauge nichrome wire, generally coiled. The coiled resistor wire is placed centrally in a steel tubular jacket. The jacket is then filled and packed with insulating material under pressure. The end of the tubular jacket is crimp sealed and equipped with terminal ends (Figure 16). The electrical power is connected to the ends of the tube. The resistance of the inner wire makes the entire tubular element hot, which in turn continually heats up the entire tube. Tubular heating elements can be installed as straight rods or if desired, can be looped back or bent into any given pattern. A multitude of tubular heater elements clamped into a fixed location in a shielded oven constitutes a thermoforming oven. The

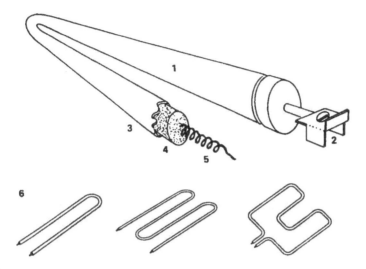

Figure 16 Tubular heater element: (1) tubular heater body (bent to U shape); (2) electrical connector; (3) steel jacket; (4) mineral (magnesium oxide) insulation; (5) nickel-chromium alloy resistance wire; (6) tubular heaters bent into popular shapes.

heater elements can be wired so that all the elements are hooked together with one controller, can be banked together and connected to different controllers, or each tubular heater element can have its own controller.

Tubular heater units are very popular because they are the least expensive. Their construction is durable and almost indestructible. This feature provides a rather important advantage, because in the thermoforming process sheet heating ovens are often subjected to heavy abuse. Tubular heater units are known to give many years of trouble-free service and very seldom burn out. On the other hand, tubular heater units do not offer the highest heater efficiencies. At least 50% of their radiated energy heats the surrounding area rather than the thermoplastic sheet. In other words, all heat energy radiation output completely surrounds the tube shape and only the side that faces the sheet produces useful heating. A reflectorized surface can be placed on the back side of the heater element to reflect back most of the escaping heat energy. If the reflector has a parabolic shape, the reflected heat waves get to be almost homogeneous. However, there will be still a hotter line under the heater because of the shorter heat travel distance (Figure 17). Reflectors are useful only when they have reflective surfaces. Neglected and uncleaned reflectors, especially once a piece of plastic has burned, will rapidly deposit a black coating on the reflectorized surfaces. When reflectors are dirty and coated with deposits

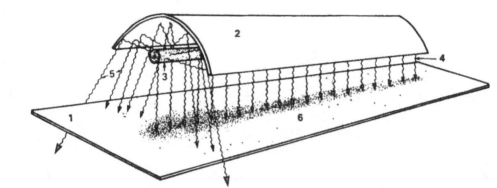

Figure 17 Reflectorized radiant heater source (with tubular heater element): (1) thermoplastic sheet; (2) reflector; (3) tubular heater element; (4) direct heat (closest distance between heater and plastic sheet); (5) reflected heat; (6) actual heat distribution on the plastic sheet.

of melted or burned plastic, they no longer provide the intended service. Therefore, just having reflectors in conjunction with tubular heaters does not guarantee a continuous reflective surface and high heating efficiencies. Cleaning and periodic replacement of the reflectors should be scheduled as part of a routine preventive maintenance program. Heating-oven maintenance is absolutely necessary not only for the efficiency of the process but as an important safety measure. Melted residue and drippings of plastic deposits could easily ignite and contribute to a dangerous oven fire.

Tubular radiant heaters' other shortcomings are that hot spots can show up in a smaller section of its full length. Hot spots do not occur often, but do show up especially on areas that have been subjected to bending. A tubular heater that is bent into shape may receive slight damage at the point of bending which will increase the resistance of the inner wire core. These sections or bends will glow at a higher rate than the rest of the tubular heater element. The higher-glowing areas will have shorter wavelengths which generate higher heat levels. In the heating of thermoplastic sheets in thermoforming, the hot spots on the heater elements will expose the thermoplastic sheet to higher levels of heat at the same pattern or location. To remedy the cause of hot spots, the tubular heater element must be replaced, or a piece of wire screen must be placed or hung between the heater's hot spots and the thermoplastic sheet. The wire screen patch will shield and pick up the excess heat generated by the hot spot on the tubular heater element and diffuse the concentration of heat. In an oven area that has a few unwanted hot spots, this remedy is ideal. However, when the hot-spot phenomenon shows up in a multitude of heater elements, change is the only way out, unless someone cares to decorate the heating ovens with wire screens, like a Christmas tree. Further use of screen shields is discussed in Section III.E. When thermoforming equipment is quoted or ordered without the heater type specified, it is always implied that the equipment will be supplied with tubular heater elements, the industry standard.

c. **Flat strip heaters:** Flat strip heaters are very like tubular heaters in that they are made of an inner core of resistance wire which is patterned in a side-to-side (zigzag) formation. The resistant wire is also embedded in heat-resistant electrical insulating material and jacketed with a metallic cover. The strip heater is made in a flat striplike fashion, which may have a width of 1 to $1^1/_2$ in. and a thickness of $^1/_4$ to $^3/_8$ in. (Figure 18).

The heater element is available in most popular lengths and can be ordered in any specialized length. The electrical connectors can be located at the opposite ends of strip heaters, or both connectors can be located on one end of the strip heater. Strip heaters can be bent on their flat side and reshaped after purchase to conform to special thermoforming needs. Strip heaters are usually bent to create open-ended V or C shapes (Figure 19). The purpose of

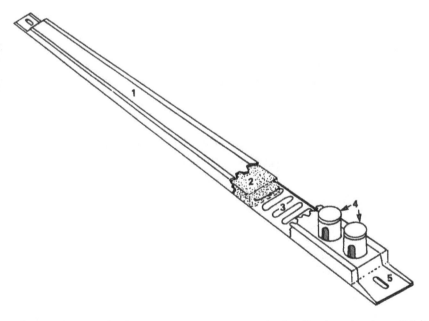

Figure 18 Strip heater element: (1) strip heater body; (2) mineral (mica or MgO) insulation; (3) nickel-chromium resistance wire; (4) electrical connectors; (5) mounting hole.

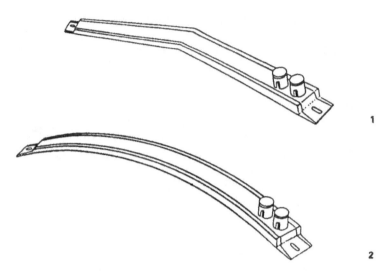

Figure 19   Bent strip heater elements: (1) V-shaped bend; (2) C-shaped bend.

bending is to obtain a greater distance between the thermoplastic sheet and the heater element at the center, while having a closer distance near the edges, where most of the heat is escaping. The shaping of heater elements is useful only when the same heater condition is needed to produce similar products. Multiple bending of a heater element is damaging to the unit and the burnout chances are considerably increased.

Strip heater elements can be placed in thermoforming ovens spaced some distance apart or close to each other. When they are placed next to each other they create an excellent radiant heat-emitting surface for thermoforming. However, the heat efficiency is not the best because the back sides of strip heaters produce nothing but wasted heat energy. Wasted energy cannot be recaptured or reflected back toward the sheet; the heat just heats the rest of the equipment or space and eventually escapes to plant air. The escaped heat may make the operating area rather comfortable in the winter but can make it unbearable in the summer. At times when excessive heat conditions persist, operating personnel seeking relief may be forced to create drafts by opening doors, windows, or any available ventilating source. Drafty plant conditions can affect the thermoforming outcome when establishing "negative heat" forces. Many consider the strip-heater element to be the most durable, indestructible heater element, which increases its popularity. These heater elements provide the user many years of trouble-free service.

d. Ceramic heaters: The ceramic heater element is also very popular and its application is often specified. Its popularity is second only to that of tubular heater elements. The cost of ceramic heater units is rather high compared to tubular heaters, but the superior control of programmed temperature often outweighs the undesirable cost factors. Ceramic units are produced with the fine coiled heater element wires embedded directly in the ceramic material, which is glazed over. When energized, the embedded coils turn glowing red. This makes the ceramic unit glow red hot and causes the entire surface to become an emitter of radiant heat energy (Figure 20).

The heater elements are available in two sizes: 2.6 in. by 4.8 in. and 2.6 in. by 9.7 in. The ceramic heater units are fitted into the thermoforming machine ovens in rows, side by side, creating heat-emitting surfaces. The ceramic heater can also be installed with a small space ($^1/_2$ to 1 in.) between the rows to allow for a reduction in the number of heater units, with only a small loss of heating efficiency. The use of individual heater units allows thermoformers to program temperatures in their heat applications. Programming makes it easy to achieve various temperatures in specified areas, making the thermoplastic sheet hotter or colder in a preplanned manner. Naturally, to achieve different temperature emissions in the heater units, the individual heaters must

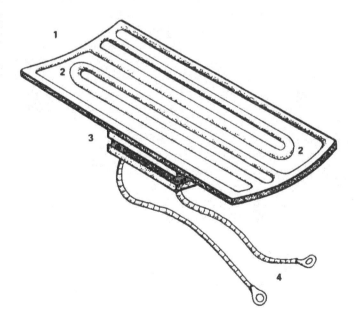

Figure 20 Ceramic heater element: (1) curved porcelain body; (2) embedded resistance wire; (3) mounting post; (4) connecting wire.

be connected to separate controls. In cases where similar heat outputs are needed in rows, the units can all be hooked up to one heat control. Ceramic heater units offer controllable heating for thermoforming and they provide the best results, especially when programmed heating is desired. Thermoforming equipment purchasers often specify ceramic heating elements for their equipment. However, to reduce initial costs, ceramic units are often ordered only for the lower heater banks. The top heating units are fitted with tubular heaters. Such arrangements provide only a slight edge over tubular heating elements. In addition, this practice minimizes the opportunity for heating capabilities.

Ceramic heater units should receive periodic maintenance and should be cleaned of melted thermoplastic residues and checked carefully for cracks. As the ceramic materials are extremely rigid, they are vulnerable to thermal shock cracking and abusive damage. Cracks can develop through the ceramic, which may not necessarily affect the internal resistant wire continuity. However, if the cracks are large enough, bending and weakening of the wire elements can increase the wire's resistance and can easily lead to shorting out. This may cause sparks to drop on the heated thermoplastic sheet, which is easily ignitable. Production and maintenance personnel have to be made aware that ceramic heaters are fragile and that extra caution has to be taken when cleaning and installing. When heater units are replaced, it should be noted that the individual ceramic heaters do not have uniform wattage. Therefore, the control units must be adjusted or connected to other control units, which should match the heat-emitting conditions of the heater element previously replaced.

The skills involved in programmed heating of thermoplastic sheet with the help of ceramic heater units is a beneficial procedure when dealing with lighter-than-30-mil thermoplastic sheets and intricate shape forming. Such a programmed heating technique allows the thermoforming operator to achieve uniform wall thickness distribution in the thermoformed parts, avoiding the natural overstretching and thinning. Also, when desired, this programmed technique permits shifting of materials to the desired locations within the perimeter of the forming area. The installation of ceramic heater units in the thermoformer is made possible by its protruding back stem, which is part of the ceramic body.

e. **Pyrex glass heaters:** These heater units, offered by Corning under the Pyrex brand name, are also used in the thermoforming process. The Pyrex glass heater units are made with a tempered borosilicate glass panel, which has an attached electroconductive film on its back side. The electroconductive film serves as an electrical resistance element and when electric current is applied, will turn hot, heating up the glass panel. The glass panel will become a highly effective heat emitter and dissipate the heat over the entire surface to a uniform level. These types of heater units are ideal for uniform heating by virtue of the Pyrex glass panel's mass, which is a good heat sink and reemitter of heat. Since the glass mass heats slowly, it also cools slowly. This results in minimal heat fluctuation, which is encountered when sudden voltage variations are experienced. The constant heat emittance of glass allows the use of less expensive control devices. The heater units are self-contained and include a built-in reflectorized back-side panel. The recommended method for cleaning these units and keeping them in good operating condition is to dry-wipe them only. Washing with water or a wet cloth should be avoided.

Pyrex units are available in many popular sizes, from 12 in. to 24 in. square and up to 6 in. by 30 in. rectangular panels. When different sizes are combined in the same oven, the same watt/square inch ratings should be used to assure uniform temperatures.

One of the disadvantages of glass heaters is their fragility. This problem can be solved by the use of protective covers when maintenance or heater replacement is performed; the covers will protect the glass from such hazards as dropped objects. Rigid or close-spaced installation of the heater panels should also be avoided, as this would interfere with the natural thermal expansion. With this type of glass heater used in a sandwich-type (top and bottom) thermoforming oven—where the heater units are facing each other—the oven must not be left on while thermoforming is not being performed. The heat buildup on these units must not exceed 660°F, as this would damage them. On the other hand, tubular and strip heaters can handle 1200°F without damage. However, with glass heater units, strong benefits are realized through their continuously uniform heat emittance. Such a performance would be quite a challenge to any other type of heater element for specific thermoforming operations and in the processing of critical thermoplastic materials.

f. Quartz heaters: In recent years, the popularity of the quartz heater element has grown faster than for any other type of heater element. Considering the high cost of electric energy, this type of heater unit is capable of providing outstanding energy efficiency. The energy cost savings can be so substantial that a changeover from existing heater units to quartz element units is readily justifiable. The uniqueness of quartz heaters stems primarily from the quartz jacketing or covering materials. Quartz itself is an excellent electrical insulating material, with equally effective heat retention and temperature stability. Quartz remains strong and rigid under high-heat conditions and therefore provides the necessary supportive structure to the glowing electrical resistance wires. The greatest advantage of quartz is that it has no significant hindering effect on the infrared energy waves. As soon as the internal resistance wire is electrically charged, it will emit infrared heat and the ensuing energy waves are transmitted through the quartz material almost instantly.

Quartz materials can be produced in a tubular shape, allowing the coiled (nichrome) electric resistance wire to be threaded through its core with electrical connectors at the opposing ends. The quartz tube ends are usually fitted with high-heat-resistance porcelain caps, which also double as installation supports for the heater elements. At the same time, the caps allow good built-in electrical connections with proper electrical insulation. Larger heating panels can be created by banking rows of tubular quartz heater elements together in multirow fashion.

The quartz heater's efficiency does not come from reduced wattage and reduced electric power demands. The true gains in efficiency are actually realized in the rapid on-and-off cycle capability of the quartz heater element. The quartz heater element can be turned off completely in the "no-heat-demand" segment of the cycle. For example, in the cut-sheet type of thermoforming process, when the heated sheet is undergoing the forming, cooling, stripping, and removal cycle steps, the quartz heaters can be turned off. Since the quartz material does not interfere with infrared wavelengths, the heater units start to emit the radiant energy as soon as electrical power is turned back on. There is no extensive heat-emitting delay with quartz-type heaters as is normally associated with standard tubular or flat strip heaters. The difference can easily be seen in Figure 21. The only drawback of quartz heater units is that they are extremely fragile in a tubular form. Quartz tubes are usually produced in $^3/_8$- to $^1/_2$-in. diameters and often longer than 12 in. in length, making them as fragile as glass tubes. Since thermoforming ovens move during the thermoforming process, they produce a rough, jarring action that could

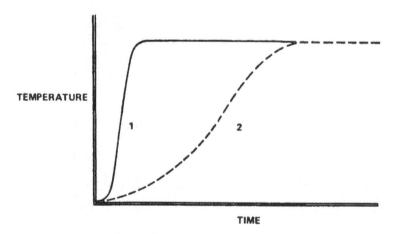

Figure 21   Charted difference between quartz and tubular heaters' temperature rise: (1) quartz; (2) tubular.

result in breakage of the quartz elements. Particular thermoforming machines that have traveling oven heaters, or machines designed with a hinged "clamshell"-type or C-type safety oven, require the installation of special oven cushions or counterweights. These will minimize slam-bang and abusive shocks, which could result in breakage of the quartz elements. A broken quartz jacket will not interfere with the heat-emitting abilities of the heater element but could possibly cause an open exposure of the resistant wire core, which might easily short out and spray hot sparks on the thermoplastic sheet. The heated thermoplastic sheet only needs to receive the smallest amount of hot sparks to ignite, resulting in an oven fire which can easily get out of control. The benefits of quartz heaters can be fully appreciated only when periodic oven maintenance programs are implemented and the quartz heater elements are routinely subjected to visual inspections.

Even though the tubular quartz heater unit is emitting radiant heat energy 360° around its body, only 50% of that heat is useful for heating of the thermoplastic sheet. Use of reflectors to recapture and redirect the escaping radiant heat energy makes the quartz heaters even more efficient. Some manufacturers offer tubular quartz heater units with built-in reflectors. Such reflectors can minimize oven maintenance because only the quartz tube has to be cleaned; there are no external reflectors to be cleaned or replaced. On the other hand, internally built reflectorized quartz heater elements require more attention at the time of installation. The installer has to make sure that the heater elements are facing in the direction intended and that they do not have the opportunity to rotate away from the proper direction.

g. **Heat-emitting panel heaters:** Heating the thermoplastic sheet in the thermoforming process can be accomplished with larger panel heaters. The size of this heat-emitting panel can be as small as 3 in. by 6 in. or as large as 12 in. by 70 in. The purpose of the heat-emitting panel is to create a uniform heat-emitting surface. The panels can be placed side by side, creating an even larger surface. Since there is neither "hot spotting" or excessive heat as caused by a rod-shaped heater or reflected heat, the surface of the heat-emitting panels will become uniformly heated. The metallic jackets spread and dissipate the internal heat uniformly over the entire surface. This uniformity of heat emittance allows this type of heater unit to be installed much closer to the thermoplastic sheet surface without the risk of uneven heating. Since the heaters can be placed closer to the thermoplastic sheet, heater temperatures can be lowered to 400 to 500°F. This is much lower than that with tubular or strip heaters. Because of this lower heater temperature, the processor will be able to maintain ideal heat transfer to the thermoplastic sheet while reducing power consumptions. The heat-emitting panels are manufactured in a manner similar to that for flat strip heaters. However, their thickness and mass are greatly reduced. The heat-emitting panel is usually made of banked strips, each of which contains a nichrome resistance wire surrounded by packed mica insulation. The whole body of the strip is very thin (under 1 in., including the built-in insulation) to minimize back-side heat losses and increase the response to heat controls. The surface jacket is made with a treated black metallic material, which provides the most efficient surface radiation when uniform heat is needed. This type of heater element is not made for programmable heating and could by no means by used in every type of thermoforming process. The heater's light weight, combined with the availability of varying lengths, makes it ideal for all types of applications. The second most popular use for these types of heaters is for preheating panels just before the thermoplastic sheet is transported into the final heating stage. Heat-emitting panels allow a wide range of creativity in the heating segment of the thermoforming process. Such uses often bring rewarding results.

   h. Infrared flat quartz heaters: The infrared flat quartz heater is produced for the same purposes as radiant-heat-emitting panels. It is aimed at producing uniform heat over the entire surface. The flat quartz heater's advantage over the heat-emitting panel heater comes from its ability to be turned off in the nonheating segments of the thermoforming cycle. Quartz in any form (tubular or flat block) is transparent to infrared wavelengths, and as soon as the embedded electrical wiring is charged, heat emittance results. The flat quartz heater is produced out of a flat block or panel of quartz which is sinuously grooved. A fine-coiled electrical resistance wire is channeled throughout the continuous grooves. The ends are properly connected to the power source. To capture and hold the electrical resistant wire in place, a cover sheet of some sort is usually placed on top of the grooved quartz panel. This cover sheet, aside from its heat resistance and electrical insulating quality, may have heat-reflecting properties that would greatly improve the quartz heater's efficiency. This type of flat quartz heating panel is available in such sizes as 10 in. by 10 in. and 14 in. by 14 in. up to 20 in. by 20 in. It is possible that they can be ordered in custom sizes as well. The panels can be ganged together, creating a large, uniform heat-producing surface. Again, flat quartz heaters are aimed at producing uniform heating to the exposed thermoplastic sheet and are not meant to produce programmed heating.

   i. Heat lamps: Heat lamps are also capable of producing radiant heat energy that would heat the thermoplastic sheet. However, the heat output rate of heat lamps is so low that they cannot be considered efficient heater units. Heat-lamp-generated heat can be used efficiently only with film materials of approximately 1 mil or less thickness. Heat lamps are useful only for surface treating, as in drying paint or other coatings. We should not consider heat lamps to be a feasible heat source for thermoforming. In my over 38-year career, I have witnessed heat lamps being used only once. I am yet to be convinced that the heat lamp is an efficient heating method.

D. Temperature Controls

Most thermoplastic processing demands closely controlled temperatures to obtain optimum production. In the thermoforming process, heating of the thermoplastic sheet requires precise temperature control. The finely controlled heating of the thermoplastic sheet ensures the quality and repetition of the forming process and guarantees uniformity in the production. All the heater units discussed previously require some type of control. When the heater elements are connected to a continuous full-line power source, their heat generation often accelerates to higher levels than most thermoplastic sheets can accommodate. The excessive heat cannot be absorbed fast enough to prevent surface scorching and material destruction. To avoid this unwanted surface damage to the thermoplastic sheet, two types of controlling factors must be implemented. The first factor is the temperature output rate of the heating elements and the second is the time of exposure of the plastic sheet to the heat—"residence time." The best condition for heating a thermoplastic sheet for thermoforming is when the heat acceptance and absorbancy rate of the sheet can be matched with the heat output rate of the heaters. At this point, the ultimate heating efficiency is accomplished. When, for any reason, less heat is generated by the heater units, a longer residence time or exposure is required. Naturally, a lower heat level makes the heating cycle safer and more controllable. At the same time, the longer residence time slows down the cycle and makes the production more costly and inefficient. Beside safety factors and improved heat controls, lower heat levels are often chosen because of other influencing conditions. Choosing the lower heat levels will provide improved forming conditions and better functioning of the final product. These conditions may not have a direct relationship to the heat absorbencies of the plastic sheet but can easily influence the production or the final product outcome. These influencing conditions often justify the slower heat cycles and greatly improve the final product outcome. These conditions, which can easily be destroyed in the rapid heating cycles, include molecular orientation, surface degradation (color, texture, shine, delamination), and foam breakdown with poor expansion.

To match the heat output rates of the heater units to the heat absorbancy of the thermoplastic sheet, the power input to the heaters usually has to be controlled. When energy inputs are increased or decreased, the heat generation of the heater units will automatically increase or decrease with them. The simplest and most common way to control any heater unit is by on/off switching. Such switching can easily be activated manually; however, for more precise

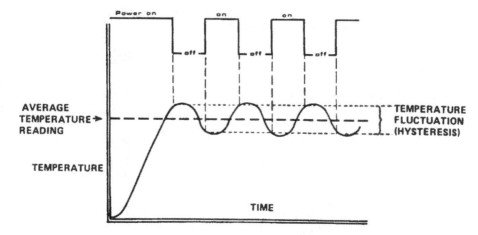

Figure 22   On-off control action with temperature fluctuation (hysteresis).

control, an automated system is implemented. The basic on/off temperature controllers will create very strong preconditional temperature fluctuation (Figure 22).

A more advanced automatic system usually works in conjunction with a timer or temperature-sensing device called a thermostat. In an automated system, achieving satisfactory control over the heat output rate of a heater unit is often accomplished with a timing relay. Such switching relays can be adjusted so that the relay is closed or opened for a predetermined time. The time ratios can be set from 0 to 100%. These switching relays are called "percentage timers" and run in repeated cycles that are applied continuously to the heating process. By changing the on/off ratio on the timer, which regulates the time period the power is on or off, the heat generation of the heater unit is controlled. This temperature-controlling method is rather simple and at the same time rather crude, as can be seen in Figure 23.

The percentage-timer control system is guided by the timer, not by actual temperatures. When controls such as percentage timers are used, constant visual monitoring is needed. The percentage timers provide no sensitivity whatsoever to temperature changes. The ambient temperature of a plant facility can greatly fluctuate not only on a daily basis but also between the morning and afternoon hours. In addition to ambient temperature fluctuation, the thermoplastic temperature can vary from batch to batch due to different storage and holding practices. The percentage timer control system is totally insensitive to such unforeseen temperature variations and therefore cannot be made totally

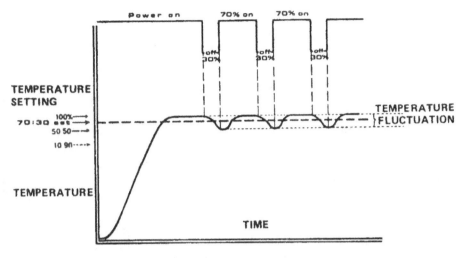

Figure 23　Time proportioning control action at 70:30 percentage setting.

reliable. The higher the percentage of power used, the less the temperature fluctuation that will be realized. This is well documented in Figure 23.

For more precise temperature controls, temperature-sensing devices have to be installed. Such devices are called "thermostats" and "thermocouples." The duties of these temperature-sensing instruments can realistically be no more than to sense the temperature at a specific location and react to a specific temperature change. When the temperature-sensing device provides only a numerical readout of the current temperature and temperature changes, the only reaction from the device is a change in the numerical readout. Its function, then, is limited to being a temperature indicator (just like a thermometer). The readout display can be as simple as a dial-type gauge or as sophisticated as a digital readout gauge. On the other hand, when the temperature-sensing device activates an on/off control or a more complicated electronic relay system, it has a full thermostatic function, although it is still capable of providing visual temperature readouts. When a thermostat is set to a predetermined temperature, that exact point is called the "set point."

Most thermostats, depending on their make and composition, have a temperature fluctuation range. This range determines the upper and lower limits of temperature fluctuation of the heating system. To trigger the thermostat, the temperature change has to surpass the actual thermostatic temperature fluctuation range. This will result in a slightly greater temperature variation. The actual sensitivity of a thermostat or thermocouple is judged by the overshoot and undershoot of this temperature fluctuation, as can be seen in Figure 24.

For better, more precise control, the thermoformer must rely on a more sophisticated level of electronic devices. Such state-of-the-art devices provide anticipated controlling by the thermostat and may rely on a multitude of "information" or sensing inputs to make the necessary changes that will compensate for the temperature changes in the process.

The next generation of controllers are the "proportional" controllers. These can provide variable power inputs, depending on the deviations from the set point. They apply full-line power up to the set-point temperature, then decrease the power input proportionally. The proportional controllers are typically precalibrated, assuming that 50% power is required to hold the temperatures at the set point. Such an assumption has a built-in offset error factor that requires constant recalibration. Although proportional controllers are aimed at minimizing the temperature oscillation, they cannot completely eliminate the overshoot and undershoot of temperature fluctuations at the set point.

More sophisticated control components are employed to narrow the hysteresis between the temperature over- and undershoot. Such components must be able to cycle rapidly without having the short life expectancy of controllers. These controller components often have solid-state relays, triacs, and thyristors

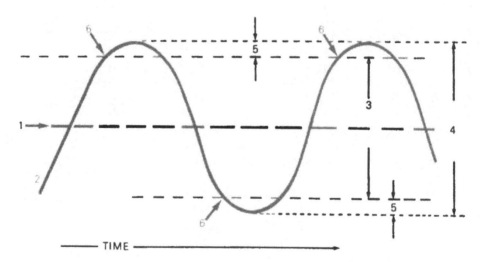

TIME

Figure 24 Temperature fluctuation curves: (1) temperature setting (set point); (2) temperature fluctuation wave; (3) thermostat fluctuation; (4) temperature fluctuation; (5) temperature fluctuation and thermostat actuation differential; (6) point of thermostat actuation.

(e.g., the silicon-controlled rectifier). Proportional-integral-derivative (PID) controls are also very popular. These controlling systems derivatively apply the temperature input signals to the integrator, which can adjust and slow down the power input ratios when the actual temperature approaches the set point. The PID controller is aimed at satisfying optimum temperature settings. However, even with the help of potentiometers, it may have some difficulty in totally eliminating the natural temperature oscillations. If and when poor rate signals are received by this system, they will generally produce above-normal overshoots.

The ultimate in state-of-the-art controllers today is the microprocessor. This device actually qualifies as a computerized control. It may have either the components or complete setup of a computer. The microprocessor system basically senses the temperature setting and temperature acceleration or deceleration deviation. From these it makes anticipated calculations and performs the necessary adjustments. The responsiveness of a microprocessor controller is very precise, within milliseconds. The function of the microprocessor controller is to successively reduce and level out the peaks of the temperature overshoot and undershoot waves, bringing them right to the set-point line. This type of instrument, as a controller, offers the most ideal responsiveness and, to date, the top-of-the-line controlling system. Microprocessor controllers, being the best temperature controls, do not guarantee faultless temperature controlling within the system. The ultrasensitivity coupled with the high responsiveness could still play havoc with the temperature control when the microprocessor is subjected to irregularly alternating conditions and process disturbances. The ultrasensitivity of this instrument cannot react sensibly to uncontrolled draft conditions or similar "negative heat"-caused temperature fluctuations. Draft and negative heat conditions must be minimized if not completely eliminated in order to obtain satisfactory response from the controlling instruments. If and when plant conditions are erratic, the controlling instrument responses can be driven, causing extraordinary responses. Microprocessor controllers, subjected to normal manufacturing plant conditions, provide the best and fastest controlling responses in use today. The electronic and electromechanical makeup of the microprocessor and computerized controllers is beyond the scope of this book. Anyone interested in pursuing the subject should consult the literature on electronic devices.

A few additional important facts should also be mentioned. In most instances the temperature-sensing devices and thermostats are positioned in close proximity to the heater elements. The actual temperatures to which these devices are subjected—and thus the readout—are far greater than the actual thermoplastic sheet temperatures. For example: a gauge readout may indicate temperatures as high as 600 to 1200°F, while actual thermoplastic sheet temperatures may be in the range 200 to 375°F. This difference between readouts and actual sheet temperature must be properly correlated, and when temperature adjustments are made, this should be kept in mind. Temperature-sensing devices and controllers do fail from time to time, giving false readings and false responses, which will result in unsatisfactory thermoforming conditions. When such a failure happens, there is no substitute for an alert and conscientious thermoformer operator making the necessary corrective steps.

E. Negative Heat

The exposure of the thermoplastic sheet to heat will cause it to soften, and that makes it formable. For the best safety and economic practices, the heat of the thermoplastic sheet should be maintained at a minimum level, which is necessary for its shaping. Any heat level above this point could increase the time needed for cooling. Excessive heat can be detrimental for thermoforming or may even destroy the thermoplastic. Thermoformers who avoid the slightest amount of overheating run a highly efficient, well-maintained operation. However, an accurately measured heating operation can easily result in negative heat effects.

Negative heat is a condition caused by many different types of heat-robbing situations. The main culprit is draft. Uncontrolled, inconstant air movement does not directly hinder either the heat output of the heater elements or the travel of radiant heat through air space. The air movement actually causes cooling or chilling effects on the heated thermoplastic sheet. The draft condition removes heat from the sheet, causing areas of lower temperatures than is intended. The unpredictability of the draft may affect all the thermoforming cycles or may irregularly affect some of the thermoforming cycles in a somewhat controllable way.

1. Draft

The customary manufacturing plant layout generally consists of a large door or doors at the back end of the plant which are used for access for the material supply. The same plant will have a door or doors at the other end for product carryout to shipping or warehousing. When the two sets of doors are open at the same time, a draft will be created that could reach "hurricane" proportions. The temperature difference inside and outside the plant creates the negative heat effect. When it is cold outside, such as in the winter, the effect is even greater. If and when a thermoforming operation produces inconsistent results, where a series of satisfactory parts is broken intermittently by unsatisfactory parts, the cause is probably a draft.

Draft, or unpredictable air movement, can also be caused by fans. Fans are used to improve the comfort of the operating personnel, usually in the form of portable floor models or duct-type coolers. Both comfort-improving cooling systems have to be positioned in such a way that they neither face nor force air movement toward the heating or heated thermoplastic sheets. It has often happened that the air duct outlets have been positioned in such a way that air is directed right onto the thermoforming machine. Each time the comfort cooling system is turned on manually or cycled by automatic controls, the thermoforming produces a reject or poorer-quality part. When checking for drafts, not only should violent or strong air movements be considered but also all changing air currents that may result from on/off air blowers or oscillating fans. Negative heat effects caused by a draft basically have two types of variations, both resulting in the same unwanted irregularities in thermoforming.

 a. Cut-sheet thermoforming: In cut-sheet thermoforming, when the carefully heated sheet is being transferred to the forming area, it is usually exposed to the possibility of draft. During the shuttle operation, when the oven and the heated sheet are separated, the sheet can easily be exposed to open air and drafts. Similarly, on rotary-type thermoforming machines, each time the machine cycles, the heated sheet leaves the heating station and rotates unprotected in the open air to the forming station. The vulnerability to draft damage is much higher in the below-40-mil gauges, due to the lack of heat retention by the thinner materials.

 b. Continuous-sheet thermoforming: When thermoforming on continuous-sheet formers, the equipment is usually well protected from draft on the sides. However, the front and back ends are, in most cases, left open for feeding of the material and to allow the thermoformed part to exit. This two-ended opening through the equipment readily presents itself as a wind tunnel. To unknowingly increase wind tunnel effects, this equipment is often positioned in the manufacturing facility in-line with the back and front plant doors. When these doors are opened, providing an opportunity for rapid air movement, the thermoforming machines in line with the draft present no resistance. The draft will penetrate well into the oven tunnels, taking heat from the heated thermoplastic sheet and chilling irregular areas. It is easy to understand why a cycle can produce satisfactory parts when the conditions are ideal, while irregular and unsatisfactory parts are produced when the draft is interfering with the heating process.
 The most important countermeasure for all thermoformers is to draft-proof their plant. A constant airflow would not cause the same consequences, so the equipment heating system can easily be adjusted to match or neutralize such a draft. Erratic and unpredictable air movement is the culprit. To eliminate draft conditions, windbreaker walls or curtains should be installed. Comfort cooling fans and air outlets must be turned away from the process and carefully directed. In severe-winter areas, double-door systems must be utilized

and regulated so that none of the opposing doors can be opened at the same time.

2. Surface and Trapped Moisture

Thermoplastic sheet materials used for thermoforming in cut sheet or roll form can be exposed to rain or high humidity while being transported or stored. This exposure will probably affect its thermoformability. The surface moisture, given time, will dry up. However, moisture seeping between the layered stacks or between the coils of the rolls from the open edges can remain in place and would not dry out for several weeks. When the thermoplastic sheet are separated (or the rolls unwound) for loading into the thermoforming process, the trapped water, in the form of droplets or beads, will remain on the surface of the sheet. In the heating cycle, the moisture beads will dry off the sheet and will absorb the heat energy that has been intended for the plastic sheet. This action of heat robbing creates colder temperatures directly beneath the droplets, which can show up as spotty "fisheye" markings on the finished thermoformed parts.

Certain types of thermoplastic materials have a moisture pickup and holding tendency. Materials such as ABS and high-impact styrenes have a moisture pickup tendency. The water molecules migrate and get trapped between the polymer molecular structures. Materials like this, when tested, will measure out with higher weights when the humidity conditions are high, and lose weight when the humidity conditions are lower or when placed in dehumidifying rooms. The effect of the moisture entrapment is blistering of the sheet in the heat cycle. This blistering is the result of the moisture turning into steam. The actions of moisture entrapment are not directly related to negative heat effects; however, by virtue of its unwanted destructive effects, moisture entrapment deserves to be placed in this category.

Minimizing or completely eliminating moisture entrapment can be accomplished with proper preplanning of a particularly sensitive thermoplastic sheet. Materials should not be exposed or stored where wet or high-humidity conditions prevail, which will encourage moisture pickup. For both transportation and outdoor storage, proper coverings should be used to provide protection from rain. In areas where continuous high-humidity conditions prevail, the sheet stock materials should be stored for a time in the warmth of the manufacturing area, providing an opportunity for "dryout." In exaggerated conditions of moisture and humidity, or when demand for material does not allow longer time for drying, dehumidifying rooms are necessary. The simplest but possibly most costly way to eliminate moisture entrapment is to shrink wrap the moisture-sensitive sheet materials with a moisture-barrier film. The wrapping should cover and protect individual stacks or rolls of material right after the sheet-making process. The protective film will be removed just prior to feeding the material into the thermoformer. Such wrapping will also protect the sheet stock materials from other types of contamination.

3. Negative-Heat (Heat Sink) Cooling Effects Caused by the Clamp Frame or Chain Rails

Any type of thermoplastic sheet in the thermoforming heat cycle is subjected to a controlled amount of heating. The exposure to heat permits the sheet material to absorb the measured heat levels to become more formable. Since the thermoplastic sheet material is held by the clamp-frame or chain-rail system, the same systems can act as heat sinks. The close proximity of these clamping devices will show heat-robbing tendencies. The carefully measured heat levels put into the sheet are subjected to cooling by the cold metal frames or rail contacts. These will channel the heat to the machine body by the natural heat conductivity of the metal structures. The heat-robbing tendency of the clamping mechanism and chain rails can be eliminated by allowing sufficient spacing between the actual forming area and the undesired cooling zone.

IV. MOLDS

In the thermoforming process the actual shaping and forming of the heated thermoplastic sheet are done with molds. The heated and softened thermoplastic sheet material is forced into or drawn over a mold. In the thermoforming molding cycle, the shape and contours of a mold transfer its form to the softened plastic sheet. When the heat is absorbed by the mold and the plastic is cooled, the plastic's newly acquired shape will be set into it. The mold itself is the determining factor in the final shape of the formed plastic. Occasionally, the forming can be done without the use of actual molds, as where "free-blow" bubbles are formed (e.g., in skylight windows). The bubble is formed with air pressure alone. Regulating the air pressure inside the bubble will determine the size of the blown bubble. As soon as the plastic is cooled by the ambient air, the forming is completed and the bubble becomes self-supportive. This type of moldless thermoforming technique is very limited in both product adaptation and production. About 95% of present thermoforming methods require the use of molds. There are three basic types of molds: female, male, and matched. The use of these three basic mold types permits shaping of the plastic sheet with some inherited differences. These differences not only affect the molding techniques but create fundamental differences in the outcome of the products formed. The products will usually provide clues as to what type of mold has been used to form its shape. This inherited difference carried by the thermoformed product allows both the type of mold and the molding technique to be pinpointed. Each mold type will have distinctive characteristics implanted into the formed part. This could serve as an advantage or a disadvantage for the final product. This inheritable characteristic may affect the decision on what types of molds will be used when a particular product is made. The material makeup of molds varies a great deal and the composition is often dictated by the twin factors of intended use and life expectancy of the mold. Temporary or experimental molds can be made out of less costly but easily reworkable materials. For other purposes, the molds may have to be made to long-life-expectancy, high-volume production. However, no matter what purpose the mold is built for—from temporary short-term molding conditions to long-life expectancy—it has to withstand the thermoforming conditions without dimensional changes or failure. The construction and the various materials from which molds are made are discussed in Chapter 5.

Molds in the thermoforming process have two functions: to provide a basis for the heated thermoplastic sheet to receive its shape, and to cool down and remove the heat from the formed plastic. In both cases the ultimate goal of the thermoformer is to achieve maximum surface contact with the mold. Having the utmost surface contact ensures the most detail transfer in the molding, plus the highest efficiency in cooling. Molds within the three basic categories come in a variety of sizes, from as small as $1/8$ in. by $1/8$ in. to as large as 20 ft by 30 ft. Molds can be used in the thermoforming process in singular units as "one-up" molds or in a group of identical configurations called "multi-up." Multi-up molds can have as few as two units to as many as 81 units or more grouped together. Molds may contain a group of different mold configurations, called "family molds." To combine several different configuration molds into a family mold, it is important that there be some similarities in their dimensions or configurations where the thermoforming method does require the same conditions. It is also impòrtant that the various products produced by the family mold have the same consumer demand. If that is not the case, an imbalance in production will result, leaving an unwanted surplus of items of less demand.

A. Female Molds

In the thermoforming process, a mold made with a cavity configuration is called a "female" mold. The thermoforming of a heated thermoplastic sheet with a female mold is always made with the sheet forced into the cavity. The actual forming is accomplished with either a vacuum force or air pressure applied from the outer sheet side, or by employing both forces at the same time. The simplest thermoforming method uses a vacuum as the force to pull the heated thermoplastic sheet material into the mold cavity. The use of a female mold cavity always leaves characteristic evidence of the mold on the thermoformed part. The female mold is usually shaped in an open, flared cavity form. This typical female mold configuration and the resulting thermoformed part wall thickness configuration are shown in Figure 25. The actual forming of the plastic sheet is accomplished by stretching a specific area of the sheet into a much larger surface area of the mold cavity. Basically, the area of the mold cavity opening is stretched out to cover the entire mold cavity surface. The actual stretching of the thermoplastic sheet material is usually negligible or limited at the outside of the circumference of the cavity. This is due to the chilling effects, self-sealing, and sheet-holding ability of the mold. At the same time, almost all the stretching in the sheet will happen within the circumference of the mold cavity. This will result in substantial gauge reductions. The gauge reduction will cause thinning of the formed part. When the lip area (flange) of the formed part is measured for gauge thickness, it can be assumed that the measurement is very close, if not identical to the original thickness of the sheet material. The reduction of gauge, due to the stretching in the forming cycle, is in direct relation to the "depth-of-draw ratio" or the depth of the mold cavity. The deeper the mold cavity, the greater its surface area that the plastic must cover in thermoforming. A larger surface area will cause more thinning in the side and bottom wall thicknesses. Ultimately, the depth of the female cavity can reach a point (maximum depth-of-draw-ratio limits) where the thinning is so extensive that the stretched plastic sheet has the appearance of a film membrane. In the most extreme case the stretched plastic will rupture completely. With simple female mold types of forming, the ideal depth-of-draw ratio

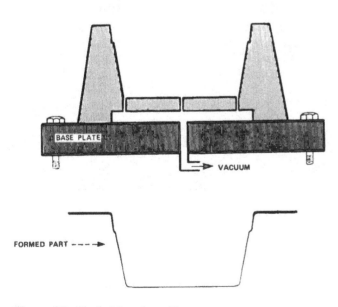

Figure 25 Typical female mold.

should be limited to 1:1. It is possible to achieve ratios of 1:3 or 1:4; however, substantial thinning of the side and bottom wall thickness will accompany the forming, and such extensive thinning may prove to be detrimental to the product. To minimize this accompanying poor wall thickness distribution, there are many thermoforming techniques that can be adapted. The use of prestretching techniques and the employment of plug-assist formings can provide unmeasurable thickness variations. With today's state-of-the-art technology, the depth-of-draw ratios have been pushed to the limits of 1:7. The various methods of improved and specialized thermoforming techniques are discussed in Chapter 3.

The heated plastic sheet when formed into a female mold and cooled by the mold will shrink in size. The various types of plastics will have different shrinkage rates. The shrinkage will make the formed part smaller than the actual mold. This is a definite advantage in the part removal or stripping operation. The reduction in size actually pulls the formed part away from the mold sides. This phenomenon allows the use of much smaller sidewall angles without the adverse effects associated with male molds. Female molds, due to the effects of this shrinkage, can have very small sidewall angles: as small as 1° without any major removal problems. Naturally, when very small sidewall angles are chosen, it is important to consider their interdependence with the overall depth of the forming and the intricacy of the mold design. Last, but not least, the number of "ups" in the mold is another influencing factor that has to be investigated. In multi-up molds, when the part removal is implemented, a curvature or bow may develop in the overall panel. Such a bow will result in a nonparallel movement in the stripping, causing the formed parts to cock and change sidewall angles in relation to the angles of the mold. If and when cocking occurs, strip bars must be used. It is the ease of part removal from female molds that makes this type of mold so popular.

B. Male Molds

A male mold is essentially a completely reversed form of the female mold. Instead of a cavity configuration, these molds are in the form of a protrusion. Male molds are often selected over female molds because the manufacturing or shaping of the mold is much easier and less costly. In most cases, the shape a male mold is less costly than creating a cavity (female mold), which would require cutting in, milling out, or removal of material. Thermoforming with a male mold is often referred to as "drape forming." This phrase arises from the fact that the heated thermoplastic sheet material is draped over the protruding mold surface and then forced down to take its shape. Thermoforming with the male mold produces a complete reversal of material distribution from the female mold. The typical male mold and its resulting material distribution are shown in Figure 26. The top portion of the formed part will best retain a thickness

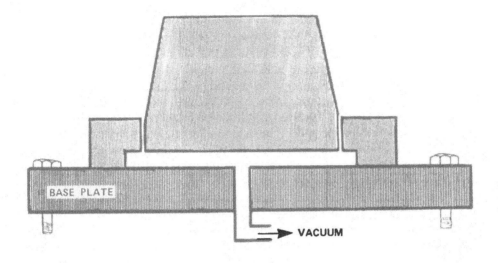

Figure 26 Typical male mold.

closest to that of the original sheet material. Most of the material's stretching and thinning will occur on the side and flange areas of the formed part. The minimal stretching at the top portion is due to male mold contact with the preheated sheet, causing it to cool and set up before substantial stretching can be accomplished. Using male molds with larger radii along their upper edges together with warmer mold temperatures will provide a much improved stretch to the center portion of the formed part. At the same time, when the stretching is increased, material stretched and moved out of the upper area will be allowed to move into the side area. The material stretched out of the upper area could well enhance the sidewall thickness to a point where nearly even material distributions is realized. Achieving ultimate material distribution is often limited to a specific mold shape and to the overall mold height.

The use of male molds is very popular with blister packaging manufacturers. The forming technique involved in making a male mold is more economical than that for the female mold and allows all sorts of opportunities for quick alterations. On the formed blister, the highest strength is obtained along the uppermost surface area of the bubble, with minimum thickness at the flange area. The slower strength along the flanges rarely weakens the package because of the added support of the cardboard blister card. The reduced blister flange is actually a blessing in the heat-sealing procedure, where it demands less heat sealing dwell time to seal the blister to the cardboard. Male mold usage is also very popular with sign manufacturers, who have only to position the male letter molds to create raised lettering on their signs. Male molding, or the actual act of drape forming, is not limited to just a very few thermoforming methods. Almost all forming can be accomplished with either a male or a female type of mold. However, the key to material shifts and thinning may dictate the type (male or female) of mold that is required to obtain the desired product criteria. With male molds, the thermoformer will experience difficulties when multi-up mold patterns are used. The spacing requirement between individual male molds must be appropriate. If less than adequate spacing is used, unwanted webbing or wrinkling will develop between the molds. The space between male molds must be looked upon as a female cavity and therefore can fall under the depth-of-draw-ratio limitations. The webbing actually develops because the sheet material is overstretched, exceeding the actual surface area. Increasing the spacing between the male molds, or using "helper bars" or any type of assist pushers, will eliminate the unwanted webbing effects between male molds. Another drawback of male molds is their sensitivity to release of the formed part. Deep dimensions and small taper angles are definite culprits in male mold forming. It is a well-known fact that plastic parts shrink in size as they cool. The actual dimensional shrinkage will vary from thermoplastic to thermoplastic; however, all of them will shrink. With male molds it should be obvious that this shrinkage will cause the formed part to tighten on the mold. With the lack of sidewall tapers and the deep dimensional configurations, the cohesiveness could be so severe that the formed plastic part is no longer easily removable from the mold. The degree of difficulty in removing the formed part is in direct proportion to either the depth of draw or to the male mold height and sidewall angles. Adding to the situation is the problem of design intricacy. Usually, angles of less than 10° present stripping and removal difficulties which rapidly increase as the taper angles are decreased; this condition will reach the point where formed part destruction will result during attempted part removal. In less severe conditions, part removal from the male mold is accomplished with the help of mechanical pushers or strip bars. Pressurized air is often used to blow off the formed part from the male mold. When pressurized air is used, careful consideration has to be given to the way in which this procedure is accomplished.

Using the same plumbing and holes for the blow-off that are used for the vacuum often results in poor or partial blow-off, which could result in a cocked (angled) part. The blown air will seek the patch of least resistance and concentrate on the vacuum holes that are the least beneficial for part lift-off. In this case it will have a lot of noise (air rushing out from under the plastic part) but will not help to strip the part from the mold. Having a separate blow-off hole system at the very tip of the mold will greatly reduce the chances of poor lift-off and stripping. The level of air pressure also has to be regulated when it is used for stripping the formed plastic parts. Strong blasts of air can often distort or even rupture the formed part. Of course, all of these unwanted effects are strongly dependent on mold configurations and sidewall angles. (This is explored in Section V.)

C. Matched Molds

Forming thermoplastic sheet into various shapes can be accomplished by borrowing a technique from steel sheet stamping. In this method the heated thermoplastic sheet is clamped and forced into a female mold by a matching male mold. The mold gap between the male and female molds is predetermined according to the specific product and is there for the thermoplastic sheet to fill. The forming is performed identically to steel sheet stamping, or can duplicate coin-making techniques. In both cases, the forming is done with a female mold, which has an opposing counterpart male mold. The two together press the heated thermoplastic sheet between them and by shear force induce the plastic to take the shape of the mold gap. Naturally, there are limits to the shapes and depths of draw that can be accomplished by matched mold thermoforming.

There are two types of matched molds: completely matching contoured molds and partially matching contoured molds (Figure 27). Each type of mold has its own place. Each makes products for different purposes, and each is made specifically for different types of materials. Both types of match molds require an absolutely perfect mold alignment between the male and female counterparts. Because this alignment is so important, the molding procedure must be made on equipment that has properly parallel platen movement.

1. Completely Matching Contoured Molds

In this type of match mold, the male side of the mold follows the respective contours of the opposing female mold. There is a size difference between the male and female molds, creating a measurable gap between the two. This gap, often referred to as "mold gap," is the space that will be filled by the heated

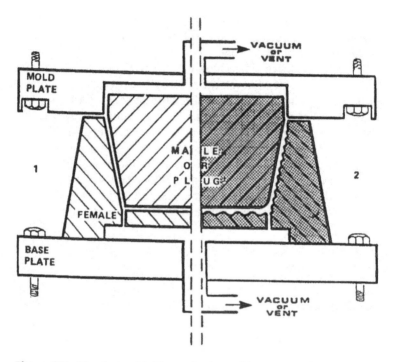

Figure 27 Matched mold: (1) completely matching contoured mold; (2) partially matching contoured mold.

thermoplastic sheet. Actually, the male mold mechanically forces the heated flat thermoplastic sheet into the female mold. By the squeezing actions of the two opposing molds, the plastic sheet will fill the mold gap and attain its surface dimensions and patterns from the respective mold sides. The implementation of this type of match mold is generally done with solid thermoplastic sheets. However, these molds are well accepted for thermoforming of foam thermoplastic sheets or even for the "twin-sheet" thermoforming methods.

2. Partially Matching Contoured Molds

This group of molds has also been well established in the field of thermoforming. Here the mold sides do not match the contours of the opposing side. The resulting discrepancy in the configurations of the molds can vary greatly. Naturally, there are limitations to the dimensional variations between the male and female molds; this should not interfere with the thermoforming operation. Partially matching molds are most popular for the thermoforming of foam thermoplastic sheets and also for twin-sheet forming. To achieve well-detailed dimensional surface on both sides of a thermoformed foam product, a match mold must be used. With molds that do not share the same contour, it is possible to thermoform different designs for the inside and outside of the formed product. For example, a small container can be thermoformed out of a foam sheet which may have a smooth inside finish and a ribbed design on the outside. In this case, the male mold will have a smooth surface and the female mold will have the ribbed pattern. Cross sections of the wall of this container will display foam density differences; the least density will occur at the ribs; these being at the thickest points. At the same time, the foam area between the rib pattern will have to be squeezed and therefore display the greatest density. Egg cartons are produced the same way. To create the hinge, the foam is squeezed into solid material where the lid folds over the base. This is accomplished using opposing male and female molds. The mold gap is reduced so much that the original foam material is squeezed to little more than a flexible membrane. At the same time, the male and female molds create larger mold gaps at the egg cavity portion of the egg carton, allowing the foam to form into soft, cushiony cups. The lid of the foam egg carton is only partially compressed between the male and female mold halves, giving the lid a perfectly smooth surface and extra rigidity for the printing of a label.

There are times when a matching mold is purposefully made and built in such a way that certain parts or areas of the mold have no clearance between the two mold halves. In this situation when the two mold halves are made to close against each other and have a heated plastic sheet in between, the plastic sheet will interfere with the mold closing. Of course if the mold closure is done with sufficient force it will squeeze the plastic first and then tear it, making a hole where there is no clearance located between the mold halves. The squeezed plastic usually ends up in the area where there is extra space left between the mold halves.

This particular phenomena can be adopted for a beneficial use, when foamed thermoplastic material is thermoformed. It can be adopted to produce a hole or holes in a predetermined location and size holes in a thermoformed product instead of punching holes out in a secondary operation. This method will and can produce holes accurately and repeatedly without any problem. As the mold halves engage it will squeeze the foam material, then as the "no clearance" comes into play will tear the material in a most controlled manner. This creates a specific size hold matching to the "zero" or no clearance area, while the material which is squeezed away will end up in the adjacent area by increasing the foam's density there.

Such a hole producing method is most popular with those manufacturers who produces egg cartons, fast-food carry-out trays, and hinged-lid containers. These items do have a built-in closing mechanism made of a tab on its lid and a latch hole on the base. The two engage and lock into each other to facilitate the closing and at the same time allow easy opening of the package. In the past, to produce such latching mechanisms, the tab holes were punched out by mechanical punch and die sets. This was not an easy task and required very costly equipment installation that had to perform such a task after the molding procedure was completed. The punches failed, resulting in damaged or unpunched units, intermittently. The entire production had to be rejected.

The above hole-making procedure with the zero clearance mold design is covered by patents whose validity often has been challenged in court.

It is interesting to observe such innovation as it is found and made, to be covered by patent as an intentionally developed purpose. The very same result is often produced by accident, when someone makes a poorly aligned mold setup. Each time when the molds are placed into a thermoforming machinery where the male and the female molds are so badly in misalignment it can and will produce the same results of a torn hole. I am most certain that this type of a finding was the basis of this hole-punching invention.

A mold with engraving or raised lettering qualifies as a partially matching mold. The markings on an engraving are usually not matched in the opposing mold surface, and this creates a larger gap at the engraved area than at the rest of the gap. In the case of raised markings (e.g., raised lettering), the mold gap is reduced by the height of the pattern design or letters. The reduced gap requires some compression of the material, which has to be squeezed out of its way.

Entirely different shape molds opposing each other are frequently adapted in the thermoforming of twin sheets. In this type of thermoforming process the two opposing molds, in their extreme, may have entirely different configurations. Twin-sheet thermoforming starts with two sheets instead of one single sheet. The two sheets are on top of each other, clamped together or carried by the chain system. Because they are placed together, heating the twin sheets naturally takes longer than heating a single sheet. When the twin sheets are properly heated for the forming procedure, the two opposing mold halves close on them. Between the two sheets, syringe-like needles are placed. These needles are open to the atmosphere or can be attached to pressurized-air lines. When vacuum is applied to the molds, it will pull the respective side of the twin sheet to form against them. The needles will allow air to enter between the two sheets, filling the newly created space. If pressure is applied through the needles, they will create the same result, forcing the plastic against the respective mold sides. The only difference between the two forming techniques is that the pressure-forming approach requires a tight squeeze and strong platen forces to hold the two sheet materials together and airtight.

Today's modern thermoforming techniques do not rely solely on the foregoing simple female/male mold configurations. Thermoformers are not limited to choosing one or the other type of mold because of its basic inherited product benefits or cost-effectiveness. Today's molds are generally far more complicated in shape and configuration. Forming may be done on a female mold which has either male protrusions in its cavity or additional female cavities within itself. Female molds with engraved lettering, for instance, should be treated as a female mold in a female mold. Also, mold combinations can be made in complete reversal of the molds listed above: They may come in male and female portions within. When the process of thermoforming is made with such combinations of molds, the practitioner must be aware of the possibilities and for optimum results, make the necessary adjustments in heating, platen height, and movement. Often, a minute adjustment is enough to greatly improve the formed part quality. In a few instances, when the mold configuration has been chosen in the wrong form, a complete reversal in the mold's shape should be prescribed. The initial examinations prior to mold making should help to avoid such costly mistakes, and most practicing thermoformers can choose correct mold configuration patterns. A truly serious error in the tooling patterns occurs only when the prominent features are not very distinguishable or the choice of the better-detailed side of the thermoformed product is disregarded. The principal or prominent side of the mold is the one that comes in full contact with the heated thermoplastic sheet and therefore transfers its own configuration onto the forming plastic. The other side of the sheet may not totally come in contact with the mold, or just momentarily contact its surface, resulting in less detail transfer. When aiming for high-quality detail on a particular side of a thermoformed article, choosing male or female mold configurations for the prominent mold side can be a key factor.

V. FORMING FORCES

In the thermoforming process of forming the heated thermoplastic sheet into a useful product shape, the processor has to coerce the sheet to assume the actual mold configurations. To shape a flat sheet form into a different form and compel it to follow the contours of the adjacent mold, outside forces must be employed. The energy levels of this force should be adjustable so that the plastic sheet can either be eased into the coerced shape or, if necessary, explosively thrust into its new shape. This shaping and forming force should have a range great enough to allow it to be applied from a gentle force through gradually increasing levels up to a powerful forming energy. Naturally, power introduction is limited to its energy source, as well as the mold's and equipment's ability to apply and withstand such elevated forces.

The common forming forces used in the thermoforming operation are: vacuum, pressure, matched mold, and combinations of those three. The selection of a particular forming force for a specific thermoforming process is usually guided by the product size, number of parts to be made, and the desired cycle speeds. In addition to the criteria above, the following factors should be considered: distinctive thermoplastic sheet material limitations, mold material strength, and the available thermoforming equipment. Any of these can make a difference in the selection of a forming force.

A. Vacuum Forming

The oldest method of shaping plastic sheet into a useful form is with the use of vacuum force. The original description for this thermoforming process was "vacuum forming." Since other methods have been adapted for forming forces besides the vacuum, the general term "thermoforming" has taken over. The process of vacuum forming still dominates the industry, and there are manufacturers who use nothing else to shape the heated plastic sheet into useful products. As a source of forming force, it is easily accomplished, manipulated, and controlled.

The basic principle of the vacuum-forming process relies on the heated plastic sheet's self-sealing ability and trapped air space evacuation by vacuum. As the air is removed from the enclosed cavity, it causes a pressure reduction on that side of the sheet, thus allowing the natural atmospheric pressure to fill the cavity and force the heated plastic sheet into the evacuated space. The principle of this basic type of forming can be seen in Figure 28. As this drawing

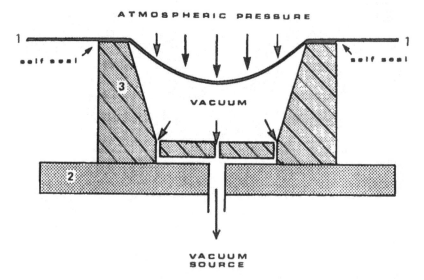

Figure 28 Vacuum forming: (1) preheated thermoplastic sheet; (2) mold base plate; (3) mold body (female mold).

reveals, vacuum forming can basically qualify as pressure forming, using the natural atmospheric pressure to accomplish the forming task. Naturally, this forming technique depends on a tight seal being created on the vacuum side of the sheet. Any premature actuation of the vacuum prior to the obtainment of a good seal is not only a waste of energy but can prove to be detrimental to the quality of forming. The rushing air movement created by the suction could very well cause chilling of the heated thermoplastic sheet, which would affect its stretching—and ultimately, forming—results.

Thermoforming using vacuum is easily accomplished. First, the heated and softened thermoplastic sheet will usually have good self-sealing capabilities. Second, creating vacuum is an easy task, thanks to the availability of a variety of relatively inexpensive vacuum pumps. The only thing the thermoformer has to watch for is that sufficient force and volume of vacuum are available at the mold. Small-volume-capacity pumps may produce high-enough vacuum ratings, but due to their lack of capacity, may not provide enough volume to evacuate all the air. The vacuum pump volume capacity (cubic inches/second) must match or be greater than the needed evacuation area of the mold. Often, smaller vacuum pumps can be accommodated if attached to a reservoir tank ("surge tank"). This larger tank is completely emptied by the vacuum pump, which has to work constantly to replenish the vacuum source during the off-cycles. If the level of pumped air runs even slightly under the total volume of the tank in a continuous cycle, the vacuum will, sooner or later, show a drop in its force. Lack of vacuum force can also be caused by inefficient plumbing. Undersized plumbing lines or couplings between vacuum pumps and the mold can cause vacuum force loss at the mold. Lack of vacuum force and volume capacity hinder both thermoforming quantity and product detail. In addition to the above, another serious result is lengthening of the cycle speeds of the thermoforming operation. In fact, since vacuum forming is dependent on the atmospheric displacement pressure, it cannot be improved on. When maximum (close to 29 inches of mercury) vacuum is applied at the molds, the highest forming speeds and best detail in forming will be obtained.

Having the highest possible level of vacuum in the beginning of the forming cycle will not guarantee that all the details of the mold will be transferred into the formed plastic. As the forming procedure takes place and the air is evacuating from the mold, the vacuum levels naturally decrease. This is evident much more with a larger sized part forming than with a smaller article forming. As a matter of fact, this should concern only the large part thermoformers and not those who produce the smaller sized thermoformed products. This will happen in spite of a sufficient sized surge tank employed to evacuate all the air from the mold cavity. It is easy to understand that as the forming begins a continuous drop in vacuum levels will result because the air from the mold will be sucked into the surge tank. This drop in the vacuum level will be continuous as more air is pulled from the mold into the surge tank. Such a drop in vacuum levels is a normal result of the forming and the results can be monitored easily on the vacuum gauge dial. When the forming is completed the vacuum drop should stop, unless there is a leak in the system. If a leak is present the vacuum levels completely diminish and the gauge will indicate zero on its dial. Normally the vacuum drop should stop before going all the way down to zero, and should start to increase as the vacuum pump regains some vacuum. Of course all this scenario only holds when the plastic sheet was heated to a proper forming temperature and it is not suffering from lack of proper heat.

Now let us assume that the forming of the plastic sheet starts with the proper forming temperature and with the highest available vacuum levels. It is a good idea to establish at this point of the discussion what is the ideal and highest level of vacuum for good forming. In my own opinion as close as to 29 in. of Hg is the optimum. However, a slightly lower level, like 27–28 in. of Hg is still an outstanding level to perform thermoforming. Anything less than that may not be as suitable and as the levels get smaller the forming quality speed and details diminish as well. But by no means is 22 in. of Hg an acceptable maximum level to begin thermoforming, as will become more evident by the following discussion. So as the vacuum forming begins a natural phenomena occurs, starting with the drop of vacuum levels. As the forming progresses a $\frac{1}{3}$ to $\frac{1}{2}$ drop in vacuum force will occur from the original levels. The fact is that the last portions of the forming are always accomplished with a greatly reduced vacuum force. It also the very fact that the final details and the finished quality of the forming is achieved when the plastic comes in full contact with the mold. This most crucial moment happens in the last stages of the forming. Due to the reduction in vacuum levels, we are definitely asking too much from the reduced vacuum force to accomplish.

To remedy this long time neglected situation, we should be able to add a boost to the vacuum levels in those critical last stages of forming. By adapting a two-stage vacuum system this can be accomplished. The first system should be left the same as the original and existing system, large enough to match the mold cavity volume and to be operating at a reasonable high level of vacuum. The second stage system should consist of a smaller and separate vacuum system, without connection or any chance of bleeding into the larger system. Both systems, the large and the small, should be working one after the other and have maximum vacuum levels. The plumbing and valve setup should be made in such a way that the first and larger system will accomplish the main air evacuation. As that system is turned off, the second smaller system instantly switches on and with its higher level of vacuum force, completing the final stages of the forming. To implement such valves of this somewhat complicated system and to make it work well is critical. If the smaller system is opened too soon, its vacuum force will bleed into the larger tank. The technique would not work. If too much time elapses before the actuation of the second system, precious time is lost and the partially formed article can cool off, hampering the forming quality. But if it is done correctly, the quality improvement is so elaborate that the results are nothing else but crisp and sharp definitions formed into the plastic parts.

To reemphasize the above mentioned two-stage vacuum system, the main source of vacuum should be large enough to evacuate most of the mold's air volume, while the secondary system's purpose is to finish the forming with a higher vacuum level. The two systems should be separate, only valve connected, and never activated at the same time. The plumbing and valve mechanism should be made and connected to the mold's air evacuation line in such a way, that at the last portions of the forming the large system is switched completely off, and just then the secondary smaller system should come on. This must be done almost simultaneously, without creating any opportunity of an accidental connection between the two systems; otherwise the whole effort is wasted. The purpose of the second smaller vacuum system is to provide maximum vacuum levels for the final forming segment, and with that, gain the optimum detail transfer from the mold to the thermoformed article. By using this two-stage vacuum system the quality of detail transfer can be so improved that it will closely resemble the details of pressure forming, and this is gained without the cost of implementing pressure forming. Of course if that is still not providing satisfactory detail or the engineer is seeking higher speeds of thermoforming, it will be necessary to implement the pressure forming technique.

B. Pressure Forming

To obtain higher forming speeds and clearly defined detail, a greater force than gentle vacuum forming can offer should be applied. In addition, certain plastic sheet materials, such as the oriented polystyrenes (OPS), cannot be formed with the slow vacuum forces. They require rapid speeds and higher forces in order to form. Vacuum forming takes too long for this type of material and would permit deterioration in the molecular orientation of the plastic before full forming could be obtained. Pressure forming is actually accomplished with higher air pressures and, depending on the needs, can be tuned from 10 psi to the maximum of full line pressures of the plant's compressed-air system. To achieve pressure-forming capabilities for thermoforming, the first requirement is to use molds that are capable of withstanding the pressure force applied. The use of molds made from weak mold materials or not constructed strongly enough to withstand the applied pressure should not even be attempted in the process of pressure forming. Poor mold construction under the pressure of pressure forming can actually act as a bomb and explode. For pressure forming, a temporary or short-term run mold should always be built like an actual production mold. Cast aluminum or machined metal tooling is a fine choice for achieving the strength required. Tooling made of wood and plaster castings should not be used with pressure forming.

The second requirement of the pressure-forming technique is to create a sealable pressure chamber. Pressure chambers can be attained between the mold lips, which are closed with a pressure plate. The addition of a sealing material may be necessary. The softened thermoplastic sheet material itself can act as a seal. The basic principle of this pressure-forming technique can be observed in Figure 29. In multi-up molds, it may not be necessary to build pressure chambers for every mold cavity; the entire mold can be secured with a common pressure box. These boxes create a complete pressure chamber for all the cavities within the mold. They have a pressure seal built in the perimeter of the mold and surround the mold. The entire pressure box and the surrounding seal should be capable of withstanding the applied air pressure of the pressure-forming force.

The third requirement is that the thermoforming machinery hold the platens strongly enough to be able to withstand the pressure forces between

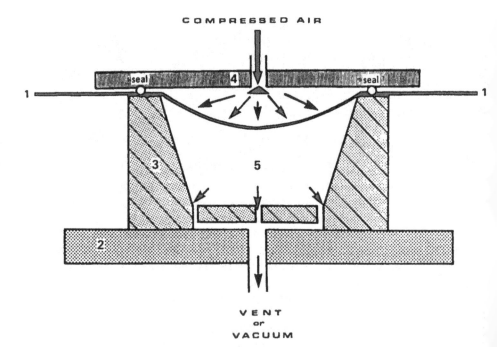

Figure 29   Pressure forming: (1) preheated thermoplastic sheet; (2) mold base plate; (3) mold body (female mold); (4) pressure plate; (5) trapped air.

the pressure box lid and the mold base. If the platens are not held against each other with sufficient strength, the pressure buildup may force them apart. In extreme cases, this may totally hamper the thermoforming; in less critical situations, this will only create a high-pitched whistling sound. This pressure loss could affect the thermoforming quality and make it more difficult to obtain superior detail. Ironically, the introduction of higher pressures to improve the detail of the formed part does not remedy the situation because the higher pressure actually increases the platen separation. For best results in pressure forming, the thermoformer should start with the lowest possible pressure force and gradually increase the forming pressure. As soon as satisfactory parts are produced, there is no need to increase the pressure further. By this method, the operator not only minimizes the chances for tool destruction but eliminates the possibility of overpowered pressure forming. Overpowered forming occurs when the heated thermoplastic sheet is forced, because of excessive air pressure, to form a female cavity. Because the air inside the cavity cannot exit quickly enough and the air pressure force is so rapid and intense, the overpowering effect results. The forming sheet, pushed into the mold in the interim, will take a bottlelike or pearlike shape. The bulging section of the particular pearlike shape will form tightly against the bottom part of the mold cavity, plugging up all the evacuating or venting holes. At the narrow neck of this bulge shape, the trapped air will have no chance to escape. This phenomenon, in a less severe form, will show up only as a slight distortion or loss of detail at the upper portion of the formed part. However, the cause is the same: The trapped air will form a flat, pillowlike space. As soon as the air pressure force is reduced and its overpowering nature is eliminated, this unwanted effect will completely disappear.

The most practical advantage in the pressure-forming technique is its forming speed. With higher air pressures than the atmospheric conditions would normally permit, the heated thermoplastic can actually be shot against the mold with a blastlike force. Pressure forming not only can achieve the forming with a powerful blast, but can also continuously hold this energy force when the plastic sheet is forced against the mold surface. Pressure forming is usually done with the cavity side openly venting its trapped air into the atmosphere. However, for best results in definition and faster air evacuation, the application of a vacuum is highly advantageous. Use of vacuum for the evacuation of air from the cavity can be accomplished with little effort and minimal cost. The use of pressure-forming techniques combined with vacuum evacuation can consistently guarantee the best uniformity and detail resolution to part after part. Today's high-volume, high-speed production can only be achieved with pressure-forming forces. To maintain good pressure-forming conditions, the manufacturer concern should have available an ample supply of pressurized air with uniform pressure levels. The air compressor providing the pressurized-air supply should have an excess of essential air volume. It is often debatable whether a thermoforming plant should depend on a single large compressor or two smaller compressor units coupled into the same pressure line. To pick the correct compressed-air system with the proper size and numbers, the type of operation and various business criteria must be considered. A change in the direction of business can leave the plant with the wrong type of air-compression system.

One large single compressor will provide a single source; however, when breakdown or routine maintenance occurs, a complete shutdown will soon follow. With two smaller compressor units, when one of the units is out of order, only a partial equipment shutdown will have to be implemented. The other compressor will be able to supply compressed air to at least a portion of the equipment.

In the use of pressure forming, it is also absolutely necessary to have a dry compressed-air supply. Moisture in the air supply plays havoc with the plumbing system, causing it to corrode. Additionally, moisture can accumulate in the pneumatic cylinders, similar to a condenser effect. In time, the accumulated moisture will collect in larger quantities, which, in effect, can change the stroke of those cylinders. The stroke change will limit the travel of the piston within the cylinder. This, in turn, will affect the sheet advance or platen movements. When the predetermined strokes are changed or limited due to an entrapped moisture layer, the thermoforming machine functions are altered, rendering improper operating conditions.

The moisture in compressed air can also have unwanted effects on the molds. The volume of air traveling in and out of the mold can cause corrosion of the metal mold parts. Condensation, which is pure distilled water, combined with airborne contaminants, can be highly corrosive. This corrosiveness is especially active on aluminum molds, resulting primarily in the production of aluminum oxide, a white powdery substance. The constant air movement through the mold and its small vents and vacuum holes will trap this corrosion residue. In addition, airborne lint, oil, and dust will become packed in the small holes. In fact, the clogging can be so thorough that within a short time, many of the original holes are completely plugged up. The substance in the holes is so well packed that it gives the appearance of concrete. Hole redrilling is close to impossible in this situation, and many expensive drill bits can be spoiled before new hole locations are chosen. Prior knowledge of this unwanted side effect will encourage preventive measures, such as extra hole placement in the molds, periodic steam cleaning and treatment of the molds, and the installation of air drying equipment spliced into the compressed-air circuit. The use of pressure force in the thermoforming process has made this manufacturing technique what it is today: an accurate, high-volume, highly competitive producer of many useful products.

C. Matched Mold Forming

In the thermoforming procedure called "matched mold forming," a heated and softened thermoplastic sheet is formed between two opposing mold halves with somewhat similar contours. In matched mold forming, the thermoformer closely mimics the long-established sheet-metal stamping operation. As the two molds close onto one another, their contours will force the sheet to take up an identical shape as the gap is created between the two mold halves. Any male protrusions on the side of a mold will mechanically force the plastic to form into the counterpart female side. The real difference between sheet-metal forming and thermoforming is that in thermoforming, the depth-of-draw limitations are much improved. This improvement in the forming depth is due to the higher stretching qualities of the heated thermoplastic materials versus the malleability of most metal sheets. When thermoforming is made by the matched mold forming technique, the forming is done by the mechanical forces of the platens. There are three basic criteria that have to be met to achieve satisfactory thermoforming with this technique. The first criterion is that the platens must have enough mechanical energy forces to accomplish forming of the intended plastic part. The platens' driving forces, whatever their source (pneumatic, hydraulic cylinder driven, mechanically cam operated, or moved by a toggle mechanism) must have sufficient power to induce the plastic to form. Naturally, the larger surface area or more intricate mold shape requires higher platen forces. When both of these conditions are present, the operation will demand even higher platen forces. Sufficient power to form and squeeze a plastic sheet into its shape is particularly important when foam materials are thermoformed. A foam thermoplastic sheet, when thermoformed, usually requires some squeezing of its thickness to achieve the intended shape. Most closed-cell thermoplastic foam sheets, when subjected to the heat cycles, will expand and increase in thickness from their original gauge. This increased thickness allows the foam material to be squeezed and forced to fill the mold gap. However, when the foam material thickness is substantially small, due to limited expansion rates, or the thickness is reduced by natural stretching, the mold gap will not be filled. When this happens, a partial or total loss of detail will occur to at least one side of the thermoformed foam article. By allowing the mold halves to travel and come closer together, the mold gap will be reduced, improving the formed part's surface detail. As the mold gap is reduced, the wall thickness of the thermoformed foam product will diminish. It is well known that the reduction of wall thickness always creates more flexibility and less structural strength within the finished thermoformed foam product.

When a remedy is introduced to solve an existing problem, the remedy may cause adverse effects on other facets of the same process. The use of up-to-date equipment should prevent this situation from arising, since today's thermoforming machines have sufficient platen force tonnage for their size and can handle all types of thermoforming.

The second criterion for matched mold forming can be met by providing proper escapement for the entrapped air. When the matched mold halves close on the heated plastic sheet and the male protruding counterparts of the mold squeeze the plastic sheet into the female cavity, the air should be allowed to escape. Lack of air escapement causes trapped air pockets, which not only will hinder the thermoforming but will not allow the mold closure to be completed. The entrapped air will compress to a point where no further compression can take place. Heated and softened thermoplastic sheet material provides an excellent seal medium—or gasketing—when placed between the molds. The air escapement between the closing mold surfaces is usually nominal and undependable, not at all like in metal stamping. To ensure cavity air evacuation, relief or venting holes should be added to the cavity sides of the mold. The venting is usually channeled to the open atmosphere. However, to obtain more definite results in air evacuation, a vacuum can be attached to the air venting channels.

The third criterion in matched mold thermoforming relates to its depth-of-draw limitation, which results when the only forces employed in the thermoforming process are those coming from closure of the opposing molds. It can easily be recognized that only limited stretching can be achieved between opposing molds. This subdued stretching can be successful only when the mold has large tapered angles. Molds that have rather steep detail angles (close to 90° to the sheet), will curtail the thermoplastic stretching, even to the point where tearing will result. With mold details that do display the steepest possible angles, tearing will occur every time forming is attempted. With less steepness in mold angles, the tearing may be just occasional.

With the matched molds, the thermoformer can adopt the techniques of compression forming. In this type of thermoforming, the thermoforming practitioner basically duplicates a compression molding. This is accomplished by heating the thermoplastic sheet to the point where the plastic material, under the molding pressure, is capable of oozing within the mold cavity. Just as it does in compression molding, the flowing material will fill the mold gap and will form into a complete reproduction of its surface details. This type of borrowed molding technique is very useful and allows the production of surface details that normal thermoforming will not accomplish. This specialized thermoforming method does not have to be chosen for the entire mold configuration; it could be limited to use for only portions of a given mold. In such cases, an absolutely flat and uniform thickness of specific areas is needed: for example, container flanges where the container must have a hermetically sealed lid, and the flange must be perfect; or clear laboratory-type plastic containers, which must have optically distortion-free bottoms. Compression forming of those specific areas is the only way to achieve these desired end results.

Matched mold forming is an ideal way to thermoform high-volume, thinly walled thermoformed containers. With a multitude of variations in matched mold designs, a great diversity in thermoforming can be accomplished.

D. Combination Mold Forming

Today, thermoforming with matched molds need not depend solely on the platen forces. Usually, this type of thermoforming method is combined with a vacuum, air pressure, or both of these forces. Consequently, the matched molds do not have to have closely matching contours at all. The male plugs can be produced in smaller sizes and substantially different shapes than the female cavities. When they are made smaller, the male plugs can act as assists or pushers for the heated thermoplastic sheet material. These assists (called "plug assists") mechanically push the softened thermoplastic material into the female mold. The purpose of the assist is to prestretch the thermoplastic sheet for the final forming. The final shaping is made with the introduction of vacuum and/or pressure forces. Using plug-assist methods in thermoforming provides an outstanding improvement in wall thickness distribution over any of the molds discussed previously.

In any thermoforming process where the combination mold has been chosen, the processor has nearly unlimited opportunities to achieve the best thermoforming results. With the combination matched molds, a multitude of deviations can be obtained. Such deviations might be a change in the vacuum or pressure force levels, the timing of vacuum and pressure introduction, mold closing speeds, or mold sequencing timing. In addition, adjustments can be made for the travel of independent mold halves and mold parting line location. With tuning opportunities such as these, the thermoformer can alter the conditions that would allow better forming techniques for a particular thermoplastic material.

VI. TRIMMING

After the forming cycles are completed, the formed parts usually have to be trimmed out of the surrounding panel. Seldom does the finished thermoformed part not require trimming. Occasionally, a portion of the area that has been used for sheet clamping is retained as part of the thermoformed product. As stated earlier, more than 90% of the thermoformed products undergo some type of trimming operation. The whole panel from which the specific product is made is usually larger than the actual finished product. Normally, when multiple parts are formed out of the same panel, some spacing is allowed between the parts. This spacing is also trimmed away when the finished thermoformed product is trimmed out. As thermoformed parts are trimmed out of the original formed panels, they leave a "skeleton"-like trim. These leftover trim skeletons are divided into two categories. The first category consists of an "edge trim," which comes from the area surrounding the finished product. This edge trim is actually those sheet areas used for clamping or sheet transport as well as the extra spacing that has been allocated for buffer between the actual forming and clamping areas. This allocated space is dependent on depth-of-draw ratios and design intricacies. With shallow depths of draw and wide taper angles, less buffer area, or spacing, will be needed between the actual forming areas and the sheet clamping area. The use of minimum edge spacing is a definite advantage for the processor. The ratio of product versus edge trim is an important factor in the economy of thermoforming. Excessive use of edge trim could easily affect overall manufacturing costs and may produce an uncompetitive situation.

Often, when multi-up molding is done, a given distance is allowed between the individual molds. This spacing also needs to be trimmed away and treated accordingly. Obviously, minimizing this type of trim skeleton is in the interest of the processor. The size of the edge trim is heavily influenced by the types of thermoforming the processor is performing. The deeper the draw ratio or the steeper the product sidewall angles, the wider the spacing has to be between the multi-up molds. On rare occasions, thermoformed products can be produced where either no spacing allowance is needed or the spacing distance is incorporated into the finished product as a flange. If this is the case, no "space trim" will result in the trimming operation.

The edge trim and space trim together make up the entire trim skeleton. The area of the trim skeleton versus the actual product area gives the basic trim ratio. The trim that is cut away from the product is commonly referred to as "scrap." The trim-off/scrap could amount to a sizable percentage of the total product weight. For example, when round objects are produced out of a rectangular sheet panel, close to a 40% trim-scrap factor can result. A trim-scrap factor between 10 and 20% is average in thermoforming, which is the reason this process is always treated as a high scrap producer. On the high side of scrap factors, there could be products thermoformed with as high as a 65% trim-scrap factor. To dramatize this trim-scrap factor ratio, a 1000-lb thermoplastic sheet material will provide 650 lb of trim scrap and only 350 lb of product. Thermoformer operators are constantly trying to improve the trim-scrap ratios. It has been well recognized that any reduction in scrap has the opportunity, after the control of energy consumption, to save the most money throughout the entire process.

A. Hand-Held Knives

Trimming the actual thermoformed product out of the surrounding area has to be done without allowing damage, distortion, cracking, or tearing. The trimming must be dimensionally acceptable and done quickly enough to meet production requirements. The type of trimming apparatus can range from a simple hand-held knife or scissors to the most sophisticated laser beam equipment. Naturally, the simplest instruments are the least costly and therefore the most popular (Figure 30). Hand-held knife cutters are most often used when only a few parts need to be cut or even when a limited quantity of production must be trimmed. For best results, hand-held knife trimming is done with the formed plastic part placed on a softer (wooden or rubber-padded) table. The knife blade tip is forced through the plastic, and guided either freehand or by a jig. The knife is pulled to cut the plastic trim away from the part. The knife can be guided by drawn markings or even by thermoformed markings forced into the actual panel. In improved versions of hand-held knife cutting, a secondary jig is built that will hold the formed panel. Where the cutting lines are located, a U channel is built into the jig for knife edge guidance. There is no other limitation to hand-held knife cutting other than that of sheet thickness. Naturally, there is a point in thickness and plastic stiffness beyond which a hand-held knife will not be effective. It is also recognized that hand-held knives, due to their popularity, overwhelming availability, and uncontrollable safety features, have one of the worst injury records. Hand-manipulated cutting knife injuries caused more lost-time accidents than those caused by all other hand-held tools combined. The simple tool should be well respected and should be used in a well-planned manner, as with any other perilous cutting instrument.

B. Electrical Power Tools

The next group of cutting instruments are the electrically powered saws and routers. This type of cutting equipment can be used and guided by hand or can be adapted for use with mechanically guided apparatus. Electrically driven saws may come with circular or reciprocating straight cutting blades. Routers always come with rotating router bits. The saw blades and router bits must be designed and made especially for cutting plastics. The high speeds of these power tools usually create high-friction areas at the cutting surface. The blade or router bit has to have the proper chip-throwing capability. If the cutting

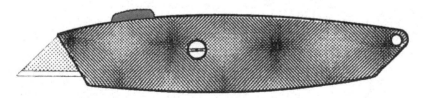

Figure 30   Hand-held knife.

blades or router bits do not have this specialized cutting edge, friction will heat up the cutter and the plastic will melt and adhere to it. The cutting surfaces will not be smooth; rather, they will be ragged or "whiskery" and the trimming actions will constantly be hampered. The melted, sticky residue can easily re-deposit and readhere to the edge of the cut—or elsewhere when it gets thrown out from the cut—on the plastic part surface. Just as critical as the proper cutting-edge designs for the particular plastic is the speed at which the tools travel along the material. Power tools moved at slower speeds tend to create smaller chips or shavings. As the chip size gets smaller, the pieces will eventu-ally become powdery and dustlike. Add this dustlike condition to the natural physical phenomenon of constantly present static charges, and the thermo-forming processor will encounter another processing problem. The static en-ergy charge acts like a magnet in pulling the fine plastic particles directly onto the plastic parts being worked on. The more wiping that is attempted to remove this contamination, the more static energy will be generated, making the effort practically useless. Even the use of a vacuum cleaner proves to be unsuccessful because the moving air creates more static charge in the plastic. To clean off the statically adhering particles requires a great deal of effort, using specially treated sticky rags (e.g., tack cloths), antistatic wiping materials, antistatic sprays, or even antistatic air blowers. Even with this specialized effort, the cleanup is rarely totally successful. For optimum cleanliness and dust-free products (or at least to minimize the problem), the cutting tool must have sufficient cutting bite and travel speed to produce larger chips that will not cling as easily to the plastic parts. Last, but not least, in the war against self-generated plastic par-ticle contamination is the need to keep trimming operations with power tools away from thermoforming operations and the accumulated finished products. Dust- and powder-residue-creating trimming operations should be done out-doors, or at least in a separate room where the particles cannot readily become airborne and contaminate any of the prethermoforming or postthermoforming areas.

C. Automated Cutting Tools

The next group of trimming equipment is used where a higher level of auto-mation is required. In this case, the router apparatus is actually guided by mechanical arms that follow a preset pattern. The guidance could be provided by a jig-following cam mechanism or a more sophisticated, electronically con-trolled guidance system. The cutting pattern is set into a memory bank of the guidance system. The cutting tool will receive commands from the memory as to which way to travel and when to change directions. This type of system is not limited to single-level cutting. The cutting can be programmed to change to other designated levels and can follow very intricate patterns involving multiple levels of cutting. For example, a cutting tool can be directed to cut a complicated pattern on one level, then receive a command to switch to another level for further cutting, and then possibly switch back to the original level to finish the cutting. Some computer control cutting guidance systems can provide five axes of controllable cutting. The cutting may be made simply by an electrically driven router bit or could involve more advanced cutting methods. The actual trimming of the plastic part out of the formed panel, as opposed to using a mechanical cutting blade or bit, can be accomplished with a high-velocity water-jet stream. The water-jet cutter applies high-pressure water with high velocity through a small orifice to pulverize the plastic material under the needle-sharp water-jet stream. The water-jet system is guided through a preprogrammed pattern. The actual cutting head will follow the pattern and make the intricate and multilevel cutting with ease. The edges of the cut on the plastic will not provide a perfectly smooth finish. The pulverizing water jet will leave fine-grooved perpendicular markings on the edge of the cut. The use of high-velocity water-jet cutting is limited to trimming one part at a time and is also limited to working with sheet gauges from 0.010 to 0.500 in. Most of this type of trimming is done as a completely separate step following the forming. However, in recent years, a four-station rotary thermoforming machine, which in the fourth station provides in-line trimming with a water-jet system, has become available from at least one manufacturer.

Thermoformers seeking higher speeds of trimming and better trim surfaces will find that working with a laser light beam will do just that. The actual cutting speeds are not overwhelmingly high, but multiples of formed parts stacked atop each other can be trimmed in a single pass. The laser cutting head is guided by computer and will follow the most intricate patterns. The laser cutter can be constantly turned off and relocated elsewhere to be restarted and ready to cut again in a new location with a different pattern. This type of cutting tool will provide practical cutting because of its ability to trim uniform parts. Additionally, cutting lines can be achieved with more complicated inner cutaways and cutouts, and even multilevel cutting can be accomplished. There are, however, inherent drawbacks to the use of laser trimming in these processes. First, the considerably high price tag of laser cutters makes most jobs economically unfeasible. Second, ideal positioning of the thermoformed panels is very important. Any misalignment will result in an off-grade trimming of the particular thermoformed parts. Laser beams require proper focusing of the work and have to be slowed down to cut thicker material. Laser beam-pulverized plastic in the trimming operation tends to produce some smoke and therefore requires vacuum or venting to produce ideal working conditions. Good safety practices must be followed with any of the cutting instruments. Processors must follow the safety instructions supplied by the equipment suppliers.

D. Steel-Rule Dies

To trim thinner-gauged thermoformed products, steel-rule dies can also be employed. Trimming and cutting with steel-rule dies is a technique borrowed from the well-known procedure of die-cutting paper and cardboard. Thermoformers using less than 100 mils of solid thermoplastic sheet material, or higher gauges of foam materials, can easily adapt this method of trimming. Steel-rule dies perform best when cutting 0.001- to 0.025-in.-thick sheet materials. Steel-rule dies are made out of strips of spring steel that have been prehardened and presharpened. The premanufactured strips may come in uniform 30-in. lengths or be supplied on continuous coils. The steel strips are available in widths from $^{15}/_{16}$ in. minimum to 2 in. maximum. The thickness of these strips is determined by "points" of intervals, where each point is measured at 0.014 in. thickness. The most popular thicknesses are 2 point (0.028 in.), 3 point (0.042 in.) and 4 point (0.056 in.). The steel strips are usually prehardened to a preferred hardness: semihard (45 to 48 Rockwell), hard (49 to 52 Rockwell), or extra hard (53 to 56 Rockwell). The hardness and point thickness are the determining factors in the minimum radii to which the steel rule can be bent. The lower limits of radii for hard rules are: 2 point ($^{1}/_{32}$-in. radius), 3 point ($^{1}/_{8}$-in. radius), and 4 point ($^{3}/_{16}$-in. radius). The strip of steel-rule knife stock is usually bent into the necessary configuration for cutting or trimming. The die cutting may consist of an intricate cutting line with many types of turns and loop-backs. The configuration of the cut may be made out of a one-piece knife strip bent to follow the intricacy of the cutting pattern, or it may be pieced together from several knife strips. The piecing together of steel-rule cutters must be precise to make the cut continuous without any interruptions or misalignments. The steel rule is sharpened after hardening in the strip form. To produce a knife edge on the steel strip, the sharpening or grinding can be made from one side or from both sides. When the blade's cutting edge is ground from one side, the steel rule has a single-beveled cutting edge. When the blade is ground from both sides, the steel-rule knife will have a center-beveled or double-beveled cutting edge (Figure 31).

Using steel-rule dies to trim thermoformed parts from plastic panels is done just as in the die cutting of cardboard. The cutting must be made in the same plane (level) and against an opposing surface. The opposing surfaces are usually made of a hardened steel plate. The formed plastic panel to be trimmed is placed between the steel-rule die and the flat hardened plate. When the two are forced close together, the knife edge will penetrate the plastic and be cut.

As has been mentioned, the steel rule is made and bent to follow the desired contours of the cutting lines. It is of utmost importance that the bends be made accurately: exactly 90° transversely to the length of the steel rule. The slightest deviation from the bending angle causes the knife blade to move out

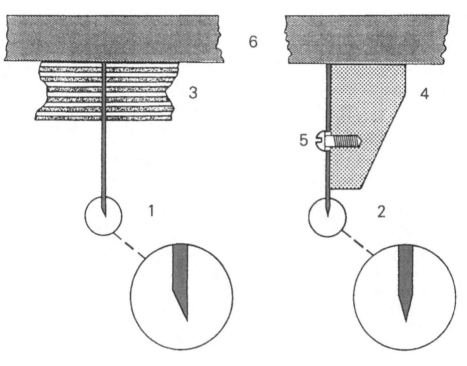

Figure 31 Steel-rule knives: (1) single-beveled cutting edge; (2) double-beveled cutting edge; (3) plywood mounting; (4) machined knife support or mold body; (5) mounting screw with larger mounting hole; (6) platen or backup plate.

of the same cutting-plane height (Figure 32). Any slight variation would not be an ideal situation since we aim to have the trim die made so as to trim in the same plane (level). Ideally, there should not be more than a 2-mil variation. However, some die manufacturers will only accept die orders of up to 10-mil variation. A slight discrepancy in the cutting-edge level will usually be overcome by the cutting press platen force as long as pressure is being applied. However, due to the nature of spring steel, the blade will return to its original shape as the pressure is relieved. If the discrepancy is of a higher order, the act of cutting will require higher cutting forces, may be harmful to the steel-rule cutting edges, and can affect the outcome of the trimming. When softer steel-rule dies are used, excessive force on the cutting edge can cause the edge to curl; when harder steel is used, excessive force will cause the edge to wear down to a flat surface and ultimately will damage the cutting plate surface (Figure 33). A curled or burred cutting edge can result in the grabbing and holding of the plastic parts or the trimmed scrap skeleton. Curling of the cutting edge

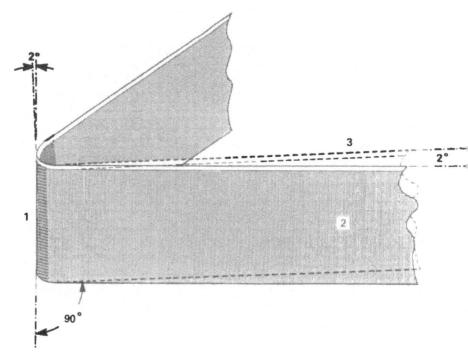

Figure 32 Formed steel-rule knife: (1) bending line; (2) steel-rule knife bent 90°; (3) steel-rule knife bent with 2° error in 92°.

should be avoided, as it could cause unwanted interruptions in the trimming cycles. Curling is more likely to occur on single-beveled blades than on center-beveled blades, usually on the nonbeveled sides. Depending on the knife's bevel placement within the die, the blade will grab either the scrap skeleton or the thermoformed cut part.

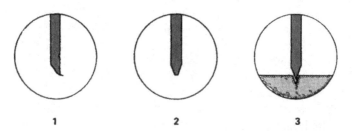

Figure 33 Damage in steel-rule die cutting: (1) curled cutting edge; (2) worn, flat cutting edge; (3) cutting plate damage (soft and grooved).

A flat surface on the cutting edge may also result from normal knife edge wear, which is solely dependent on the type of plastic being trimmed. Some plastics lend themselves to easy cutting, needing only to be scored to fracture at the cutting lines. Other plastics not only require a complete cut through but will resist any knife edge penetration. Premature cutting-edge wear can also be caused by poor knife bending, as has been pointed out. This causes a higher discrepancy than is ideal in the cutting-level tolerances. The cutting-height discrepancy will be overcome with each cut, as the platen forces cause the knife to conform to a single level. In other words, the cutting will be made complete only when the entire knife blade comes into contact with the opposing plate. Naturally, the edges that are in a higher position (before the blade is leveled out) will be the first to come into contact with the plastic and possibly cut through it before the rest of the knife reaches that position. To facilitate complete cutting, the higher knife edges are mechanically forced by the platen forces to line up with the rest of the cutting surface. This action will cause lateral (highly frictioned) movement of the knife edge on the opposing cutting plate, which will promote premature cutting-edge wear. This is the leading cause of a flatly worn cutting edge.

Besides poor trim tool assembly, an improper steel-rule knife cutting press setup can result in premature knife edge wear and cause cutting surface damage and ultimately, the inability to trim. The platens that hold the cutting knife assembly (the steel-rule knife and its support, a plywood structure or machined metal frame) and the opposing cutting plate constitute the entire cutting mechanism. The platens must have absolute parallel closing abilities with adjustable closing travel limits. Exerting pressure during cutting is acceptable, but at the moment of cut-through, any pressure between the knife edge and the opposing cutting plate must be avoided. Usually, stop blocks or adjustable screw-type platen travel limiters are employed to alleviate this problem. Having a good cutting tool setup and accurate platen travel will ensure good-quality trimming and long cutting-edge life.

Some multilevel cutting can be accomplished, but some tricks may be required. The actual cutting is always made with the knife cutting against an opposing plate. To achieve multilevel cutting, the opposing surface must be made to match the particular cutting levels. Because of the maintenance of tolerance and discrepancy variations in the process and the equipment, this type of multilevel cutting operation is difficult. But this does not mean that such cutting and trimming cannot be done. In fact, there are configurations and product designs that cannot be made any other way and the thermoforming processor occasionally has to fight through the difficulties of multilevel cutting.

For the popular single-plane (same-level) trimming operation, steel-rule knives are mounted into a plywood board. The plywood is cut with a bandsaw in the same pattern as the configuration of the trim. The saw blade cutting width is made to match the point thickness of the steel-rule knife blades, which are forced into the cut completely through the plywood. To implement the cutting, the plywood-mounted steel-rule die is mounted to the platen or in the case of a nonsolid platen, to a backup plate, then to a platen. This mounting provides the necessary backup to prevent the knife from slipping through the plywood support under the pressures being exerted between this plate and the opposing cutting plate.

The steel-rule knives are also mounted against either a mold body or a separate knife mounting assembly. For this purpose mounting holes have to be placed in the steel-rule blade, but they cannot be drilled in hardened steel; they must either be ground or punched. It is necessary to make oversized holes to allow some knife seating movement. When these holes are not made large enough—or elongated—it is possible that the knife, as it is beginning to seat, will lean on the screw shaft and shear it off (Figure 31).

In large cutting patterns or in multi-up dies, it is possible that the continuity of the cut will not be maintained throughout its entire cutting line. The knife's cutting failure has to be analyzed. If the cut interruption is constant and repeatedly shows up at the same locations, the explanation can be found in the following two areas: First, if the die is new, unsuccessful complete cutting will be caused by the excessively wide variations of cutting-edge heights. Second, if the die is an older tool that has performed satisfactorily in the past, the inability to produce a complete cut can be ascribed to the loss of a cutting edge. In either case, an increase in the cutting platen pressures may renew the die's ability to produce a complete cut. The improvement is only temporary, however. To obtain a longer-lasting improvement, placing shim stock under the back sides of the cutting blades will elevate the blade's height to the proper cutting levels. Although shimming will restore cutting ability to an irregularly worn steel-rule knife, it will extend its cutting life only up to a point. Sooner or later, the knife will have to be replaced. When steel-rule knife edges are worn and no temporary remedy will restore their ability to cut, most thermoforming companies have the die re-knifed. Only a very few manufacturers attempt steel-rule knife sharpening themselves. Even with the best sharpening skills, resharpened knives rarely match the cutting qualities of a new knife.

In the process of cutting plastic parts out of a thermoformed panel, it is often required that the trimmed part and the scrap skeleton exit together from the trim station. This procedure will help empty the trim die and ready it for the next cycle of trimming. It will also allow easier receiving and handling of the trimmed thermoformed product. This would be a problem in the trimming operation if the trimming is made so perfectly that when the cutting platens open, the completely trimmed product falls out of the die and onto the bottom plate. To have to reach into the machine to retrieve the cut parts presents a safety hazard to the operator and threatens destruction of the formed part. To eliminate these hazards while maintaining the flow of the trimming operations and implementing automatic thermoforming procedures, it is imperative that the scrap skeleton carry the formed and trimmed part out of the trim station. To accomplish this, the continuity of the steel-rule knife trimming edges must be interrupted. This interruption in the cut is made in a few strategic points to allow the scrap skeleton to remain attached and hold the cut part and at the same time, permit it to be separated from the part with a minimum of effort. The interruption in the cut is made by notching the blade with a delicate file. Where the notch is made, the blade, instead of cutting cleanly through the plastic, will leave a fine tablike connection between the cut thermoformed part and the scrap skeleton. Such connections are called "tear tabs." Obviously, too wide a notch may make too strong a tab that cannot provide an easy breakaway connection. The notch should be made delicately and with some planning. It can always be increased, but once made cannot be decreased.

1. Angel Hair

When steel-rule knives are used to trim thermoformed plastic parts out of entire panels, plastic fragments called "angel hair" are often produced. These hairlike slivers not only make the cut edge unattractive, but when the parts are used in food packaging, could contaminate the food being packaged. All plastics used in the packaging of food do have FDA approval as packaging material but not as a "food additive" intended for internal consumption. In any case, the minimization or elimination of angel hair fragments is of utmost importance during the trimming of thermoformed food containers. Cleaning the angel hair fragments from plastic parts has the same difficulties as cleaning powdery particles from parts. The same static problems will be present and even the best wiping methods will not guarantee total success. The best way to minimize the problem is through an understanding of the cause. First, hard plastics are more vulnerable to fracturing than are the flexible type. Second, the angel-hair fragments are produced by the steel-rule knives. A brand-new and a dulled knife will both produce an overwhelming amount of angel hair. The very sharp edges on a brand-new knife, combined with the rapid platen closing, may produce only a partial cut, then a bounce, and finally a completed cut. This will create a double cut, the first cut causing a fracture and the second making the separation a hair's width over. Between the two, angel hair is produced.

The other notorious cause of angel-hair fragments is a worn and dulled knife edge which will no longer penetrate for cutting but will pinch the plastic, creating the severing. Most thermoformer practitioners find that when trimming is hampered and angel hair increased, sharpening or re-knifing is necessary. With sharpening, only limited sharpness can be achieved; rarely will total sharpness be regained, and even then, the new sharpness life expectancy is often much shorter than that of the original sharpness. Sharpening steel-rule dies requires skill; not everyone can do it. On the other hand, when re-knifing is chosen, the chance of the newly built steel-rule knife producing angel hair is initially as high as, if not higher than, that of the replaced knife. This will often temporarily upset the practitioner. Brand-new knives need to be broken in and may require a period of dry cycling before satisfactory trimming will result. At the time re-knifing is decided upon, or even when the first steel-rule dies are made, the steel-rule cutting edge must be carefully selected. Between center-beveled knives and single-beveled knives bent with the bevel placed to the inside or outside, parts of different dimensions will be produced. This is especially critical when tight-fitting parts and components are produced. Choosing the wrong cutting edge or positioning it improperly may produce a different-sized part, which will be larger or smaller according to the bevel thickness.

There is also a limitation in the cutting of thermoformed plastic parts relative to the maximum available knife height, 2 in. If the thermoformed product is more than 2 in. in formed depth of draw, such cutting can be accommodated by provisions made in the die not to crush the formed part while trimming. If the height of the thermoformed plastic parts exceeds the steel-rule knife's maximum height, the toolmaker has two choices: build a machined metal frame assembly that will elevate the steel-rule knife blades to proper heights, or manufacture an entire knife out of forged steel stock.

E. Forged Knife Trim Tools

To overcome the 2-in.-height limitation of the steel-rule knives, many tool and die makers produce their own cutting blades. These types of blades are either machined out of steel blocks or formed out of a thicker steel stock strip. After the shaping and forging, the steel is annealed to give the necessary hardness and then sharpened to a cutting edge. Since these knives are fabricated from stock steel, they respond better to shaping and their cutting ability can be made much more precise. If the knives are made to ideal tolerances, their cutting life is also greatly improved over that of steel-rule dies. Forged knives initially cost more than steel-rule knives. However, their longer service intervals and resharpening qualities may offset some of the price difference. This type of cutter is usually mounted to the die plates or to molds with welded tabs that are attached to the base of the cutters at given distances. Naturally, forged knives are no less vulnerable to abuse or improper cutting setups and can be damaged just as easily. The use of forged knife cutters in thermoforming is most popular with the manufacturers of cookie trays, especially where a high volume is manufactured with volume trimming performed, and the tray depths are over the 2-in. limit of steel-rule knives.

Most of the cutting procedures with forged blades, and their inherent problems and defects, are similar to those used with steel-rule dies. The improvements and remedies associated with steel-rule dies (e.g., shimming and notching) are also beneficial to this type of cutting tool. Forged knives are rarely replaced when their cutting abilities are lost. Even the most severe cutting-edge damage is repaired and welded (filling the damaged area) and then the edge is resharpened to the specific high-quality tolerance. The decision to use steel-rule dies rather than forged knives is entirely up to the thermoformer and is dictated primarily by the product line. The justification to use one rather than the other may be dictated by the availability of supply.

F. Matched Punch and Die Sets

The greatest speed in trimming and accuracy can be only accomplished with matching trim die sets. This type of trimming is made on high-quality, machined matching punch and die trim tooling. The material from which the punch and dies are made is high-quality tool steel that has been properly stress-relieved before being machined into cutting equipment. Equipment made of high-grade tool steel is designed to render exceptionally long service life and precision cutting of most plastic material.

The principle behind this type of trimming is a technique borrowed from the steel fabrication industry. This method closely duplicates steel sheet trimming and punching procedures. Just as with steel sheet products, thermoformed and relatively cool plastic paneling is properly indexed between the punch and die sets, which when forced to close together, will stamp out and trim the plastic part. The punches are machined out to match the contours of the opposing dies. The clearance tolerances between the punch and the die are made to a minimum of 0 to $1\frac{1}{2}$ mils. In the actual cutting mode, the male punch side of the trim tool forces the plastic against the female die cavity. The plastic yields to the pressures between the punch and the die and is cut against their moving edges. The cutout is forced into the die by the punch.

There are two separate criteria in this type of trimming. Since a zero—or very close—tolerance limit is made between the male punches and the female dies, the opposing punch and die must mate accurately for cutting. In trying to ensure positive and absolute mating of the punch and the die, the platen mating accuracies will usually be found unacceptable on the thermoforming machine. Therefore, the punch and die assembly should be built with its own independent die set. The slightest wear of the die-set bearing or bushing can afflict unwanted damage on the cutting edge or even cause complete "wipeout." Most punch and die sets are built to the closest possible tolerances with absolutely no metal-to-metal clearance between them. Of course, such closely fitting trim assemblies must have precise alignments each time trimming is made. A minimal clearance can be acceptable, depending on the type and thickness of the specific thermoplastic material. The average trimming tools are usually built with as much as 1 to 11/2 mils of spot clearance. It is possible that tooling with larger clearances may still perform to a degree, but with the increase in clearance, trimming can be achieved with only a few thermoplastic materials. For example, thermoformed foams usually allow trimming with larger clearances because the foam thickness is squeezed down just prior to the actual cutting.

As the punch and die wear against each other and succumb to the abrasive plastic, their sharpness is lost. The same wearing effects also increase the punch and die clearances. Both the dulled cutting edges and the increase in clearances gradually lead to diminished cutting quality. The cutting ability is lost gradually, except in the case of misalignment, which causes instantaneous wipeout. The first sign of punch and die wear is a slight "fuzziness" on the trimmed plastic edges. If neglected, the fuzz or irregularity will increase and eventually pinch the plastic instead of cutting it. Mistrimming may not be constant and may show up at various spots of the trim line. When occasional mistrimming appears, it is a good indicator of far too much neglect of the trim tool and the need for immediate sharpening or rework of the equipment.

Matching punch and die trim tooling can be sharpened and its cutting qualities can be renewed such that it will trim as perfectly as when new. Only dulled cutting edges and/or those with minor damage can be resharpened. When the cutting edges have been damaged severely, the trim tool may be beyond help, and complete replacement will be necessary. The other criterion for sharpening is the metal hardness of the trim tooling. Thermoformers may request various hardnesses in the makeup of their trim tooling, depending on the type of plastic materials they are attempting to trim. Commonly chosen hardnesses are 45 Rockwell for trim tool dies and 65 Rockwell for punches. At 45 Rockwell hardness, the steel remains malleable enough to accept the attempts of sharpening, while steel of 65 Rockwell hardness no longer accepts peening for resharpening. If both the punch and the trim die are made with the same elevated hardness, such as 65 Rockwell, the cutting abilities may last much longer, provided that the setup of the assembly is ideal.

The actual sharpening of this type of trim tool is always made on the softer die portions. The "lips," or leading edges, of the dies are peened. The procedure is made with a hand-held hammer, or in a more sophisticated setup, with an air-operated peening tool. The malleability of the softer steel will allow for redistribution of the metal by hammering from the leading face into the cutting area (Figure 34). The peening will bulge the steel into the cutting edge and will usually bulge farther than proper clearance between the trim die and the trim punch would allow. Overbulging is not a problem, however. Following the peening of the trim die lips, the trim tool is slowly closed once without

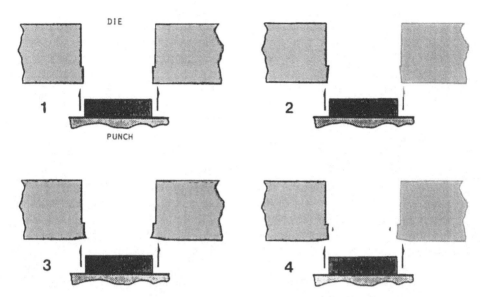

Figure 34 Punch and die trim tools: (1) new punch and die tooling; (2) worn die lips; (3) peened die lips; (4) renewed cutting lips with excess steel sheared off.

trimming. The harder steel punch edges will shear into the peened female die edges and remove any overhang of steel. This creates a zero clearance between the punch and die. After the initial closing of the trim tool, the tooling has to be opened again. Before trimming is attempted, the trim tool should be wiped clean of any metal shavings or debris that resulted from the excess peening. The newly created edge may be slightly recessed, but this is normally absorbed by the flexible plastic and will not affect the trimming. The sharpness of the die-lip cutting edges is perfectly renewed by the peening, and the trim tool cutting life expectancy is now just as good as when the tool was new. Extremely hard tool steel trim tools will not allow sharpening by peening; the only way to resharpen these is by grinding and removing the worn surface. Toolmakers generally leave extra material in trim tools to allow grinding and sharpening.

Both types of sharpening methods are far less expensive than replacement of the trim tool and result in less downtime. For peening of the trim die, tool removal is not required. However, for the grinding procedures, removal of the tool from the trim press and access to proper surface grounding equipment are required. The number of resharpening attempts is also limited since trim tools repeatedly subjected to this procedure will eventually reach a point where replacement is inevitable.

Trim tooling in matched punch and die sets is often machined so that the cutting edges of the two are parallel. The thermoformed plastic placed between the punch and die will be cut at the exact moment the punch and die engage. Trimming of the thermoformed part is instantaneous. There is no major difficulty in punching and trimming all around the full circumference of a thermoformed part all at once. However, the difficulty increases when there are a number of parts to be trimmed at the same time or when the circumference of the part is especially long, more than the customary linear inches. Increased difficulty in the punching and trimming operation can be encountered when the thermoformed part is made of higher-gauge thermoplastic sheets. The power requirement of a trim press to produce punching and trimming is in direct proportion to the thermoplastic thickness, type, and total length of the cut. When the combination of these trim criteria is well above the power-delivering capabilities of the trim press, the thermoformer has two options: to make adjustments or alterations to the trim press, or to concentrate efforts on trim tooling.

If the trim press is actuated pneumatically, increasing the input pressure can increase the cutting power of the machine. If this measure does not provide the necessary force for trimming, an increase in air cylinder size may have to be considered. Use of a larger-diameter cylinder will increase the trimming capabilities of the trim press. The consideration of increasing the size of the air cylinder should be made with the understanding that the resulting increase in power must be acceptable by the machine frame. In borderline cases, the frame may hold up under the increased pressure loads, yet bend or distort enough to provide less-than-expected cutting powers. In the most severe cases, the frame of the machine can be permanently distorted or rupture. Trim presses with oversized framing are often equipped with hydraulic boost features. In this type of machine the pneumatic cylinder operates a toggle mechanism, and when the toggles are locked into closed trim position, the actual cutting is made with a hydraulically driven movement in one of the platens. The entire punching or trimming operation can, of course, be made with a completely hydraulic system that operates both platens. Hydraulic trimming operations are usually far more forceful than pneumatic forces. Their power levels and platen travel speed are fully controllable and follow very closely the physical rules of the pneumatic forces. It should be pointed out that the increase in power favors the use of hydraulically operated systems over that of pneumatic machines. However, like any moving machinery under continuous loads and high levels of pressure, the hydraulic press will sooner or later experience leaky fittings and joints. Loss of air in the pneumatic system will only result in a little noise. But hydraulic fluids drip and leak, create puddles, and easily contaminate plastic parts. Hydraulic fluid contamination should not be allowed on any plastic part even if it is to be used in a nonmedical or nonfood-packaging item. The fluids can act like a solvent on most thermoplastic materials and permanently etch the surface.

Trim-press operations can also be made by mechanical forces. The movements of a customary trim press are supplied by an electrically powered, constantly rotating heavy flywheel. The continuously turning flywheel's deadweight provides the energy for the punching force, and adjustments to the cam will regulate the stroke length. A trim press can be engineered with either a vertical or a horizontal stroke format. The horizontal trim press is the more popular; with this format, trimming of single or multiple rows can be accomplished.

Trim presses engaged in high-volume production can be incorporated into the thermoforming machine as an in-line trimming station. The trim press can also be placed as a separate machine, in line with the thermoformer. A setup like this will require that the continuously formed parts remain in the form of a web, which must be rethreaded and reregistered in the trim press for proper trimming.

Automated high-speed trim-press equipment is capable of running in excess of 120 strokes per minute. In addition to its trimming duties, the machines are capable of automatic stacking, counting, and even predetermined stack separations. Some equipment is also available for automated packaging and boxing.

The other option used to ease the punching and trimming of thermoformed plastic parts is to make alterations to the punch and trim tooling. When multiple holes and cutouts have to be made on a thermoformed plastic surface, the simplest method is to stagger the punching action. This method eases the punch force requirement. The punches will not meet the plastic being cut all at once. Some will have cut through the plastic as others are making the cut and still others are at the prepunching stage. With this method of staggered punching, a 50 to 70% reduction in the power requirement can easily be realized.

When the punch and the die meet the plastic at a 90° angle and the entire trimming is made at once, an extreme force will be needed to cut through the plastic. This type of cutting may be damaging to the plastic. Exerting pressure on the plastic can cause distortion in the cut and may force the shearing to shift or cause crazing on the plastic near the cut. The crazing may not show immediately, but in time, can make the product undesirable or nonfunctional. Today, most punch and die trim tooling is no longer built by this outdated broad punching design, but rather, with the newer shearing action (Figure 35). When the punching or trimming is made on a plastic part, the punch side of the trim tool is built to a slight angle. Due to a 0.015- to 0.060-in. tapered angle, the punch will meet the plastic only at its most extended point, where the cutting will begin. As the punch penetrates the material, the trimming will be made like a shear, making a cut only at two points on the circumference of the cut, which eventually meet and finish the cut on the opposite side. Naturally, shear-

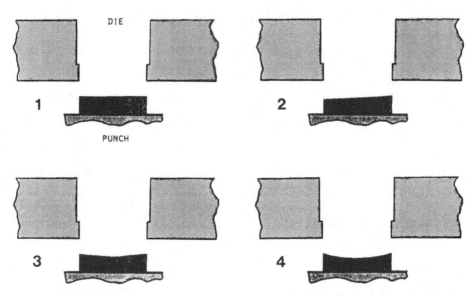

Figure 35 Punch and die trim tools with different types of punches: (1) broad punch; (2) shearing punch; (3) double-angle shearing punch; (4) hollow-ground punch.

ing-type cutting will require much less platen force and will allow handling of the heaviest thicknesses with ease without damaging effects to the cut or the plastic. The only shortcoming of tapered shearing action is that the punches may push the cut out of correct registration as they penetrate the plastic. To minimize the one-sided push of the shear punch dies, the punches can be alternately located in a multi-up tooling so that one punch may start the shearing cut on one side while the other punch starts on the opposing side. This will equalize somewhat the unwanted pushing forces. In some trim tools the shear angle is placed into the die side rather than the punch side; either way, the resulting shear action provides the same result.

To improve trimming further, the operation can be conducted with "hollow-ground" punch dies. With this type of cutting tool, the punch side of the cutter carries a slightly curved pattern (i.e., hollow ground). Instead of meeting the die portions square on, the cut will be made simultaneously at both ends, due to the concave radius shape of the punch. Since in the beginning of the cut the plastic is penetrated at the opposing end, the force of the shear will not exert a distorting pull on the plastic. With this type of hollow-ground trim tool, trimming of multiple parts can be accomplished with lighter-duty equipment and greater accuracy. Making hollow-ground punch dies for this trimming setup is one of the most costly toolmaking efforts. The punches are usually ground after hardening, and generally only for tools rated 65 Rockwell hardness or better. The grinders must follow a preset patterned radius, which depending on size, can provide 0.015- to 0.060-in. shearing action. The surface finish of the steel must make the cutting edge extra smooth, to give flawless shearing.

3
Thermoforming Methods

I. BASIC FORMING TECHNIQUES

A. Preforming Sheet Behavior

To form a thermoplastic sheet into a final product shape, the sheet has to be held by all four sides. The captured thermoplastic sheet is then exposed to heat, softening it and making it stretchable for the shaping. The actual forming of the thermoplastic sheet is almost always made by subjecting the heated sheet to the forming forces. After the forming, residual heat is removed from the formed part, setting it into a permanent shape. With these steps, the forming cycle is considered complete.

The heating cycle begins as the secured thermoplastic sheet is exposed to the heating apparatus discussed previously. There is no difference in the reaction to heating between the precut sheet or rolled material forms. The plastic material will react the same way in both cases, provided that the material source and thickness and exposed sheet panel sizes match. Naturally, sheet materials from different sources can be compared only if their raw resin type and sheet manufacturing techniques are the same. A calendered sheet will not react the same way as extruded or cast sheet materials even if they are made of the same resins. On the other hand, resins having the same chemical composition but different manufacturing sources and different polymerizations may not behave in the same way. Theoretically, the same thermoplastic materials produced in the same manner into sheet form should provide identical behavior in the heating and forming process. Such similarities or slight differences can only be pinpointed by experts. In all cases, a comparative study must be made and analyzed for choosing the most ideal thermoforming condition. A thermoplastic sheet placed into a thermoforming machine will provide a consistent sequence of clues as to its behavior in the heating process. Most thermoforming practitioners fail to observe the behavior changes that take place in a sheet as it is being heated but before an actual sag is noticed. Between the time of the cold sheet exposure and the resulting sag, there are changes in the sheet that can be observed. Such changes are not necessarily slow and obvious; often, they occur in just a fraction of a second. The rapid changes could take place so quickly, in fact, that even with the best skills of observation, they may pass unnoticed. That is the reason that for most thermoformers, the first clue to the change is sagging of the sheet. The sheet's response to heat is most likely to follow the sequence shown in Figure 36.

1. The cold thermoplastic sheet is placed into a sheet-holding apparatus and its exposure to heat begins.

2. As soon as the plastic is exposed to the heat, various temperature levels are created within the plastic and the first changes can be observed. Due to a higher temperature on its surface than in its inner core and possible temperature variations on the surface itself, the partially heated plastic will go through a wavelike movement. This movement closely resembles a ripple spreading across the surface of a body of water. The reaction of the plastic sheet to the heat may come in such pronounced levels that the sheet shows an immediate but temporary sag. The sag rapidly disappears with further heating. This temporary condition should not be confused with materials displaying an immediate and permanent sag condition, which are discussed in a later section. It is possible that thermoplastic sheet materials will react only minimally to heat. In reacting to heat, some thermoplastic sheets never display anything as defined as a ripple-like expansion movement; they may show only a slight color or surface sheen change. Some plastics provide such subtle clues as gaining more clarity or getting either shinier or duller. Either way, a change will take place that can be observed and recognized.

3. As the sheet is further exposed to the heat, its normal reaction causes it to tighten up. This tightening gives the sheet the appearance of well-stretched-out material, almost like the skin surface of a drum. In this state, knowledge of the changing temperature conditions is vital to successful heating.

4. During the heating cycle a definite sag will appear in the heated sheet. At this point, the plastic will be heated and softened to the point where its self-supportiveness is reduced and the sheet will yield to gravity. Actually, the softened plastic can no longer support its own weight and will stretch out of its original shape, creating a sag. The sag is a well-defined change in plastic behavior, and being so noticeable, most people recognize it as the first reaction

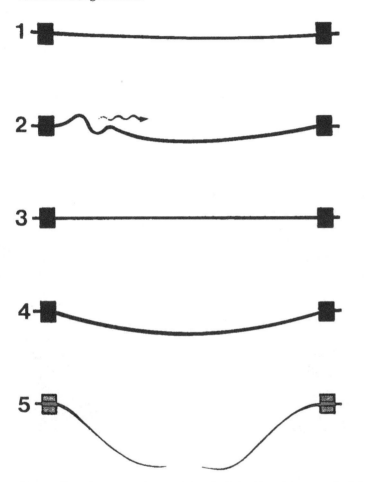

Figure 36 Thermoplastic sheet behavior in heating: (1) clamped cold sheet; (2) rippling reaction to heat; (3) heat-tightening sheet; (4) sagging sheet; (5) overheated sheet.

to heat. With many plastics, this is somewhat true. If the plastic sheet does not have any orientation—or even if it does but its weight, due to its gauge or panel size, cannot hold itself up—a sag will develop. This may be the first observable reaction in the sheet when exposed to heat. Many thermoformers automatically use sag as the indicator for readiness of the thermoforming process, although this is not a completely reliable method. Some thermoforming equipment builders even offer an electric-eye sensing device which will trigger an automatic forming cycle as soon as sag interrupts the light signals.

A heated sheet that develops sag is not necessarily unusable. However, in most instances, sag should be avoided if possible. Using sag as an indicator of readiness for thermoforming is only slightly better than using measured time. The development of sag in the heated sheet will obviously be a more accurate indicator than the use of timers, which are completely unaffected by ambient and sheet temperature changes. For example, a manufacturing area could have wide temperature variations between the early morning and afternoon hours and even greater temperature changes between the day- and night-shift hours. If this is the situation, it is easy to see why sag would be a better indicator than timer units.

Developing sag in a sheet and using it for prestretching of the sheet is also acceptable. This is a common practice in moldless thermoforming or when deeper, well-tapered cavity shapes are formed. In fact, the use of developing sag will help to achieve better wall thickness distributions than its nonuse. Controlling the development and size of the sag is not an easy task. Allowing sag to develop in the heated thermoplastic sheet for thermoforming may not prove to be advantageous at all. Creating a predetermined sag size, no more and no less, is one of the most difficult procedures. Since the sag is produced by the weight of the plastic sheet, it is an ongoing, continuous process. At the point where the sag is judged to be of correct size, it is still expanding and will rapidly stretch out of control. At the same time, the plastic may not be stretching evenly, resulting in a nonidentical bulge shape, hindering uniformity in the shot-to-shot process.

More important, and to the dismay of thermoformers, the developing sag in the plastic sheet alters the distance between the heated sheet and the heater units. Such a change in a given distance, especially when the sagging sheet drops closer to its center to the bottom heater element, will prove to be detrimental. This could cause runaway heating of the plastic, and the overheating will accelerate greatly as the sheet comes closer and closer to the heater units. It is easy to see that a fraction of a moment on the timing controls may mean the difference between forming success and forming failure. In more cases, a sagging plastic sheet will be hotter than is necessary for forming and shaping.

5. Of course, the plastic sheet can be so neglected that it will reach the point where the sag radically overdevelops and the sheet melts and flows apart, making it no longer useful for thermoforming. At this point, not only is the thermoplastic sheet lost in the overheating, but the molten plastic can fall into the heater elements and result in fire. By no means should neglect of the heated sheet continue this long. Most thermoplastic materials derived from petrochemical and fossil-fuel sources, when exposed to extreme heat, will constitute a potential fire hazard. Precautions should be incorporated in thermoforming equipment which enables it to prevent ignition of the plastic material, or even

to put out the resultant fire. More important, these precautions should include good control of the sag and perhaps even its complete elimination.

Thermoplastic sheet materials used in the thermoforming process to make various products are produced from different types of resins and in a multitude of gauges and sheet sizes. Their reaction to heat is just as wide ranging as their sources and forms. Some materials demand the most precise control range in which to perform and not get ruined, while others are especially "lenient" regarding overheating. Such differences, shown in Figure 37, should be recognized and, when known, well respected. For example, biaxially oriented polystyrene sheets display one of the most narrow heat ranges between the formability temperature and the loss of orientation, which comes about at the slightest overheating points. When high-impact polystyrenes (HIPS) and acrylonitrile-butadiene styrene (ABS) plastic sheets are heated, substantial overheating is possible without damage to the plastic. Working continuously with the various materials, thermoforming practitioners will soon discover the specific sensitivity and "criteria windows" of the individual plastic sheets, as well as their behavior under different heat ranges. Knowing where the data may fall on the chart will help to achieve the most advantageous forming conditions. Each time a change is encountered in thickness, color, resin material, heaters, or even ambient temperature, the reaction to heat will be altered.

B. Moldless Forming

Thermoforming a bowl or dome shape can be accomplished quite easily. In fact, such shapes can be obtained easily without the use of a mold. There are two basic ways to form the plastic moldlessly. The first and less complicated way consists of producing a sag in the clamped sheet. When a bowllike shape is approaching the desired size, it should immediately be removed from the heat source, for cooling. This allows the weight of the heated plastic sheet to form and stretch it. With this method, producing a bowl shape will take the longest time and the shape produced will never have a uniform curvature. Since the bowl shape is the result of a sag that may produce uneven stretching of the sheet, the curvature of the bowl shape may result in inconsistent placement of the center point. These variations in shape are easily found from one part to the next, giving the product a poor nesting quality.

The second way to produce bowl- or dome-shaped configurations without molds is to build equipment that resembles a pressure box. One of the walls of the box is actually formed by the clamped and heated plastic sheet. When the plastic sheet starts to show some softening, the box should be internally pressurized. The pressure will force the softened sheet to bulge upward to form a dome shape. At low levels of pressure, the softened plastic can be kept from sagging and thus held straight and flat. As the pressure level is increased, bulging

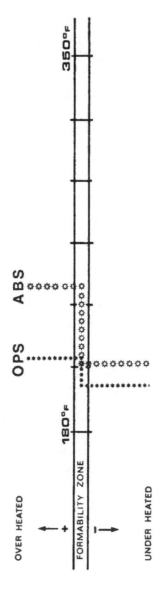

Figure 37 Thermoformability heat range chart. Note the narrow formability range of OPS and the wider formability range of ABS.

will appear; as the pressure is increased further, the size of the bulge will increase accordingly. With the force from the air pressure, the plastic will stretch and form into a domelike shape. When the desired shape is reached, further heating and any increase in pressurization must cease. Consequent cooling will take place or, if desired, fan-forced cooling can be introduced to set the newly acquired shape into a permanent form. Forming domes with this method of pressurized air is much more rapid than producing them by sag, and the comparative shapes are more uniform. It is well known that this type of forming, despite improved qualities, will never produce as uniform a shape as would a mold. However, for the purpose of specific products, this type of forming technique is well accepted and routinely used. Most skylight windows and window fixtures are produced in this manner, for example.

C. Basic Female Forming

Thermoforming a heated plastic sheet with a female mold is one of the simplest methods of forming. Using a preshaped female cavity as a mold makes this thermoforming possible. As soon as the properly preheated plastic sheet is positioned over the mold, the forming cycle is ready to begin. The entire forming procedure takes only a moment to complete. To fully understand and appreciate the intricacy of this brief forming procedure, it is necessary to break this short time span into even smaller segments. As if using stop-action photography, the rapidly created forming can be broken down into six major time segments for examination. Each time segment will have well-defined actions that individually can affect the overall forming procedure. The time segments occur consecutively and always in the same order. The forming procedure, however, can stop short of completion of all the forming segments, thereby causing incomplete forming. The forming procedure can easily be divided into six segments, as shown in Figure 38.

    a. The properly held (clamped) sheet heated to a correct temperature level. The well-prepared sheet is positioned above the female mold cavity. The mold and the sheet will move toward one another. Any vacuum force actuated prematurely will be useless. In fact, vacuum actuation may be harmful since due to the heavy loss of high vacuum energy, the forming may not receive its maximum force. In addition, the heated sheet may be chilled by the movement of air sucked through vacuum holes.

    b. At the point where the sheet and the upper horizontal surface of the female mold make contact, the vacuum should be activated. The softened plastic sheet, having made contact with the mold's upper surface, provides outstanding self-sealing action. Naturally, the vacuum introduced to the cavity will pull this seal even more airtight. At this point it should be clear that the sheet and the mold together create an enclosure, with the sheet stretched over

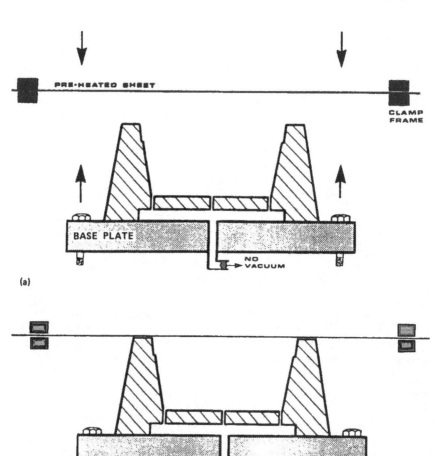

Figure 38 Female molding.

its opening acting as a flexible membrane. The softened sheet will have no resistance against the vacuum force, which will pull it into the cavity.

    c. As the membranelike softened sheet is pulled into the cavity, its thickness will be reduced by the stretching. There will be extremely little stretching where the sheet and the upper mold surfaces are joined to create the sealing effect. Most of the stretching and thinning of the material will occur at the membranelike portions of the sheet. If the sheet should not receive sufficient heat to provide the necessary softening for the thermoforming, its chances of

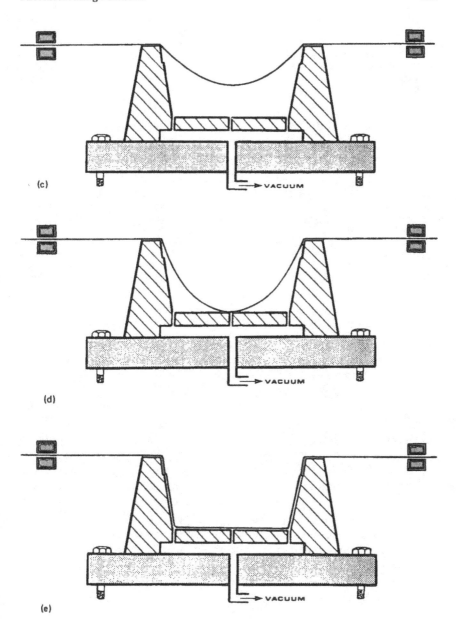

(c)

(d)

(e)

Figure 38 (continued)

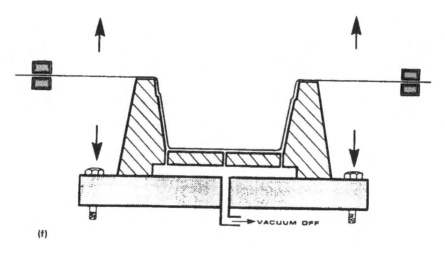

(f)

Figure 38   (continued)

stretching will be curtailed. The sheet material will firm up before full detailed forming can be accomplished. Depending on the sheet's preheated temperature, the incompleteness of forming can be either extensive or slight. In the most severe cases, the forming is so incomplete that the plastic sheet only develops a bulge. On the one hand, the forming will stop just short of completion and full detail obtainment. Recognition and determination of the cause of this condition are confusing and difficult. The slight loss of detail may indeed be caused by the underheated sheet, but other culprits—lack of vacuum forces and an overly chilled mold, to name two—could result in the same symptoms.

d. The location of the vacuum holes is also critical in forming of the plastic. When the vacuum channeling holes are placed too centrally in the cavity, the partially formed plastic can lean against the holes and plug them up, rendering further forming impossible. Air pockets will be trapped in the cavity. Properly distributed vacuum holes places in the deepest corners of the mold cavity can eliminate air pockets.

e. When the thermoplastic temperatures are at their proper levels, sufficient forming forces are being applied and fully detailed forming will take place. Reduction of the sheet thickness from its original gauge is a normal physical reaction. To create a three-dimensional shape out of a flat plastic sheet, its surface area must be increased and stretched. To estimate the amount of gauge reduction in the material in order to choose the proper gauge material for the end product, the intended mold surface area must be established and its ratio to the original sheet dimensions correlated. The established increase in surface area provides a good averaging estimation of the reduction in gauge thickness. Of course, this estimation is only good for approximating the material thinning. In the actual thermoforming, substantial variations can be realized based on the particular thermoformed part configurations. The reduction in gauge could be so severe that certain areas of the formed part will display undesirable thinning. if this is the case, many possible alterations can be implemented in the overall thermoforming procedure before such parts are deemed impossible to make. The programmed heating method previously discussed is one of the options available to remedy this thermoforming situation. Other improvements may become more obvious in later segments.

f. If and when the thermoforming is made with all the proper conditions and its outcome is satisfactory, the formed plastic will be in full contact with the inner mold surface. At this point, the mold's cooling effect takes place. The mold, being cooler than the forming plastic sheet, will extract heat out of the plastic through surface contact. The cooling of the formed part not only sets its shape into permanent form but results in a normal shrinkage in size. This shrinkage will cause the part to pull away from the mold surface, facilitating easy removal. Natural shrinkage is a welcome phenomenon with female molds and happens with all plastic material. The shrinkage rates may, however, differ from one plastic type to another. The establishment of shrinkage rates is usually worked out in a range from the lowest to the highest point. The range for a family of plastic will be representative of all members of that family. Substantial variations in shrinkage can be found from one plastic family to another, allowing no substitution of materials if final product dimensions are critical.

D. Basic Male Forming

Thermoforming a heated plastic sheet with a male mold follows closely the "stop-action" sequence of female forming. The same criteria will hold for each sequence. However, certain aspects of the outcome may result in some variations. The actual steps of this sequence are shown in Figure 39.

a. The preheated plastic sheet is placed over the male mold and the two elements are brought together. The clamping mechanism holding the sheet will provide the necessary holding—or even draping—action that is needed when male molds are used in the thermoforming. The particular thermoforming is often referred to as "drape forming" because the heated plastic sheet is actually draped over the mold. The results of the forming will be the same provided that only the clamped sheet moves to drape the plastic over the mold, rather than the mold moving against the sheet or both moving toward each other simultaneously.

b. Unlike the case of the female mold, the vacuum system should not be actuated at the moment when the heated sheet comes in contact with the mold

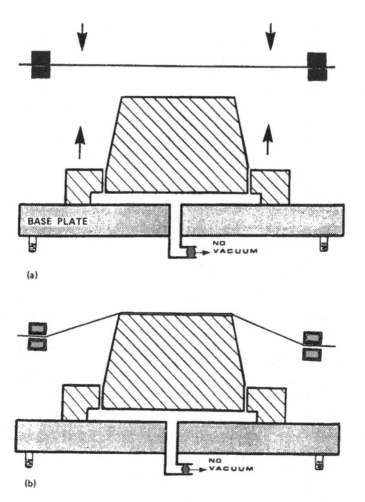

Figure 39 Male molding.

surface. At this stage, the thermoforming sequence is not timed to match the
female mold condition. Vacuum actuation at this point is premature. If the
vacuum is actuated, there will be no plastic forming. When the mold is much
colder than the forming plastic sheet, the mold will chill the plastic upon contact. If this cooling is extensive, the contacted material will be unable to stretch
for the forming.

    c. The clamp mechanism and the male mold surfaces must close further
upon each other to create a sealing effect at the lower surface edges of the

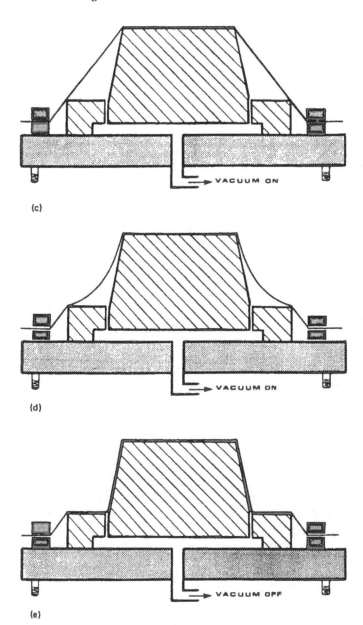

(c)

(d)

(e)

Figure 39 (continued)

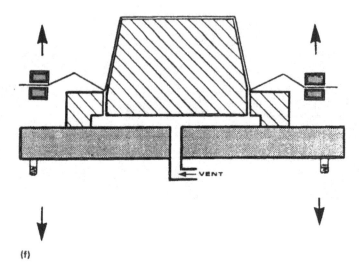

(f)

Figure 39   (continued)

mold. As soon as a seal is created between the plastic sheet and the outer surface corners, actuation of the vacuum forces will finish the forming procedure. If the natural sealing effect is incomplete due to the loss in vacuum, the overall forming of the plastic will be affected.

d.   When a good seal is achieved between the softened plastic and the mold surface, the forming forces of the vacuum will enhance the sealing qualities. At this stage, the process of forming with the male mold will mimic many characteristics of female mold forming. The softened and draped plastic sheet will trap air between it and the mold. This air is evacuated by the vacuum. This will force the membranelike plastic directly against the mold surface. Another similarity of male forming to female forming is that when the plastic sheet does not receive sufficient heating for softening, the forming may not be fully produced. If the vacuum holes are not placed in the deepest parts of the mold configuration, the sheet will cover and plug them up as the sheet is stretched, leaving air pockets and stopping the forming short of full detail. Underdetailed forming can be caused by any of the conditions noted above as well as with low levels of vacuum forces. Such a variety of problem-causing conditions necessitates a careful examination. Their elimination one by one will ensure the success of thermoforming.

e.   When all the elements of forming are successfully achieved, the thermoplastic should be fully formed over the male mold. The heat of the plastic sheet will be absorbed by the mold throughout its entire surface, resulting in a permanent shape being set in the plastic. At this point, the vacuum forces must be turned off.

f. As the cooling of the formed part is accomplished through contact with the cooler mold, the formed plastic should shrink, just as it does with female forming. Since the mold is male, the shrinkage of the formed material, with its normal reduction in size, will tighten it onto the mold. The general shrinkage of the formed plastic part over the mold may not present an overwhelming problem. However, there are occasions when the formed part and tooling do not have sufficiently tapered angles. In addition, ribs and ridges may be present on the sides of the part, or extreme depth-of-draw ratios may be insisted upon. In such cases the shrinkage on a male mold easily proves to be a problem. The formed part will be difficult to remove from the mold. As shown in Figure 39, the components do have sufficient draft angles and the drawn flange area distortion is exaggerated. However, with substantially smaller tapered angles, at mold separation, part removal will cause the flanges and even part of the sidewalls to distort. In the most extreme cases, the shrinkage and grabbing by design-related flaws can be so severe that part separation from the mold is nearly impossible; to attempt it can result in damage or rupture of the formed part.

When the designs of a part do contain slightly tapered angles, which would cause part removal difficulties with its shrinkage, it is always best to choose the female molding method over the male. The choice of female molds versus male molds should be guided by the ease of forming and by the material and wall thickness distribution factors. Female mold forming tends to stretch material in different locations and proportions than an identically shaped male mold. The thermoforming practitioner has the option to choose whichever forming method will produce the best result.

E. Basic Matched Mold Forming

Thermoforming of a heated thermoplastic sheet can also be accomplished with only male and female mold stamping forces. In this type of thermoforming, the male counterpart of the female mold will push the heated sheet into the female cavity. In most instances, this stamping force will act similarly to compression molding techniques. The limitations of this type of matched mold forming are the dependence of the mold configuration and the thermoplastic material stretching and flowing characteristics. Using the matched mold method to force material into the contours of the gap left between the male and female molds is not an easy task. With limited taper designs and substantial depth-of-draw ratios, the heated thermoplastic sheet can only form up to its limits of stretching; after that, it will develop tears and holes. Some materials have greater tear resistance than others. The two limiting factors—mold configuration and thermoplastic material features—are so intertwined and interacting that their specific limits must be tested under actual working conditions. Due to the unknown factors present here, you can only judge the approximate failure limits of specific combinations. This condition is aggravated even further by design patterns containing elaborate configuration combinations, such as male and female elements existing within each mold. Mold configurations that display an open, well-tapered design with smooth surfaces and clean lines are good candidates for matched mold forming. In fact, all thermoplastic foam materials are formed using the matched mold forming technique. Using a matched mold, whether on completely or partially matching contours, is the only way to produce satisfactory details on both sides of the thermoformed product. Matched molds, as they close together, will squeeze the softened thermoplastic foam sheet between their surfaces, forcing it to flow and fill the predetermined mold gap. For example, examination of an egg carton cross section will demonstrate such foam forming results. At the hinge line between the cover and the base, the foam material is squeezed into a solid plastic to allow it to hinge. The egg-holding cells, on the other hand, are produced with a much larger mold gap, giving the foam a softer, cushiony effect. The foam thickness variation is evident throughout the entire egg carton design and gives us a clue that it is thermoformed using a partially matched contoured mold.

There are two basic variations to thermoforming with a matched mold:

1. Having the male mold mounted above the female mold and forming the softened thermoplastic sheet downward into the female. In this case the actual act of forming can be combined with a purposely produced sag. The sag may provide prestretching, promoting better material distribution and resulting in a more uniform wall thickness of the formed part. In this particular case, the not-so-desirable sag can be justified, because it will both improve the forming and benefit the thermoformer. It should be emphasized that developing sag in a sheet does create a slightly overheated condition and therefore extra cooling time will be required.

2. Affixing the male mold to the bottom platen for pushing the heated thermoplastic sheet upward into the female. With this method, the male mold can prestretch the heated plastic sheet by moving upward, causing a tentlike stretch in the sheet. Normally in this process, the male mold's travel is curtailed, as it should be limited to satisfy only the prestretching purposes. Excessive stretching may be more damaging than helpful. Overstretched material may develop wrinkles when the female mold closes onto the male. As the material, overstretched through lack of control, develops more surface area than the mold surface requires, it will be forced to fold over itself as a bona

fide wrinkle. To eliminate this unwanted flaw, the thermoformer has only to limit the prestretching and the wrinkle will disappear.

In both types of thermoforming with a matched mold, the forming success is highly dependent on the flow characteristics of the particular thermoplastic material. These characteristics, in turn, are highly contingent on the sheet temperature, which softens the sheet and ultimately allows the forming. With some of the more sensitive thermoplastic materials, just a few degrees' decrease in heat levels could result in major forming difficulties. Individual testing and understanding of a specific material's limitations are very important to success in thermoforming. The particular temperature sensitivity and forming criteria are discussed further in Section II.H.

Matched mold forming is not limited to the use of molds that are closely matched in size. In modern thermoforming processes, using male portions of the mold to facilitate forcing the sheet into the female mold is a common practice. Such "plug-assist" forming, which could qualify as a type of matched-mold thermoforming, is discussed in the following section.

II. THERMOFORMING METHODS

The thermoforming process need not rely solely on the three basic mold types and their simplest forming methods. In many instances, the actual forming of the plastic sheet needs to be managed with some variation of the methods discussed previously. These variations tend not to change the basic thermoforming concept; rather, they improve either the forming technique or the finished product. In some cases, a specific deviation from the basic forming method makes it possible to thermoform an otherwise unformable plastic sheet, or at least to provide better control of its shaping. Some of the ultrasensitive thermoplastic materials cannot even be considered thermoformable without implementation of specialized thermoforming techniques. The main purpose of carrying out any thermoforming method is to improve the outcome of the forming and to achieve the best possible material distribution and stretching. This is accomplished when the finished thermoformed article has been produced with an even wall thickness and excessive thinning is not permitted at any location. Product weakness and failure always take place at the thinnest areas of a product.

Variations in and combinations of forming methods are attainable in numerous ways. Some variations are chosen strictly for their obvious benefits and may serve a specific purpose. On the other hand, some thermoforming methods are implemented to satisfy the inventiveness of the processor rather than for their usefulness. The methods should be examined individually and broken down into segments to be judged for appropriateness. We discuss in detail below the most common thermoforming methods in use today: trapped sheet, plug assist, form and trim in place, pressure bubble plug assist, pressure bubble snapback, snapback vacuum, trapped air assist, solid-phase pressure, and twin-sheet forming. There are other, less popular variations or combinations that can be used successfully. As the industry pursues better results, additional variations will be developed, and there is no limit to the inventiveness that we should see in the future.

Variations in thermoforming methods usually begin with one of the basic methods and then proceed to combine methods borrowed from other styles. Other variations include the altering of forming force levels, changing platen speeds and sequencing, incorporating clamping mechanisms, and resetting vacuum and air-pressure switching and timing.

A. Trapped Sheet Forming

The trapped sheet thermoforming method should be viewed as a specialized type of thermoforming. However, the technique has the simplest components of any of the forming procedures. Since the thermoforming is carried out on a captured and trapped area of the sheet, the method is called "trapped sheet forming." At the same time, because the three basic sequences of the thermoforming (heating, forming, and trimming) all occur at the same location, this type of thermoforming can also be called "heat-form-trim-in-place" thermoforming. The first is more correct because trapping of the sheet is the dominant factor in the process. To understand the inner workings and motion sequences within the thermoforming cycle, Figure 40 should be followed. The thermoforming cycle in this forming method always begins with the unheated sheet and can easily be broken down into four basic segments. There are easily identifiable advantages and shortcomings to the use of this process. The shortcomings will be discussed first because these provide the factors that limit the opportunities and applications of this method. The two major limiting factors relate to thermoplastic sheet heating and exclusive female mold use.

To implement thermoforming by the trapped sheet method, the thermoplastic sheet is heated by a contact heater. Contact heaters provide excellent heat-level controls but slow down the heating procedure because contact can only be made with the underside of the sheet. This sheet heating method limits the use of this process to the lighter-gauge sheets. More than a 25-mil material thickness requires prolonged contact heating, making the heavier or thicker materials uneconomical to run.

The second major shortcoming of the process lies in its restriction to the use of strictly female molds. Typical female mold use curtails the depth-of-draw ratios to 1:1, and parts with the maximum draws will suffer from substantial material thinning. From a product manufacturing standpoint, trapped sheet

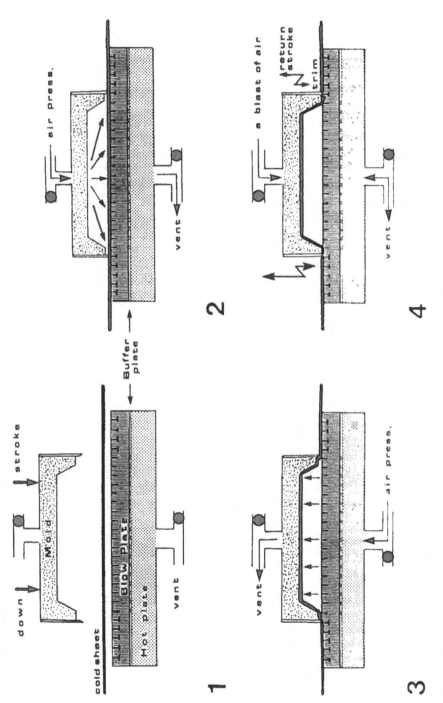

Figure 40 Trapped sheet, contact heat/pressure forming: (1) sheet clamping; (2) sheet heating; (3) forming; (4) trimming and stripping.

forming should be limited to products that have shallow part designs and do not have critical specifications for even wall thickness.

There are definite advantages to the use of this process which make it ideally suited to high-volume production. The simplicity of the process makes it so easy to carry out that many products fall well within the limits. The use of contact heating, with its bulky heating plate, is not only easy to control but is not subjected to temperature fluctuation. In fact, this type of heating is not vulnerable to ambient temperature changes or drafts.

Finally, and most important, the type of mold used, with its own wrap-around cutting knives, makes this style of tooling complete and comparatively inexpensive. New tooling, tooling changes, and reknifing can be made reasonably cost-effective. Most production tooling for this type of thermoforming is made by using a heavy $1\frac{1}{4}$- to 2-in.-thick surface-ground stress-relieved base plate. The mold cavities are sand-casted aluminum bodies with a machined back surface and the surrounding cutting knives are either wrapped around steel-rule knife blades or are one-piece forged knives. When a change in the thermoformed product is to be implemented, an alteration in tooling can also usually be made, thereby avoiding purchasing a new tool. Of course, this alteration of tooling does have limits, and major changes can be accommodated only by building a new tool. This type of thermoforming process does not place stress conditions on the molds, and the only part of the tooling that will wear out is the cutting knife edge.

The four-process sequence shown in Figure 40 illustrates how this thermoforming method works.

1. The unheated sheet is trapped by the cutting knife edges. In this type of thermoforming, the knife has a double duty to perform. Through most of the operation, the knife edge acts like a clamp frame, capturing a predetermined area of the sheet and firmly securing it around each mold's circumference. Naturally, the final duty of the knife remains the trimming of the formed part. In the trapped sheet thermoforming process, each mold cavity is completely surrounded by its own knife. Consequently, each forming is made from its own captured sheet area. The upper platen moves downward with the mold and the knife forces the unheated plastic sheet against the heated cutting plate. The cutting plate is mounted on a stationary platen and is made out of a hardened surface-ground plate, which has built-in blowholes at 1-in. increments. The blow plate is mounted on the heat plate, which is usually heated by embedded cartridge heater units and thermostatically controlled. Between the heat plate and the blow plate, a thin buffer blow plate is placed. This plate carries the same hole pattern as that of the mold's outline configuration. The purpose of the buffer flow plate is to plug those blow-holes that fall outside the mold's knife area, leaving only the inside ones active. All three plates are sandwiched together and connected to a two-way valve system, which is cap-

able of either delivering air pressure or opening itself to the atmosphere for venting.

2. As the mold closes on the sheet, its knife edges dig slightly into the plastic sheet, providing a sealing action and simultaneously holding the sheet in a trapped position, just as a clamp frame would. A valve opens and sends air pressure through the mold, squeezing the sheet against the heated cutting plate. All the trapped air underneath the sheet is vented out, providing absolute contact between the sheet and plate. The heat of the plate transfers to the plastic sheet, making it thermoformable. The heating cycle is regulated by timers that repeat accurately from cycle to cycle.

3. As the sheet is heated to ideal forming conditions, the valves change flow direction and air pressure is provided through the blow plate. At the same time, air pressure on the mold side is cut off and the valve opens to its venting mode. The switching of air pressure will force the now-heated plastic sheet to form against the mold cavity. As the plastic is shaped and formed up against the mold cavity, it not only receives its new shape but will be chilled by the cavity's colder temperatures, setting the newly acquired shape into permanent form.

4. At this stage of the thermoforming cycle, the upper platen will move downward with sufficient power to force the knife edges against the cutting plate and shear the plastic sheet between them. Cutting against a heated plate makes the cutting less difficult. As soon as the cut is made, the upper platen will move upward with the mold and a blast of air will be introduced via the mold. This will provide the necessary stripping forces to get the formed article out of the mold cavity.

To avoid having the formed part drop back onto the heated cutting plate and to facilitate orderly removal from the molding area, the continuity of the cutting edge is interrupted by making several notches in it. The strategically placed notching on the knife edge will leave uncut tabs between the thermoformed part and its surrounding scrap. When the scrap is advanced out of the molding area, it will carry the formed part out with it. At this point, scrap and part separation can be made without interference. The formed part needs only a slight lateral movement for the narrow tabs to break away. Close examination of light-gauge thermoformed parts often provides a clue that this type of thermoforming has been used—fine barblike projections on the trimmed lips made by the broken tabs.

When this type of trapped sheet thermoforming is produced on a continuous basis and the thermoplastic sheet material is fed from roll stock, the scrap advancement not only helps to remove the formed part from the molding area but expedites movement of new sheet material into the forming area. Formed part removal and stacking can be implemented manually or in a more sophisticated manner using automatic stacker units. These units are capable of sliding or pushing each trimmed part into a magazine rack and even automatically counting the number of parts in the nested part stacks.

The tooling for this type of thermoforming is not limited to single-cavity format but can be made into a multicavity format. The number of "ups" is determined by the thermoforming machine platen size. Smaller machines come in 10 in. by 12 in. platen sizes, while larger equipment may have platens as large as 42 in. by 48 in. The number of thermoformed products that can be formed out of a specific platen area—with proper allowance for minimum spacing and edge trim requirements—is the guiding criterion for the number of "ups" for proper tooling format. In the production of rectangular or square products, it is not unfeasible to build the tooling with common knives between them. This method eliminates the spacing between the individual thermoformed parts. The elimination of spacing reduces the combined width of the overall sheet, allowing the use of narrower sheet widths. This, in turn, reduces the final material cost.

Products produced by this thermoforming method are used mostly for packaging, particularly for food-handling purposes. Most cookie trays and food service trays are produced in this manner. The combined advantage of cost savings for cheaper tools and lighter-gauge material and the limited depth-of-draw ratio required makes this process one of the most ideal thermoforming methods for these types of products. At the same time, the process is under constant scrutiny with regard to both quality and economics by manufacturers using other methods of thermoforming, who may be capable of improving both material distribution and cycle times using their methods, thus creating strong competition. The trapped sheet thermoforming method does run slower cycles, making it vulnerable to competition. A 6- to 15-second cycle run is normal for this process. In the past, economic pressure has been so intensive for this process that most of the original machinery suppliers have phased out their new machine lines. However, there is much machinery still operating in various plants. In fact, when a trapped sheet thermoforming machine is placed on the market, it is not only sold rapidly but carries a high price tag.

B. Plug-Assist Forming

Plug-assist thermoforming is the most widely used technique at present. The majority of deep-drawn containers are thermoformed by this method. The assist is used to help stretch and push the heated thermoplastic sheet into the female mold. With the use of an assist, a draw ratio deeper than 1:1 is possible. If all available technological innovations are implemented, a 1:5 draw ratio can be realized and even extended to 1:7 ratios. Forming with plug-assist tooling is basically a type of matched tool forming, which should be categorized in the partially matching contoured tool classification. The plug assist is somewhat similar in configuration to the female mold, but is much smaller. The assist may be only seven-eighths or even three-fourths of the size of the female mold. This size difference results in a pluglike appearance of the assist in relation to the female cavity—hence the name "plug assist."

The use of a plug assist in the thermoforming process offers definite advantages in addition to providing deeper draws. It can provide clearly visible material distribution improvements. The material distribution can be made so precise with the help of the assist that hardly any measurable difference in wall thickness can be found in the thermoformed container. The key to successful use of a plug assist lies in two of its basic characteristic features: (1) assist shape, and (2) assist temperature. Both features can be altered and made to fit a specific thermoforming condition that, in turn, will result in an improved product. By not understanding the inner workings of plug assists, the thermoforming practitioner can actually negate its benefits and make it work against the process. The assist shape and temperature both require careful consideration and some preplanning in order to benefit the process. For the purpose of better wall thickness distribution, the plug assist's single duty is to provide suitable prestretching. Inadequate prestretching results in a nonuniform thickness of the walls of a thermoformed part. To examine the plug assist's inner workings, we can break down the complete cycle into four individual segments, shown in Figure 41.

1. The properly preheated plastic sheet is carried between the two opposing mold halves by a clamp frame or a chain sheet transport. The two opposing molds are mounted such that their alignment is properly maintainable and their centerlines are matching. Mounting can be made with the plug assist facing either downward or upward, as discussed earlier in the section on matched molding. The mold halves can travel against each other at a same time, or one can remain stationary while the other moves; it is best when the plug assist does the traveling, but the generally preferred method is to have both move toward the sheet. The thermoforming practitioner must choose the particular setup and platen travel speeds to accommodate the particular thermoforming process.

The preheated plastic sheet can be clamped and captured by either the plug assist's base plate or an independent clamp mechanism, as shown in Figures 13 and 14. This independent clamp mechanism can be activated prior to the plug assist's motion in order to capture a predetermined area of the heated thermoplastic sheet. Independently operable clamps are very important in continuous roll-fed thermoforming operations. The clamps should eliminate any sheet "pull-in" from stretching by the plug assist upon the already formed or subsequently heated areas adjacent to the forming. The plug assist can exert so much stretching force on a thermoplastic sheet that the two opposing surfaces of the molds or clamps cannot hold the sheet securely. If such is the case,

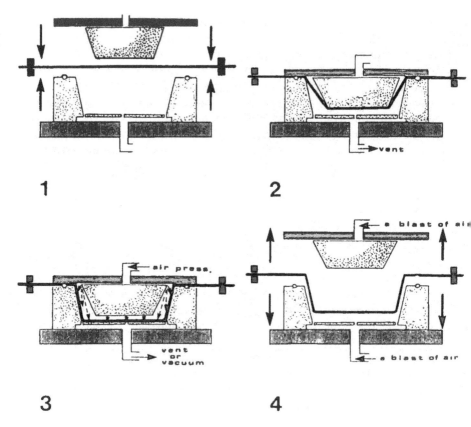

1   2

3   4

Figure 41   Plug assist thermoforming: (1) sheet clamping; (2) prestretching; (3) forming; (4) mold opening and stripping.

a machined V-groove pattern must be placed into the two opposing surfaces in such a way that the mold side will carry a V-groove while the clamp side will be made into a bladelike V shape. This will pinch and crimp the sheet material between the two V-shaped mating surfaces and will not allow pull-in or sheet slippage.

2.   In this stage of plug-assisted forming, the heated thermoplastic sheet is stretched and pushed into the female mold. The plug assist does the actual stretching; it is easy to recognize that the size of the plug assist is influential in the amount of prestretching. Larger plug assists will provide more prestretching of the sheet, and with steeper or milder angles and different-sized radii on its leading edges, the plug can produce variable results in the subsequent stretch.

It is at this point in the forming procedure that the actual size and shape of the assist become critical. The shape and size of an assist can make the difference between satisfactory and poor results in the evenness of wall thickness distribution. Poor assist designs can also interfere with the attainment of full detail transfers.

The temperature of the plug assist can also cause variations in the forming. With the normal escape of heat into the equipment, the plug assist will usually have lower temperatures than that of the preheated plastic sheet. Due to its lower temperature at the moment when contact is made, the plug assist will absorb heat from the sheet at the converging area. At the same time, the female mold's lips, together with the plug-assist base plate or clamp bars, have the same sheet-cooling effect. When enough heat is taken from the preheated sheet at either location, the sheet becomes less stretchable. Where no contact is made with the sheet by the sides of the plug assist, the sheet will retain its heat and will remain easily stretchable. The use of cold mold lips and a cold plug assist will force all stretching to be made out of this narrow and limited area. This will result in a thermoformed product that will display the properties of a much heavier flange and bottom areas and probably, surprisingly thin, weak sidewalls. The results of this thermoforming failure can often be observed on thermoformed soft-drink cups or sour cream and cottage cheese tubs that display an extra-strong and heavy bottom area with extremely thin sidewalls. Poor material distribution will render products unusable in the most severe cases. Better uniformity in material thickness will reduce container failure rates and improve the overall strength of the container.

To minimize or completely eliminate the plug assist's chilling effects, the plug's temperature must be elevated. For the best results, the ideal temperature of a plug assist should match that of the softened thermoplastic sheet. When a plug assist has the same temperature as the sheet, the sheet's stretchable state cannot be affected. In fact, as the plug assist pushes and stretches the sheet, it will allow material stretching along the area where it makes contact. As the mold and plug assist close together, the sheet material can be stretched evenly within the captured area.

Bringing the plug-assist temperature up to match the preheated sheet temperature is one thing; keeping it there is quite another. The plug assist can be heated with electric power, using cartridge heaters and incorporating thermostatic controls. With this type of heating system, the temperature, just as with the heater units, will fluctuate quite frantically. This fluctuation will render unwanted temperatures for even stretching of the sheet. To provide better temperature control for the plug assist and to match the temperature of the sheet, the thermoforming practitioner has two options. First, a mold temperature controller can be used. This unit controls the temperature of a reservoir of fluid that is circulated and pumped through a plumbing system coupled to the plus assist. The reservoir fluid temperature is easily managed and will, in turn, provide good temperature control to the plug assist. The second way to provide outstanding elevated temperature control is to make the plug assist out of an insulating material. The idea behind this approach is very simple. The insulating material can, by nature, pick up and maintain exposed temperatures. By repeated contact with the heated sheet, the plug-assist's temperature will be identical to that of the sheet. The thermoforming practitioner has already made an effort to heat the thermoplastic sheet within a well-controlled temperature zone. The same temperature will be transferred automatically, without deviation, to the plug assist. To heat the plug-assist insulating materials, several thermoforming cycles are required. Through these cycles, contact between the heated sheet and the plug assist will bring the temperature of the plug assist to the same level as that of the sheet. Until the temperature of the plug assist has reached the levels of the sheet, the products will be improperly formed and probably unusable. The number of discarded products and necessary nonproducing cycles is dependent on the composition and insulating abilities of the material from which the plug assists are made. A densely made material requires more repeated cycles and contact with the heated sheet before reaching the same level of temperature than does a less dense material. In today's sophisticated thermoforming, we have adopted a material to use for that very purpose. Called syntactic foam, it gives the needed temperature control to the plug assist. Syntactic foam materials are discussed further in Section III. This material gives exactly what is needed at this stage of the forming and can be made into a variety of shapes. A plug assist made entirely from this material can be formed into almost any shape. If required, syntactic foam can be added to a metallic plug-assist body as inserted pieces or full layers. When the plug assist is made out of differently composed materials within the same mold body, varying face temperatures can be achieved, bringing more sophistication to the technique. As we have already learned, programmed heating is obtainable in the sheet-heating process; with pieces inserted in the plug assist, we can manipulate different temperature zones within the same mold.

In most plug-assist thermoforming practices, the use of rigid plug assists dominates. The rigid materials offer superior dimensional stability and therefore greater accuracy in the forming. The unsuccessful cooling efforts of such plug assists are also well recognized, and metallic material is often substituted.

The use of flexible, cushiony foams for a plug-assist body is not unacceptable and is occasionally useful when everything else fails. In the thermoforming process, there is often a need for some type of mechanical pushing force, a force that could, almost like magic fingers, squeeze the formable plastic into every corner of the mold. Use of a block or large piece of flexible foam material, such as a urethane foam block, may provide the solution. The foam must be larger than the actual female cavity so that it is capable of ramming the heated plastic sheet into the female mold. Since the foam possesses a cushionlike resiliency, it will conform to the female cavity shape and push the forming plastic into very corner and crevice of the mold. With the foam material, the harmful effects of cooling will not be present. As an assist, the foam material has to be selected carefully and replaced frequently. Foam materials under high levels of heat and repeated compression tend to break down and lose their resiliency. Through continuous cycles, the material recovery rate will slow down and forming efficiency will quickly be lost.

Using large blocks of foam (or ideally positioned smaller foam pieces) as an assist can provide outstanding results in thermoforming. Such use of a foam assist should not be considered an inferior method. The best opportunities for adopting foam assist use can be found in smaller-volume runs and those of larger articles. Use of this material will provide surprisingly favorable results.

3. In this segment of plug-assisted thermoforming, the actual forming takes place. Within the female cavity, the plug assist will be fully stretching the preheated sheet. As the two mold halves are closed together, a circumferential seal is created. At this moment, the forming force should be activated. The forming force may consist of a vacuum force, air pressure alone, or both forming forces can be activated at the same time. Using just a vacuum force results in slower forming of the plastic than that obtained using air pressure. Pressure forming will not only improve the speed but will provide more power to the forming, resulting in a more precisely detailed transfer from the mold to the plastic. Depending on how strong the air pressure is on the plug-assist side, it will either gently push or blow the softened thermoplastic sheet against the female mold surfaces. Of course, the air underneath the sheet must be ejected. The air pocket is forcibly ejected by the incoming sheet, which acts almost like a plunger. The trapped air must find a sufficient escape route and provide room for the incoming pressure-forced sheet that will take over its space. The trapped air can be squeezed to vent to the atmosphere, or for more forceful evacuation can be sucked out with vacuum forces. Adding suction in the air-pocket-removal procedures ensures prompt but thorough evacuation, ultimately giving outstanding definition to the thermoforming results. Both vented and vacuum air-pocket removal require plumbing of sufficient size and a multitude of connecting holes in the female mold. Insufficient openings anywhere in the system can cause restrictions in the air-flow that can interfere with and slow the forming. If lack of definition in the thermoforming dictates increased pressure forces and these do not bring an improvement, the cause may be restricted air escape from the female mold. In the most severe cases, the pressure-induced forming is so intense that the formed sheet can trap and push some of the air up against the flared sides of the female mold. This unwanted thermoforming error is caused by improper air evacuation. To remedy this situation,

the forming pressure may be reduced and/or the air evacuation rate increased by adding extra holes. This problem was discussed in Section V.B of Chapter 2.

As the heated plastic is formed and forced against the female mold surface, the mold will extract its heat, causing it to cool and firm up. To remove the heat from the formed part requires some time and is dependent on the part's original temperature, type of plastic composition, and material thickness. To obtain best cooling results, not only can the plastic be held in the mold longer but the forming forces can also be left on longer. Leaving the pressure and vacuum on will hold the formed plastic part firmly against the mold surface, creating an uninterrupted heat sink. As cooling takes place, the plastic part will shrink. At this point, a slight pressure force leakage and weakening will take place. This is a good indication that the forming forces should be turned off.

4. In the preceding step, the forming forces were no longer activated, the plastic part was properly cooled, and shrinkage took place. The mold and the platens should not be separated. To ensure proper stripping, a blast of air is introduced at both sides. The mold will be opened in the same manner as in the mold-closing mode. Ideally, both halves should be opened to allow the formed part to exit the forming area. The opening of the mold must have sufficient "daylight" to allow the formed part to be removed freely. For the most ideal thermoforming conditions, the mold opening should not be made any larger than is required by the formed part for ease of removal. If the opening of the mold halves is allowed to be larger than that required by the thermoformed article, it will not have harmful effects on the stripping and part removal but can impair both the actual forming and timing of the cycle. The extra travel time of the platens in closing and opening will require a longer travel time and will increase the timing of the overall cycle. The increased size of the mold opening will expose the carefully preheated sheet to uncontrollable temperature conditions and result in unwanted cooling by draft. Good thermoforming practices demand accurate platen travel controls. The travel should be altered and adjusted according to the thermoformed product height. For example, forming of a $1/2$-in.-deep part does not require, and should not make use of, the maximum daylight opening of the thermoforming machine platens. Thermoforming equipment that does not offer such options must be modernized and equipped with platen travel restrictors.

C. Form-and-Trim-in-Place Forming

Form-and-trim-in-place thermoforming is a natural extension of the plug-assist method. Actually, the initial four sequences of this technique follow plug-assist forming exactly. To fully understand all the motions of this type of thermoforming, the following step-by-step sequence should be studied, in conjunction with Figure 42. The first four steps are simply a repetition of the plug-assist thermoforming (Section II.B).

1. At the beginning of forming, the preheated thermoplastic sheet is transferred into the forming area by a clamp frame or a chain-rail sheet transport system. In this operation, both sides of the mold (female and plug assist) are equipped with an independently movable and closely fitting frame system. Tooling made in a round configuration must have a ringlike frame surrounding the mold; square and rectangular molds have frames matching their shape. These surrounding frames have two functions, which are carried out in the second and fifth steps.

2. As the mold is ready to close onto the sheet, this special frame will move into place first and act as an independent clamping mechanism. There are instances when both sides of this clamping mechanism will be activated simultaneously, in an attempt to capture a predetermined area of the sheet. For best results, the female mold side should accompany the clamp frame, to simplify the operation. The purpose and results are the same in both cases. Capturing the sheet will not only ensure a uniform area for forming, but in cases of multi-up mold use, will prevent material robbing between adjacent forming areas.

3. Moments after the sheet clamping is accomplished with the tightly fitting frame mechanism, the plug assist will move in and complete prestretching of the heated sheet. Since the independently movable frame mechanism performed its sheet-capturing function prior to this move, stretching will only occur inside the trapped area.

4. With the mold closure and the maximum prestretching of the sheet accomplished, the forming forces are activated. The forming can be made by vacuum, air pressure, or both. The prestretched sheet will be forced against the female mold surface and as contact is made, cooling of the formed part will take place. At the end of the cooling cycle, the shape of the part will be permanently set.

5. At this stage of the thermoforming cycle, the independently moving frame mechanism will perform its second duty: trimming the thermoformed part away from the surrounding area. In this trimming mode, either the two mold halves or the two opposing clamping-frame mechanisms will move out of their aligned position, rendering a shearing effect on the continuity of the captured plastic flange. Since there are very close tolerances between the outside surface of the mold and the frame mechanism, this offsetting movement will result in a clean-cut edge. To ensure proper cutting, this type of tooling is not only made to close tolerance, but of high-quality hardened tool steel. Most tooling for this type of trim-in-place process will simply have a closely fit insert or ring at its shear line. The inserts and rings are either resharpenable or replaceable. There are some limitations to the maximum thickness for this type

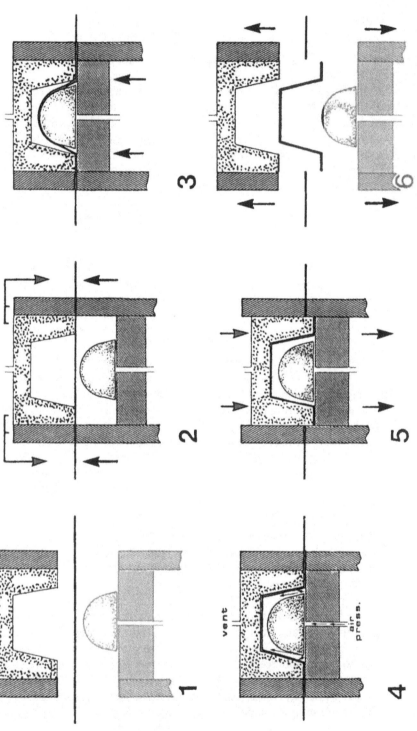

Figure 42 Form-and-trim-in-place forming with plug assist: (1) preheated sheet locating; (2) clamping; (3) prestretching; (4) pressure forming; (5) trimming; (6) mold separation and stripping.

of trimming. However, the upper thickness limits are well within the restrainment levels of this process; this is also true of its material handling limits. Some tooling may have a slight shearing angle built into its cutting face, which will lessen the demand for higher cutting-force requirements and make a cleaner cut.

6. After trimming, the thermoforming mold is opened together with its surrounding cutting-frame mechanism. As the mold bodies are moved apart, the completely trimmed part will drop out of the female mold cavity. To aid part ejection, a blast of air is shot through the female mold in unison with an air jet shot from the oven side of the mold. The air jet will help to blow the formed and trimmed parts out toward the front end of the machine. The thermoforming equipment usually has trimmed part-receiving magazines that are built to funnel the blown parts into a stack. The receiving magazine size is dependent on the number of "ups" in the mold, where each cavity will have its own magazine for receiving and stacking the blown-out product.

This type of process is very popular; however, with its benefit limitations, the method can be adapted to only a special segment of the industry. The accuracy of trim-in-place trimming is desirable. Every time a product is both thermoformed and trimmed in the same location, the trim is guaranteed to be made precisely in the same place. For example, round thermoformed parts made with this process always produce an even and concentric trim. Since there are no indexing and reregistration errors, this type of trimming never produces an offset cut. This feature is critical for thermoforming producers who manufacture tightly fitting food and beverage containers and lids. The uniformity of trim results in a trouble-free lid and container mating, the most important feature of this type of product. However, together with the benefits of utmost trim accuracy comes the negative aspect of higher price, a result of the fact that more time is required for the part to pass through the six stages of trim-in-place thermoforming. In-line thermoforming, in contrast, requires only four forming stages to produce a thermoformed product. Trimming, which is usually done in a secondary operation, can be accomplished within the same period as the forming, thereby reducing the cycle time. In the form-and-trim-in-place operation, the extra step (5) does require extra time and increases the overall cycle time. As this operation creates a significant increase in cycle time, and such increases add up in a long continuous production run, thermoforming practitioners should be aware of both the benefits and shortcomings of each method before selecting a process and equipment. Proper analysis of the product specification criteria will help in making the proper choice.

D. Pressure Bubble Plug-Assist Forming

The pressure bubble plug-assist thermoforming technique follows some of the same rules and procedures as those of the two preceding thermoforming meth-

ods that we have discussed. The clue to this special forming technique is revealed by its process name: "pressure bubble." This method is used for prestretching a trapped area of the sheet. In utilizing this process, the ultimate
goal is to create improved material distribution with maximum uniformity of
wall thickness in the finished article. Pressure bubble plug-assist forming is
used equally with light- and heavy-gauge materials as well as small or large
articles. This technique can also be broken down to four separate forming
sequence elements. The four steps are usually produced in a simultaneous
motion and combined into a quite rapid cycle. Figure 43 shows the four sequences as a stop-motion study, which allows us to examine each segment
individually.

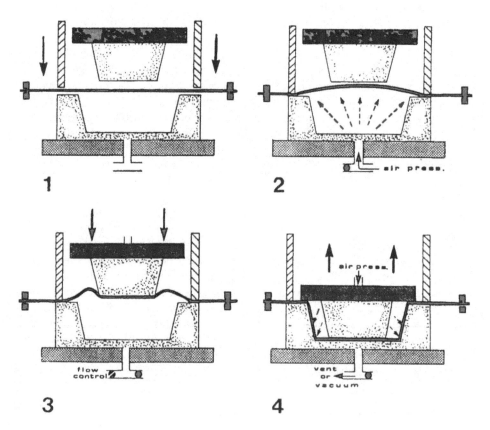

Figure 43   Pressure bubble plug-assist forming: (1) preheated sheet clamping; (2)
prestretching of sheet; (3) mold closing; (4) pressure forming into female mold.

1. In this thermoforming method, the process begins with a preheated thermoplastic sheet. An independently actuated clamp mechanism is used to trap the needed area of the sheet for thermoforming. With a small product, particularly in multi-up mold configurations, each mold comes with its own clamping mechanism; with large articles a large circumferential clamp mechanism is employed. In either case, the clamp mechanism will compress the heated sheet against the lip areas of the female mold. The crimp should be made strong enough so that there is no material movement and no escape of pressurized air. As soon as the heated sheet is located in the forming area, the clamp mechanism is activated for capture of the sheet. Pressurization of the bubble can now begin.

2. In this step, air pressure is introduced from the female cavity side. Since the clamp mechanism has created a seal to hold the sheet in this entrapment, the pressure will force the softened sheet into a domelike shape. (The same results can be achieved with vacuum force, also coming from the assist sides, provided that additional sealing is made between the plug assist and the clamp mechanism.) As the domelike bubble is formed, the result will be prestretching of the thermoplastic sheet. The prestretching can be controlled by the size of the bubble: Smaller bubbles will cause less prestretching, larger bubbles will provide the most stretching.

3. As the predetermined stretching is completed, the plug assist will move against the sheet. The bubble will be forced downward against the plug-assist surface. It is easy to recognize how using this method of prestretching will produce better results than those with plug-assist forming alone (Section II.B). A comparison should be made between the second stage of Figure 41 and the third stage of Figure 43. This comparison will provide the necessary understanding of the advantages of bubble prestretching. Some controls should, of course, be incorporated when the prestretched bubbles are produced. An overproduced bubble can, in turn, overstretch the sheet. In fact, the overstretching can be made so extensive that the sheet will bulge around the sides of the plug assist as it plunges into the female mold. This can be caused either by overpressurization of the prestretched bubble or by insufficient relief passages for air to escape and make room for the plug-assist takeover. For both reasons, excessive air entrapment will cause heavy bulging around the circumference of the plug assist. In the most severe case, this bulging can curtail full mold closure or may actually cause the bubble to rupture. In less severe cases, the bulging will produce only a slight overstretching of the sheet. An overstretch will mean that the bubble and the resultant bulge have stretched out over a larger area than that of the entire mold surface. An overly stretched sheet, with its excessive surface, will develop a wrinkle when formed. To eliminate this unwanted wrinkle, all the thermoformer has to do is reduce the size of the bubble. The reduction in the bubble will reduce the prestretch and the wrinkle

will disappear. Fine tuning of the air pressure will curtail the size of the bubble that is formed. Adjusting the travel speed of the plug assist or controlling the outflow of entrapped air will lessen the resulting bulging of the sheet around the plug assist. Limiting the prestretching to an ideal level and providing proper air outflow will make the forming possible.

4. In the last step in this process, air pressure is introduced from the plug-assist side. The pressure will force the softened plastic sheet to form against the female mold surface. As it forces the sheet to form, the air pressure will squeeze the air out from underneath the sheet. For more positive air evacuation and better transfer of detail, the introduction of vacuum forces is recommended. Using a vacuum setup may even improve the forming time.

Following the forming step, all conditions and results match those of the thermoforming techniques discussed earlier. As the forming of the sheet is completed, the mold will cool the plastic, setting its shape into a permanent form. The thermoformed article will shrink and pull away from the mold cavity walls. At this point, the molds are separated and the formed part with its surrounding scrap is removed from the molding area. Products made using this thermoforming method are trimmed in posttrimming operations, which can be implemented either in-line or on separate secondary equipment.

As has been pointed out, in this particular practice, the actual shaping of the sheet is made against the female mold surfaces. Although the actual prestretching method in this technique is capable of providing improvements over straight plug-assist forming, some of the typical female mold characteristics will prevail. The persistence of the desired female forming characteristics should be the main factor to consider in choosing this method of thermoforming.

E. Pressure Bubble Snapback Forming

The pressure bubble snapback thermoforming method proceeds identically to pressure bubble plug-assist forming (Section II.D) except for the fourth stage. The same first three stages form the pressure bubble in order to prestretch the softened plastic sheet. This method also has equal, if not greater, sensitivity to overstretching and the resulting wrinkle.

With this type of thermoforming, the plastic sheet is not formed against the female mold surfaces; rather, it is "snapped back" to be formed against the male mold surface. The male mold not only acts as an assist but actually provides the shape for the plastic and even performs the cooling duties. The female mold acts only as a pressure chamber. The forming sequence is shown in steps 1 to 3 of Figure 44, which are identical to those in Figure 43. But, step 4 differs.

4. As the male mold travels and pushes the softened plastic into the female mold, the air trapped between the plastic sheet and the mold is only partially vented or may not be vented at all. As the two mold halves close

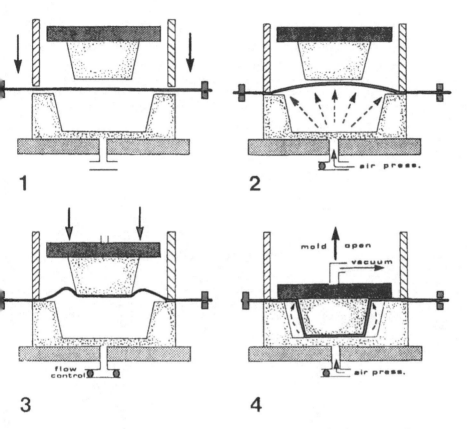

Figure 44   Pressure bubble snapback forming: (1) preheated sheet clamping; (2) pre-stretching of sheet; (3) mold closing; (4) pressure forming over male mold.

together, the pressure that has been built up will increase and force the plastic against the male mold. As the full mold closure is completed, a vacuum is introduced from the male mold side that will snap all the plastic back against itself and complete the forming. Cooling will take effect that will cause shrinkage in the formed article. For each part removal and stripping, the male mold must have sufficient taper angles. This shrinkage onto the male mold is a typical forming result. Since this forming is made with a male mold, most of its adapted features will carry the inherited characteristics of the typical male mold. Although using this type of thermoforming, with its pressure bubble formation for prestretching, will greatly improve material distribution, this forming technique can still carry and display characteristic clues of material shift or thinning typical of male molding.

The thermoforming practitioner has the option of using either of the two bubble forming methods and switching back and forth between the two. With a single mold setup, two different-sized products with varying material distribution can be produced simply by actuating and applying air pressure and vacuum forces to different sides of the mold. Armed with this knowledge, the thermoformer can choose the principal mold sides and forming modes for the product being formed. By making a switch in the forming mode, a secondary product of a different size can be produced without additional tooling cost.

F. Snapback Vacuum Forming

The snapback vacuum forming method is a very simple thermoforming process, well liked and often implemented. Its popularity stems from the fact that its use greatly improves material distribution and can produce improved wall thickness conditions. The snapback thermoforming technique may be one of the oldest and most reliable forming methods for the forming of medium- to heavy-gauge and larger-sized thermoformed products. However, the basic concept can be adapted to any of the thermoforming methods. The simplicity of this forming concept is easily seen in Figure 45, which illustrates its simplest version, in which only a female mold is required. The steps in this forming method can be describes as follows:

1. The preheated plastic sheet is clamped against the female mold lips, creating a trapped air chamber by acting as a cover for the cavity.

2. Controlled levels of air pressure are introduced to this sealed chamber, forcing the heated sheet into a dome shape that will continue to grow with an increase in air pressure. This stretching is the key to obtaining better wall thickness distribution in the final product. The size of the preblown dome will determine the amount of prestretching and will provide some control.

3. When the desired dome size is reached, the air pressure is immediately turned off and vacuum introduced. From the pressurized dome shape, the vacuum will snap the prestretched plastic sheet back into the female mold cavity. This movement finalizes the forming and subjects the plastic to the cooling phase.

4. Now the part is ready to be stripped and removed from the mold. Depending on the shape and sidewall taper angles, removal may be accomplished by simply lifting the part out or may require a slight air blast to assist. In the most difficult cases, stripping may have to be done with mechanical stripper units to push or pull the formed part out.

There are many variations of snapback vacuum forming that can be adapted for the purpose of better wall thickness distributions; all follow the basic format. One variation is shown in Figure 46. In this drawing, for the purpose of showing a different type of heating method, the heater unit is

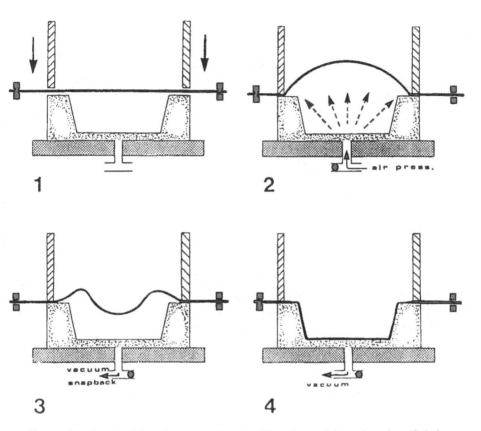

Figure 45   Snapback female vacuum forming: (1) preheated sheet clamping; (2) bubble blowing/sheet prestretching; (3) snapback of sheet; (4) vacuum forming.

positioned above the cold thermoplastic sheet, directly in the forming area. This type of heater is built like a drawer, and when proper heat levels are reached in the plastic sheet, the entire unit is slid out of the way. The heater unit is built with rollers and rails or a track assembly. For heating the thermoplastic sheet, this heating method will work as well as any other. The actual selection of different types of heating apparatus is not usually up to the thermoforming operator. The make of the thermoforming equipment and its heater unit design will determine the heating method to be used. In this version of snapback vacuum forming, the actual forming procedure can be broken down to four basic forming steps.

   1. The thermoplastic sheet is heated between the opposing mold halves. (With a different type of heater unit, the sheet would be transferred in a pre-

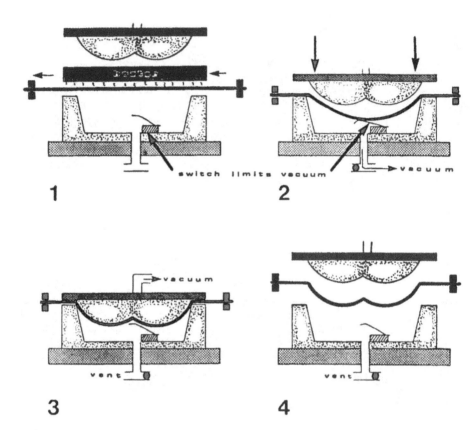

1   2

3   4

Figure 46   Snapback vacuum forming with a male mold: (1) thermoplastic sheet heating; (2) prestretching of sheet; (3) snapback vacuum forming; (4) mold separation and stripping.

heated condition into place between the mold halves.) At this stage it is easy to recognize that the development of a sag would not be harmful; in fact, sag would simply aid the prestretching effort.

2. The heater unit has been removed to the side so that the two mold faces can move toward the sheet. At the moment contact is made between the female mold upper lips and the sheet, vacuum is introduced through the female cavity. The sheet will self-seal at the contact area and the vacuum can then pull the softened sheet into the cavity. A microswitch is placed on the bottom of the female cavity. This switch is usually equipped with an extension arm to make it ultrasensitive. Additionally, the switch assembly is made to be adjustable in height. As the softened sheet is pulled into a bowllike shape, its bottom

will lean on the microswitch arm and trigger it. With the adjustable height of the switch, proper controls can be set for the amount of prestretching that is desired.

3. At the moment of sheet triggering, the microswitch will signal the vacuum valves to switch from vacuum to venting mode on the bottom female side. Simultaneously, the vacuum on the upper mold side will be activated. The rapid change in the direction of the vacuum compels the original bowl-shaped plastic sheet to snap back and form against the upper mold surface. With the full surface contact, the forming is completed and the natural cooling mode will take place.

4. In the final forming step, cooling and shrinkage take place and the molds are separated. The vacuum must be deactivated in both mold halves to facilitate easy mold separation and part stripping. A short blast of air can be introduced to both sides of the mold to aid stripping. However, the use of stripping air pressure must be used with some moderation to avoid distortion or even destruction of the formed article.

Figure 46 shows a male configuration of the upper mold with a female indentation in the center. The same thermoforming procedure can also be accomplished if this mold half is made in a female mold configuration instead of the male. Figure 46 also shows a female bottom mold that in actuality has no function other than to act as a vacuum chamber or vacuum box, one having its upper side covered by the softened sheet. The only contact that has been made between the sheet and this mold side is at the lip seating around the circumference and the centrally located microswitch arm.

G. Trapped-Air-Assist Forming

The trapped-air-assist thermoforming procedure is a questionable on because it is highly dependent on a well-functioning seal created between the clamp mechanism and the assist base plate. This key factor of this seal is so critical that its slightest deterioration will render the entire forming procedure useless. If and when good airtight conditions can be achieved and maintained, the process will function satisfactorily. However, maintenance of the sealing surfaces is usually a difficult task, particularly for large multi-up molds, creating repeated breakdowns and unwanted downtime. Usually, mold configuration and design conditions do not permit the precision necessary to build and maintain sufficient high-quality moving seal surfaces for this type of manufacturing. Since the process is keyed to this single critical factor, its use is highly discouraged. Of course, there are thermoformers who claim success using this method and have proof products of sufficiently good quality made using this system. On the other hand, there is no telling how many breakdowns they encounter while maintaining the seals. Figure 47 shows this thermoforming method broken

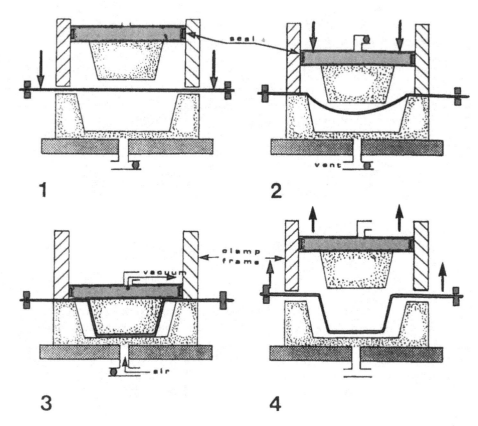

Figure 47 Trapped-air snapback forming: (1) preheated sheet clamping; (2) pre-stretching with trapped air; (3) snapback vacuum forming; (4) mold opening and stripping.

down into four stages. The following description explains the inner workings of this technique.

1. The preheated thermoplastic sheet is transferred in between the opposing mold halves. The mold arrangement can be made to function by having the female cavity located on either the lower or upper platen. Both arrangements work equally well. In this specific arrangement, the female mold cavity is placed on the lower platen so that the upper platen, with the male plug-assist configuration and clamping mechanism, will close onto it. This move will capture the heated and softened thermoplastic sheet positioned between platens.

2. For prestretching of the sheet, the plug assist will move toward the sheet, and due to the existence of the airtight seal between the clamp mech-

anism and the plug assist, this travel will condense and pressurize the trapped air. As the plug assist moves downward, just like a plunger, it will push and bow the softened plastic sheet ahead of it. With continued travel, the air compression exerts more pressure on the sheet. Regulation of the amount of prestretching in the sequence is determined basically by the plungerlike actions of the plug assist. If more prestretching is required, increased "head space" is given for the plug assist by starting it from a higher initial position to increase the air volume beneath it. Naturally, the vacuum valve is set in the closed mode in the assist mold side for this trapped-air utilization.

3. The prestretched sheet can be formed against either side of the mold. If vacuum is introduced through the female mold at this stage, the plastic will form against the female mold sides. This forming can be accelerated by air pressure introduction from the male mold side.

If preferred, vacuum can be introduced from the male mold side, with pressurized air then squeezing in from the female mold side, as illustrated. By this arrangement, the forming will be made against the male mold surfaces. Naturally, the change from one mold side to the other could not possibly produce equal-sized products unless the mold dimensions and configurations are matched with a proper mold gap. If there is a substantial size difference between the female and male molds, the differences will be duplicated on the formed plastic.

In the last sequence (as shown in the figure), the formed plastic part is cooled off by the respective mold surface. The customary shrinkage will take effect and mold separation and stripping can be done. The mold halves are separated and the part is removed.

H. Solid-Phase Pressure Forming

The process of solid-phase pressure forming (SPPF) should not be categorized as a separate thermoforming method. However, recent years of heavy publicity have generated enough interest to justify devoting the time to cover this thermoforming practice.

The clue to this type of thermoforming is contained in its name: "solid-phase pressure forming." The thermoforming procedure is made with pressure forces alone shaping the plastic from a sheet form into its final configuration. The key factor in this process is the heated status of the sheet. In this type of thermoforming, the thermoplastic sheet is not heated to normal levels where the sheet is softened to be almost flowing and in a formable state. The plastic sheet is heated, but only to levels that keep it in a solid state. Basically, any underheated plastic sheet being thermoformed could be categorized as being in solid-phase forming. However, the nature of most thermoplastic materials will not permit stretching or forming in the solid state. Most common thermoplastics will tear or rupture in the slightly heated and still solid state. They will accept shaping and forming only when they reach the proper softening temperatures. Exceptions are found only in a very limited number of plastics, one of which is polypropylene.

Polypropylene thermoplastic material has a unique stretchable quality unmatched by that of any of the other popularly priced plastic materials. In fact, polypropylene is distinctly qualified for solid-phase pressure forming. Polypropylene can be shaped quite well into a form by thermoforming just below its crystalline melting point without tearing or rupture. Since the forming is made at a temperature too low to allow material flow, the results of the thermoforming will be high molecular orientation within each thermoformed part. The resulting products will display a high degree of transparency and strength, and due to the nature of the polypropylene material, exceptionally outstanding chemical and physical toughness. In addition to these useful qualities, polypropylene has good fracture resistance and can withstand repeated flexing and comparatively low moisture-transmission rates. All of these advantages, combined with FDA approval, make this type of thermoformed product desirable for food packaging.

To be successful in solid-phase pressure forming, it is essential that the polypropylene sheet be made of a high-grade resin material and extruded with close-to-zero internal stress within the sheet. Use of improper extruder screw designs that are not made specifically for polypropylene resins may produce some internal stress in the extruded sheet. As polypropylene sheets have internal stress conditions locked in, they do not give visual clues to their conditions. However, when exposed to heat, those locked-in stress conditions will seek release, resulting in a material "wander," or shift, from one location to the other. This condition will result in variations of thickness that not only interfere with stretching and forming but also affect the outcome of the thermoforming. Whether the forming process is plagued by this problem constantly or only occasionally, it will be very aggravating to the thermoformer.

I. Twin-Sheet Forming

Twin-sheet forming is one of the many thermoforming techniques to produce double walled products. This method produces similar and competitive items to blow molded and rotational molded products. It even can replace some products which are fabricated by gluing two separately formed article portions together to create a needed product. Most of the time such products can be made by the usual thermoforming process, and after the two individual halves are produced by the two separately made moldings, then they are united together by some fabrication or gluing method to create a single item. In most cases twin-sheet forming could eliminate the two separate thermoforming moldings and definitely eliminates the secondary gluing fabrication by making everything in a single step. In this twin-sheet forming there are two independent sheets used to make the molding. The two thermoplastic sheets, whether in a precut form or a continuous web can be clamped or treaded into the feed system of a thermoforming machine, right on top of one another. The two sheets can be heated very much like single sheets are heated, preferably in a "sandwich" type of double sided oven. Heating is accomplished from top to bottom; therefore both sheets are equally heated from their respective sides. As the heat penetrates the thermoplastic plastic sheets the sheets will go through the customary heat reactions, and eventually will reach the ideal forming temperature. Just like with any of the single sheet heatings, under- or overheating often could jeopardize the thermoforming results. Careful attention should be paid to obtain optimum heating conditions, not only judging each sheet by itself, but considering that each sheet may interfere or effect the other sheet's heating results. The heating must be made in such a way that both sides of the twin-sheets receive the optimum levels of heat for the ideal forming conditions. The heater output levels must be regulated to compensate for heat output variations, which is often observed, between the top and bottom heaters. These heating differences are even further complicated when different colors or different types of plastic sheets are used in the twin-sheet forming. Naturally both sides of the twin-sheet must be at ideal forming temperature for satisfactory molding results. Special attention should be given to the fact that the twin-sheets inner sides facing each other also should be heated to the desired temperature levels for ideal thermoforming. The thermoformer must be patient in the heating of the twin-sheets, because, in spite of the fact that heating is done from both sides the inner surfaces only receives its heating by conduction from the outer surface of the sheet.

The twin-sheet is usually formed by two opposing molds, set up in such a way that each side has its own respective configurations. The molds are most likely to have a dominant female configuration; however that is not the case all the time. In fact the two sides do not have to have a mirror image of each other either. However, they must have the same outline configurations, because that outline is where the unification or sealing of the two halves will take place. As the molding is made, the two molds will create enough clamping force between the two halves that the two parts, which formed to their respective mold sides, are unified by sealing the two together (just like heat-sealing), at the perimeter of their configuration. The retained heat in the formed plastic should be sufficient to make the sealing complete. By this thermoforming molding technique the twin-sheets are formed against the two molds. Each side will resemble the respective mold sides, those creating a hollow product. Naturally, to achieve such a forming, an air passage must be allowed between the twin-sheets. To allow air movement between the twin-sheets, there must be an opening, a small void between the two crimped circumferences of the parts. Often a syringe-like device is employed for the purpose of providing the entrance for air movement. That syringe can be placed in between the two sheets or can be made to puncture and penetrate the plastic sheet as it is formed on the sides of the mold.

In case the forming technique is the vacuum forming procedure, the two molds receiving the vacuum will pull the respective sides of the twin-sheet against each mold side. In order to accomplish this forming, air must be allowed to enter to displace the space between the sheets.

If pressure forming is attempted the introduction of the pressurized air is also made through the small syringe opening. The syringe will act as an air ejector to pressurize the space in between the twin-sheets, forcing it to conform to the respective mold sides. Application of vacuum to the molds may not be necessary, but would not hurt, as it will most likely help to minimize any chance of trapped air pockets.

Both forming techniques highly mimic the blow-molding process, not only in the resulting product but in the thermoplastic material stretching. The only difference is that the blow-molding starts with a molten parison (a round hollow tube) while the twin-sheet forming uses a two thermoplastic sheet arrangement. The twin-sheet forming technique will offer better results over blow-molding, because the sheet extrusion process provides a more precise quality control in sheet thickness and a wider option of variables in the thermoplastic sheet making.

This twin-sheet forming production method often competes with the above mentioned blow molding or rotational molding procedures. Depending on the product line, it may or may not have a unique advantage over them. Since this process does not necessarily start with the same source of thermoplastic material supply, the twin-sheets can be made of a different color or even different makeup of plastics. This way twin-sheet thermoforming can produce a product which has different appearances from one side to the other side, or have different physical characteristics as long as it does not interfere with the forming procedure. This way, for example, a hollow product can be made with different colors on opposite sides. It may have a smooth finish on one side and a textured finish on the other, can be a clear plastic on one side while the other side is made of opaque plastic material. In case it is needed the two sides can be made of different plastics for accommodating a finished product with a wide range of different criteria or purposes (see Figure 48).

To further enhance the twin-sheet process and allow it to improve cycle time, the two sheets can be clamped separately or, in the web-fed situation, fed in two separate paths. By this method the individual sheets can be heated from both sides. The processor can have two independent clamp frames, top and bottom heaters, and one heater in the middle. The top heater will heat the top

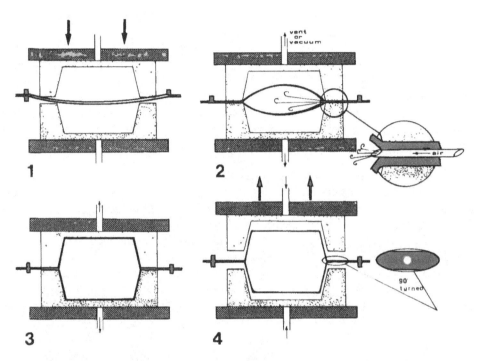

Figure 48 (1) Clamping the twin sheets. (2) Activating the forming forces. If vacuum is applied at the mold, the syringe allows air to enter between the sheets. If pressurized air is used for forming, venting or vacuum is opened to the mold. (3) Full forming and cooling take place. (4) Mold opens and a blast of air through the mold helps the stripping.

side of the top sheet, the bottom heater heats the underside of the bottom sheet. The middle heater unit will heat the top sheet underside and at the same time the top side of the bottom sheet.

Also, the forming of the twin-sheet can be accomplished by an overlapping method using dual rotary equipment, as will be described later in Section III.D of Chapter 4.

Dual-sided heating of twin-sheet also can be accomplished on the web-fed machines. In this case, there are two separate loops of pin chain tacks right above each other. Each chain will carry and index forward with its own sheet supply. There is a triple-decker oven which will heat each sheet on both of its sides. As the two sheets approach the molding area the distance from each other is diminished to almost nil, so that they run on top of each other. This closeness permits thermoforming of the twin-sheet, making it possible to form a product that cannot be produced any other way.

It should be pointed out that there are many more variations and different adaptations that can be produced with the twin-sheet forming technique, and many innovations are currently under investigation which will undoubtedly become general practice in the future.

III. TYPICAL MOLD ARRANGEMENTS

The implementation of various thermoforming techniques begins with mold installation into the thermoforming equipment. Even before installation of the mold, some decisions must be made as to the mold configuration and available equipment choices. These decisions will be tied to the production quantities desired. There are several criteria that will guide the thermoformer, the two most outstanding being product size and product quantity. The remaining criteria may not be as dominant in the decision-making process and might not be considered at all if there are sufficient funds available to deal with them. The factors that limit a thermoformer, such as equipment, type of forming force, or available know-how, can easily be overcome by expedient purchases.

Product size is a strong influence because it can limit the equipment, molds, or even the techniques that are adaptable for a specific thermoforming process. Products of larger dimensions obviously cannot be produced on smaller-format equipment and may require additional specialized installations to support the process. On the other hand, small-configuration products can be produced in multiples at the same time.

The quantity of a product to be made is also a crucial factor in the decision as to what approach the thermoformer should take. When only a few pieces of a specifically configured product are needed, the process certainly will require different planning and arrangements from a large-scale production. The forming technique coupled with the mold arrangement (one-up, multi-up, or family molds) can make the difference between a satisfactory thermoforming operation or merely an acceptable one.

A. One-Up Molds

When a single mold is used in thermoforming, it is often referred to as a "one-up" mold. When the mold consists of a female mold, it can also be called a "single-cavity" mold. There are three independent reasons for using a one-up thermoforming format. The first can be found in the thermoformed article size. Many thermoformed products are so large that the equipment can accommodate only one at a time. The second purpose of thermoforming in single units is that only a limited number of products are needed. Finally, experimentation and testing are implemented most ideally with a one-up mold configuration. Initial mold building and rework are always easier and less costly in a single-mold format. It is possible to find other reasons to use one-up molds; however, their justification should be analyzed carefully before they are accepted. No thermoformer will argue the fact that for oversized parts, due to machine-size limitations, single-unit molding may be the only way to accommodate the forming. Even if more than one part can be formed at once, multi-up molding in a larger size can create interference between the molds that will hamper the outcome.

One-up mold thermoforming is preferable when the manufactured products are not being used or sold in high quantities or when the demand for a larger volume of products is spread over a certain period of time. In either case, the product quantity demand does not justify the purchase of multi-up production molds that will be used only in short production bursts and result in production downtimes between uses. On the other side of the production scale, markets are continuously being supplied by manufacturers sticking to one-up mold equipment long after the product demand has outgrown the single mold's supply capabilities. This can create possible product shortages and invite competition from other suppliers.

All experimental tryouts of new product designs must be made in a one-up mold format. Use of single molds not only keeps the cost down but removes all interfering reactions. Besides the initial mold cost, tooling rework or alteration can be done without a major expenditure. Even if an entirely new mold is suggested, the cost is comparatively moderate. Experimental molds can be produced out of a number of materials that are suitable, inexpensive, and easy to alter or rework and then to reassemble with simple shop tools. Of course, one-up production molds must be made of materials that provide good tool life expectancy and the durability to withstand maximum thermoforming forces.

B. Multi-Up Molds

When product size and demand are not limiting factors, thermoforming producers can install multiples of the same mold in the thermoforming machine, known as "multi-up" molds. Using the same thermoforming procedure, more than one unit of the same product can be produced at the same time. Two or more parts produced in the same cycle will each have its own mold. Multi-up molds come in formats from two-up to any conceivable matrix. Multi-up molds are built with a predetermined, fixed format. The mold arrangement can follow several optional patterns, which are usually guided by the platen size of a machine and the respective sheet panel size. For example, a four-up mold arrangement can consist of four individual molds placed into a single row (1 by

4) or in two rows of two molds (2 by 2). It is not unusual to find molds made into formats as large as 48 up (6 by 8) or even larger.

The tooling for multi-up molds consists of identical individual molds with fixed spacing in between. The molds can be made out of any common block of tooling material or may even be individual molds secured to a common backup plate. This mold arrangement provides a fixed overall dimension to the final mold, which can, in turn, be tied to the thermoplastic sheet dimensions. The overall panel dimensions are determined by the individual thermoformed part dimensions, multiplied by their number in each direction plus the spacing between them and their edge allowances. The calculation and estimation of the proper sheet dimensions is most critical for successful, cost-effective thermoforming. The economic implications of panel size for each cycle are discussed in Section V of Chapter 6.

The spaces between the individual molds in a multi-up mold do not need to be identical or uniform. When two or more moldings are produced in the same cycle it is easy to recognize that molding interference can originate between the neighboring molds. The space allowed between the individual molds may have to be varied according to several factors, including mold configuration, depth of draw, and design patterns. Minimizing the spacing demand between the individual moldings can be accomplished through the use of clamping mechanisms. Such mechanisms were discussed in Section II of Chapter 2 and are illustrated in Figures 13 and 14. In the heating phase for multi-up molds, uniform heating of the thermoplastic sheet is of utmost importance. The thermoplastic sheet that is introduced into the forming cycle must have an identical temperature and soft consistency throughout its entire body. When sheet temperatures and forming conditions are not the same from one molding to another, the thermoforming will suffer. This criterion is as important in two-up mold formats as it is in very large formats. For example, in a large-format mold (that could have as many as 48 up and a panel dimension of 50 in. by 50 in.), the sheet temperature and forming conditions should be identical from one corner area to the other and from the center to the sides. To achieve and maintain temperature uniformity over such a large area is not an easy task, and to provide such uniformity from cycle to cycle requires first-rate equipment controls and thermoforming skills. Understanding the criteria limits of each type of thermoforming can help in the preplanning and final outcome.

It should be noted that relatively thick thermoplastic sheets will be required to form a multiple mold pattern producing several parts using closely placed tooling with a substantial depth-of-draw configuration. From this greater thickness of sheet will be produced an abruptly decreasing wall thickness in the finished thermoformed product. Soft-drink cups, for instance, are usually thermoformed out of 80- to 100-mil-thick thermoplastic sheet and experience a tenfold reduction in their wall thickness. Multi-up thermoforming requires multi-up trimming procedures as well. If the trimming operation does not keep up with the thermoforming speed, it causes a bottleneck in the flow of production. Trimming of multi-up-sized thermoforming can be managed all at once with a matching multi-up trim tool. Such tooling requires some type of registration capability in order to locate the fully formed panel accurately for trimming. With the use of any of the cutting instruments discussed previously, the parts can be trimmed, ejected, and gathered from the trimming apparatus. This type of trimming can be made by both post and in-line operations.

In continuous roll-fed thermoforming operations, the trimming is often accomplished in row formations in a secondary trim press. Basically, this type of trimming of a multi-up molding is made by individual rows, one row at a time. Each row of the thermoformed panel is indexed and registered before trimming. The number of rows produced in the forming determines the number of strokes the trim press must produce to keep up with the thermoforming machine. For example, a mold that displays four rows within its molding format will be four times faster in its trim-press run.

The size of a multi-up mold is often described as much by its numerical array as by the overall numbers of ups. A 48-up mold, for instance, may be eight columns wide by six rows deep, or the other way around. The number of parts in the mold's width and depth give a proper description of a specific multi-up mold arrangement.

C. Family Molds

The family mold is an extension and variation of the multi-up mold arrangement. A family mold is composed of individual molds in a multi-up format that do not share the same configuration. The setup may consist of cavities each made in an entirely different configuration or may contain groups of different mold configurations. The criteria for selecting this type of mold are based on the similarities of the formed parts. Such close dimensional comparability of the finished products allows them to be combined into the same molding. For best results in thermoforming with a family mold, the mold configurations should have the same depth-of-draw ratios. Of course, this criterion is not absolutely necessary, but it can minimize interference between the moldings. It is also best for the parts to be closely matched in at least one of their dimensions. If that is the case, such similarities in size minimize any dimensional misalignments in the combined mold configuration. However, such dimensional differences are also not absolutely essential for success in family mold adaptations.

The strongest argument for successful use of a family mold is an equal product demand on the individual articles. The actual product demand and the resulting sales ratios must match the product output ratios of the family molds. If, for example, a mold has a production ratio of 2:1 between one product item and another, the products should enjoy the same comparative sales opportunities. A constant deviation between the two ratio factors will provide either an overage or a shortage of one of the thermoformed items. Ironically, due to overaggressive merchandising efforts, the item in short supply will be the one to be brought up to its proper level, which results automatically in an oversupply of the other product. This oversupply will generally instigate "creative" marketing ideas, such as price reduction or discounting of the overproduced product, which do not solve the problem. The only way out of this predicament is to replace the family mold with newly built independent molds.

Despite the shortcomings of family molds, many products lend themselves to being good candidates for a family mold setup. Those products may enjoy somewhat similar sales volumes, to the point where an oversupply of one size will not be realized. For example, using family moldings for the production of various sizes of trays for flower pots has had good success.

The family molding technique can also be helpful where a larger area of product is blanked or trimmed out. The specific area that is cut away can be classified as normal scrap. However, with the same effort and energy consumption and a little creativity, a second, smaller product or group of products can be thermoformed out of the blanked area of the first. All that is necessary is to insert smaller molds where the blanking area will occur. The trimming apparatus may be able to handle the trimming of both products, or it is possible that the blanked-out panel may require separate trimming to cut out the smaller items. If there is no further product demand for the smaller units (which have been formed "free" out of the larger product anyway), the blanked sheet portion used for the secondary smaller products can still be scrapped and reclaimed. Opportunistic uses of family molds should not be overlooked. It can often be turned into a profitable supplier of secondary products.

D. Alternating Mold Arrangements

A quite different and innovative mold arrangement came about when high-speed, high-volume operations were demanded for the production of low-cost food containers. At the same time, for several reasons, the polypropylene thermoplastic resins came to be very attractive. Polypropylene can offer several advantages over other resins, as well as favorable costs. However, in the thermoforming process, the material demanded some extra cooling-time allowance. Extending the cooling cycle within the customary thermoforming process lengthened the overall cycle time. Such an increase results in reduced productivity and tends to destroy a large portion of the cost advantage.

To furnish extended cooling cycles for thermoformed articles without increasing cycle times demanded innovative mold arrangements. Such arrangements, which originated in Europe, even involved some reengineering of the thermoforming machines. The thermoforming equipment has been modified for each method and made specifically to implement the respective mold arrangements. Some of these specialized thermoforming methods have proprietary features that are covered by protective patents.

The basic principle of the "alternating" arrangement consists of a single set of male molds that will thermoform against a "team" of recurring female mold cavities. The male side of the mold will perform its normal function of assisting in prestretch of the plastic sheet directly into the female cavities. After the full forming procedure, the molds undergo the customary trim-in-place steps. However, after trimming is completed, the thermoformed parts are not ejected. As the male mold halves are withdrawn from the female cavities, the thermoformed parts are carried away from the molding area and a new set of female mold cavities substituted in their place. The previously thermoformed parts, being retained in the mold cavities throughout the remainder of the cycle, receive the desired extra cooling time. With this innovative method of recurring indexing of female mold cavities, the extra cooling demand is easily met without any sacrifice of thermoforming speed.

There are two basic innovative mold arrangement methods that can accomplish the same task: shuttle molds and revolving molds. Each has its own distinct approach. The results produced by both systems are equally satisfactory. The choice of one system over the other must be made according to criteria other than simply the resultant thermoforming.

1. Shuttle Molds

In the shuttle mold arrangement, the tooling has a double set of female mold cavities opposing a single set of male plug-assist molds. The principle of this method is shown in Figure 49. The female mold cavities will "shuttle" perpendicularly to the sheet advancement motions. The thermoforming is always made in a central location into which an empty and recurring set of mold cavities will index with each cycle. The shuttle movement is made in a horizontal plane, and is repeated cycle after cycle. The previously formed and trimmed parts will remain in the female cavities and will shuttle to the side. The parts will be retained and cooled as long as the center molding is ready to open. As the mold is made to open up, the parts made previously will be simultaneously ejected. As soon as the steps are completed, the mold will shuttle back again and repeat the entire process. It is easy to see that with the shuttle mold arrangement, each set of molds receives twice the usual cooling time of customary thermoforming and does not require more cooling.

2. Revolving Molds

The thermoforming technique using revolving molds follows the same principles and goals as those of the shuttle mold arrangement, but the female mold

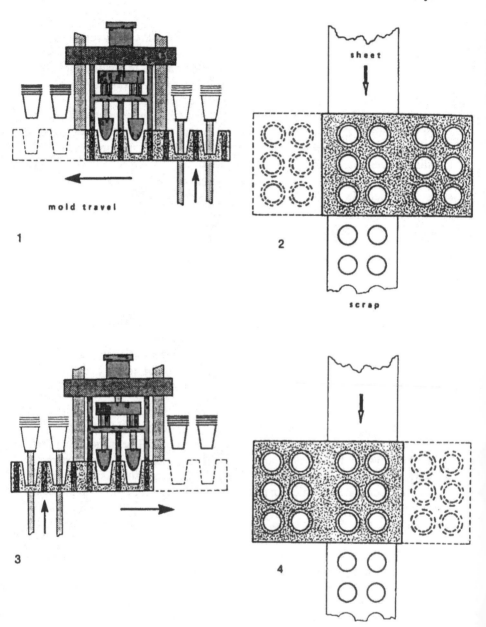

Figure 49   Shuttle mold arrangement: (1) front view, mold on right side, part ejecting to the right stack; (2) upper view, mold shuttles perpendicular to sheet; (3) front view, mold on left side, part ejecting to the left stack; (4) upper view, mold on left side.

cavities are arranged in a horizontally rotating system. This procedure still uses a single set of male plug-assist moldings opposing a four-sided "team" of female cavities. The mold arrangement and its functions are shown in Figure 50. This mold arrangement consists of four horizontally rotating and indexing mold cavity sets. The mold can contain single- or multicavity rows within the revolving mold arrangement. Each rotating index is completed with a quarter turn. The upward-facing female mold cavities are used to implement thermoforming and trimming. The next two indexing locations hold the trimmed parts

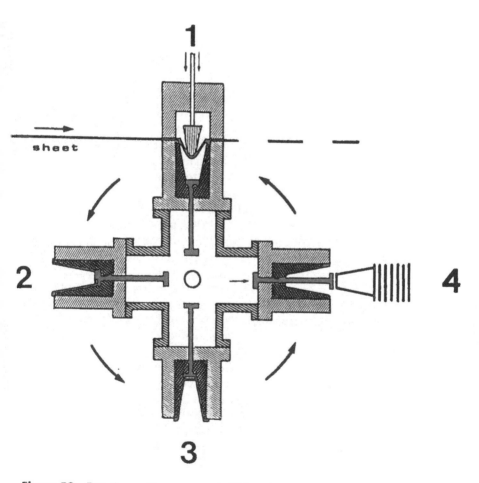

Figure 50   Rotating mold arrangement: (1) forming station; (2) cooling and lip cooling; (3) additional cooling; (4) ejecting and stacking.

within the cavities for extended cooling. In the last indexing location, the parts are horizontally ejected from the mold cavity. In the final rotational indexing, the empty female mold cavities return to the molding area for the thermoforming and trimming cycles. The ejected thermoformed parts are pushed onto a receiving magazine for nesting and stacking. For the nested stack, an automatic counter can be incorporated to help in packaging the thermoformed goods. With this rotating mold arrangement, thermoforming can be managed with 50 or more cycles per minute. At such high rates of thermoforming, it is of utmost importance to make the thermoforming procedure absolutely flawless. If errors should occur, they can easily "hide" within the stacks of finished product and slip unnoticed by all quality control efforts.

E. Rotary Mold Arrangements

In the rotary mold arrangement, a constantly rotating mold arrangement is utilized. The aim of this type of molding is to create continuous movement, resulting from the nonstop rotation of the molds. The principal shortcomings of this type of forming resides in its restriction to one-sided molds. In this type of rotary mold arrangement, it is not feasible to attempt mold-half mating, so female forming methods alone are generally used. Of course, with extended effort, such limitations can be overcome and even plug-assisted thermoforming can be carried out with this type of mold arrangement.

The thermoforming technique using the rotary mold can be approached in three ways, all of which are aimed basically at providing continuous thermoforming and rely on a continuous sheet-fed operation. The simplest version of the rotary mold arrangement has a large rotating wheel setup, somewhat similar to a Ferris wheel. The female mold cavities are placed side by side on the rotating wheel circumference. As the wheel rotates, it picks up the heated thermoplastic sheet and continuously wraps it around the rotating surface. As the sheet makes contact with the rotating mold surface, vacuum is introduced to the particular cavity or cavities. The introduction of vacuum forces will pull the heated plastic sheet into the cavity and force it to form into its own shape. As the rotation continues, succeeding cavities follow the lead of the first, one after the other. The formed parts remain in the cavities until they are stripped off. Next, the vacuum is turned off and venting takes place in the respective cavities. This permits stripping and removal of the formed parts, which remain in the form of a continuous web. The web is then subjected to trimming to separate the individual parts. In this mold arrangement, quite a number of molds can be placed into the large Ferris wheel form to produce the least number of revolutions. The larger-diameter wheels also offer long contact time with the mold after forming, resulting in better cooling.

The second version of the rotary mold arrangement follows the same principles of rotary mold setup. In this thermoforming method, 12 to 18 mold cavities are secured in a turning wheel pattern. Each cavity has a clamshell-type lid fixture that closes to capture the hot sheet between the lid and the mold cavity. This hermetic enclosure of the clamshell lid permits the use of pressure-forming forces. The mechanically closed clamshell lid will be able to hold up to 15 psi pressure and will give better forming details. This molding arrangement can produce up to $2^1/_2$ in. of depth of draw; however, it is most ideal for shallow-depth articles such as container lids. The specialized equipment utilized with this mold arrangement is discussed further in Chapter 4.

The third version of the rotary mold arrangement consists of a continuously moving mold face formed in the surface of a moving belt. The design possibilities of the belt range from a pattern of individual cavities placed next to each other, to one long pattern that concludes with each rotation of the belt, or even one that provides a continuous form. The advantage of using a belted mold system lies in the belt's ability to repeat itself. The mold system must be able to flex enough to complete the belt's rotation. Mold materials must have a rubbery consistency or be segmented in order to follow the belt turns. A single belt with its applied molding surface can be used together with vacuum forces. This type of system must have an efficiently engineered vacuum system and must manage to introduce and hold the vacuum forces when the thermoforming process calls for them. The cancellation of vacuum is also important at the end of belt travel to facilitate stripping the formed article. It is also feasible to use two opposing belt forming systems; each belt having a matched contoured design. Pressures created between the belts will result in a matched mold forming. The two belts squeeze the heated thermoplastic material to form and transmit their surface details on both sides. This method also qualifies as a compression molding technique. Naturally, the mold surface details must have enough venting holes to relieve any air trapped in the design structure. This double-belt mold arrangement works equally well with preheated thermoplastic sheets or with direct extrusion, where the material flows directly between the moving belts.

The mold arrangements discussed so far are not necessarily the only ones that can be implemented in thermoforming. There are combinations and variations that can be developed, and innovations are constantly being introduced. As new mold materials are developed, new uses will be found. Since many inventions like these have been made in the past and will continue to be made, outstanding patents on mold arrangements can limit the adaptation of some of the features. It is the obligation of the thermoformer to select the process that is free of restrictions. Patented mold arrangements are discussed here only to provide the most complete mold arrangement list available, not to promote the concept or imply its unrestricted use.

IV. LIMITATIONS IN THERMOFORMING

The thermoforming process has been very successful in producing a great many useful items. The manufacturing method, with its ideal volume-producing capability, has swept over established product lines, substituting and squeezing out traditional well-accepted manufacturing techniques. The thermoforming industry has not only replaced existing products but has created entirely new products. As the technology of thermoforming has developed and advanced, it has also grown bolder. Equipment capabilities, sophisticated toolings, and forming techniques have made it possible to manufacture products out of a single piece of sheet material where previous products required the assembly of multiple components. The design opportunities are practically unlimited for this manufacturing method. However, in real situations, thermoforming has defined parameters and limitations for which the process cannot offer any solution. It is an advantage to know and understand such limitations before thermoforming attempts are made. Being well acquainted with such limitations can eliminate any unrealistic expectations in respect to both sales and manufacturing. Through the years, many hindrances have been overcome and there is an excellent chance that a few more restrictions will be dismissed. However, some limits to thermoforming will remain constant. In fact, some of the characteristic limitations of the process will forever be inherent in the forming technique. There are shapes and forms that will never lend themselves to production by the thermoforming process. Such impossible situations should be relinquished as a thermoforming goal. There are restrictive boundaries in the thermoforming process which should be learned, respected, and worked with (and not against). Correct knowledge of what a process is capable of doing will save a lot of time, effort, and money. The thermoforming process cannot, by its very nature, be thought of as a total substitute for any other forming process, such as injection, compression, or blow molding. It cannot make products of identical configurations or dimensions or create minutely precise mold duplication. However, in many cases, thermoforming can permit similarly shaped competing products to be substituted. The thermoforming practitioner has the opportunity to design around the particular limiting factors in such a way that it will not affect the function of the product, yet will render a most cost-effective manufacturing method. The three basic limiting factors in thermoforming will remain with the process and under any circumstances, will require some effort to minimize their effects.

A. Detail Loss

In the process of thermoforming, not all conceivable details will be transferred from the mold to the formed part. From a meticulously made mold finish, some loss of detail transfer can be expected. Better transfers will be obtained with higher levels of forming forces, which can be further enhanced with temperature elevation. However, such improved detail transfers can be accomplished only up to a certain level, never 100%. In addition, where only one side of the sheet is in contact with the mold, the opposite side will probably show poor mold details. A thermoformed plastic article may show satisfactory detail transfers on one side, yet exhibit somewhat "washed-out" details on the other side.

A close examination of the forming technique will allow recognition and pinpointing of where and when such fine detail is transferred from a mold surface to the formed part. As the heated and softened thermoplastic sheet is forced against the mold walls, it will take up the contours. Where direct and substantial stretching or strong angular changes appear, higher levels of detail loss can be expected. In stretching and forming, actual material shifts will take effect and work against the obtainment of full detail. Figure 51 shows the two basic types of detail loss where even with the best conditions and the highest levels of forming forces, some lack of detail transfer can occur.

1. With the traditional male taper angles, the mold surface side of the formed article will be best equipped to obtain details. As the soft forming thermoplastic is forced against the mold surface, the mold detail will be closely transferred. The opposite side of the formed plastic article will be somewhat stretched and thinned over the mold radius. The stretching force will reduce

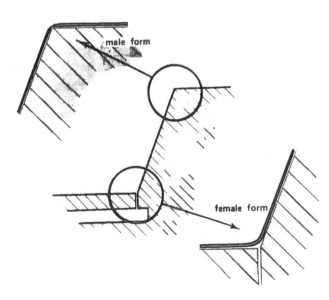

Figure 51 Detail loss in thermoforming.

the resulting radius on the opposite side from the theoretically obtainable parallel curve. As shown in Figure 51, the resulting radius can qualify as a detail loss. For example, when a textured sheet is used in thermoforming and the textured pattern side of the sheet is turned away from the mold, the results of the forming will be a stretching out of the particular textured patterns. The stretching will expand the texture design laterally while reducing its depth of detail. This will give the appearance of a worn and polished surface. In corner areas where at least three surfaces are junctioned together, the resulting stretch and detail loss can be so severe that the texturing will become almost smooth and shiny. This detail loss can be avoided or at least minimized by shunning the use of designs that require small radii. Larger-radius forms usually allow retention of the texture details. This particular sensitivity is most important in products where such loss of detail will diminish the product's attractiveness, as with suitcases, briefcases, and automobile dashboards.

2. With conventional female angles, the forming material will stop short of full corner detail. Even with the most ideal heat conditions and mold temperatures, the forming radius will be limited. This limitation is closely tied to the thermoplastic sheet thickness. Thinner sheet materials tend to form into the radius or corners better than heavier-gauged materials. For all practical purposes, using the sheet material thickness for judging the minimum obtainable radius is a satisfactory method. The thickness of the thermoplastic sheet should be used as the least formable radius in thermoforming.

When trying to obtain a smaller corner radius than the thermoforming can offer, the matched molding method must be used. This permits the mechanical forming forces to become effective and force the plastic to conform and squeeze into smaller radii than can be handled by ordinary thermoforming. Using matching molds, with their compression molding features, is often not feasible. Most thermoforming is made without the possible adaptation of a matched mold, and therefore such radius-forming limitations will often exist. Knowledge of this limitation is most important. It can eliminate overexpectations in the forming results and eliminate the requisition of tooling with unattainable details.

In the case of textured material forming with conventional female angles, the pattern details of the inwardly placed textured sheet finishes (pattern side away from mold) tend to close up. As the thermoforming is accomplished, the curvature of the inside radius will be made smaller than the actual mold radius, depending on the material thickness. This reduction of curvature, together with the tightness of the corners, will tend to compress the texture pattern so that a dulled and wrinkled surface is created. Such compressed textured patterns can be minimized or even eliminated with an increase in corner radius. To maintain ease in forming and to lessen detail loss, the adaptation of larger-radius forms is highly desirable. However, using the larger-radius design patterns can force possible size increases and even diminish product quality. Products thermoformed and made with sharp, strong design angles (smaller radii) will offer superior strength. Products that contain a fair amount of rib design pattern are testimonies to this fact. Products with sharp rib design patterns are far stronger than those with large, less pronounced design patterns. This is only valid, however, when comparing the same thermoplastic sheet materials and thicknesses.

B. Depth-of-Draw-Ratio Limitations

Depth-of-draw limitations involve the process's ability to stretch a given area and thickness material into a three-dimensional product. Certain thermoplastic materials will allow far more stretching than others and, through the stretching, some can resist ruptures much better than others. Some of the materials allow for so much stretching that extreme reduction of gauges will be accomplished without loss of continuity, Products can be thermoformed out of a 100- to 120-mil thermoplastic sheet material and formed into products that display a 10- to 15-mil wall thickness throughout the entire body.

Draw ratios of 1:1 can easily be achieved with any of the simplest methods of thermoforming. This 1:1 ratio means that the smallest entrance distance of an open cavity can be drawn into the same measure of depth. The same limitations pertain to male molds. In cases where multi-up male molds are considered, the spacing between the male molds should be treated the same as a female cavity and the 1:1 ratio limitations must be honored. For example, 2-in.-high male molds must be placed 2 in. apart to be produced by the basic thermoforming methods. If they are placed closer together, serious material thinning will occur. In fact, at some point, depending on the material, the material thinning will be so extensive that the walls will be formed membrane thin and can even rupture. In any type of thermoforming, with either female or male molds, the resulting and unwanted thinning is normally compensated by an increase in sheet gauge. Using a heavier-walled thermoplastic sheet may add material thickness and strength to the radically thinned out areas but, at the same time it will increase material thickness proportionally to all the other areas. So where one problem is being solved, another is being created in areas of exorbitant thickness. Poor material distribution makes a thermoformed part more susceptible to product failures. Most thermoformed products are judged by, and fail at, their weakest areas; no value will be given to the heavier and stronger portions. Most failures occur at the weaker portions of a thermoformed article, making the entire product useless in spite of having portions that display "tank"-like strength.

Increased draw ratios can be achieved through various molding techniques with improved molds and temperature control and planned programming of both. Depth of draw can easily be produced in 1:3 or 1:5 ratios and even 1:7. However, this ultimate draw ratio requires sophisticated process controls and equipment together with specific thermoplastic material. Extending the limits this far cannot be expected using the simplest thermoforming methods and alternating materials. In extending the upper boundaries of this thermoforming criterion, many proprietary methods and process controls are implemented; patented techniques, out-of-the-ordinary equipment, and special materials are used. Where one thermoforming practitioner may be bound by lower depth-of-draw limitations, others will be trying to extend the upper boundaries and may succeed further than is generally known today. Although there will be many advances and much success, there will be always a challenging task ahead.

In any event, the depth-of-draw limitation, even with a well-mastered process, requires constant attention together with proper controls and monitoring. Particular products will demand constant tuning and an ongoing extra effort to maintain production and product quality. At the same time, other products will not be hampered at all by this limitation.

C. Reversed Draft and Undercut Limitations

Producing articles by thermoforming often involves well-proportioned open flair configurations. However, every so often a product design may contain virtually no taper angles, or angles so small that special considerations pertain. It is well accepted that with sufficient taper angles, forming and formed part removal presents no difficulties. With smaller taper angles, the male mold's shape tends to hamper part removal and stripping much earlier than with the female mold. The interference in part removal is caused by the shrinkage, which grabs onto the male mold.

The female mold is less vulnerable to this natural physical phenomenon. The formed part shrinkage on the female mold will work for part removal rather than against it. Formed parts with natural shrinkage and reduction in size are easily removable, even from an absolutely straight walled mold cavity. In fact, parts whose designs contain a limited number of reversed draft angles or undercuts (due to size reduction related to shrinkage) can also be stripped from the mold. Slightly larger undercuts can also be removed and stripped out of the mold but require mechanical stripping assistance. The attempt at removal is affected greatly by the resiliency and deformation recovery quality of the particular thermoplastic material. Mechanical strippers are acceptable for removal of a thermoformed product from a mold, but care should be taken that the stripping force is not damaging to the article. Some materials display a rather rubbery flexibility that permits a deep undercut to be pulled out of a mold. Other materials may rupture and result in permanent damage at the nonreleasing portions of the thermoformed article.

Prior knowledge of the sensitivity of various materials to forced stripping can eliminate unrealistic expectations. It is advisable first to make an oversized drawing of the preferred undercut or reversed taper angle. Such a drawing can easily reveal stripping clearances and mold removal restrictions as well as the desired shrinkage factors. Different materials with different shrinkage rates may not allow direct material substitution without causing removal interference where close tolerances in the undercut are concerned.

Other limitations that exist with most undercuts and reversed draft angles are found in the difficulty of producing them with sharp definitions. The very purpose of producing undercuts and reversed angles is to obtain a fully detailed form. Poorly or incompletely formed undercuts will not provide this function.

Undercuts are used for two reasons. The first is to form a spacing device directly into the thermoformed article. In thinly walled thermoformed products, dense nesting will occur due to the identical configurations of parts and dimensional similarities between the inner and outer surfaces. To overcome the tight and dense nesting results, reversed draft angle protrusions in a ring or lug formation are thermoformed into the plastic product. Because these protrusions are made with reversed draft angles, the nesting articles will be hung up by the tips of their "undercuts." The undercut designs basically create larger nesting dimensions within the parts, which therefore cannot fully slip into one another. The depth of the undercut design will determine how well the stacking feature will work; its location and overall height will regulate how closely the nested part will be spaced when stacked. A typical nesting lug feature is illustrated in Figure 52.

To achieve full details in the forming of an undercut design, the reversed draft cavity must be allowed to evacuate the air trapped within it. To ensure complete forming, it is highly recommended that those critical areas be tied together, with open channels all the way to the main forming force chamber. In this way, when any combination of forming and stripping forces is used, the undercut cavity will be activated. Since undercut designs make formed part removal difficult, the reversed angles are kept at a minimum. The slightest underforming will show up first in detail loss at the undercuts. As little as 3 to 5% loss in detail can cause the production of a nonfunctioning stacking feature.

The second use for undercuts and reversed draft angles is to provide good snap fit between containers and their lids. The undercut designs in this case will give a positive locking feature that is much better than a friction fit. As already pointed out, in the thermoforming process, an undercut feature is the first place where detail will be lost. With even the smallest loss of detail, good, strong snap fits will deteriorate into a weaker snap and eventually to a point where lids will no longer hold onto the container. Lid failure from poor forming and the resulting detail loss are not uncommon. In fact, this is the main culprit of most improper lid and container matings.

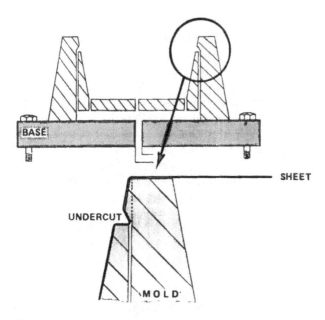

Figure 52   Typical undercut.

Improper fitting of thermoformed parts caused by reduced forming qualities and detail loss should not be confused with those caused by incorrect shrinkage calculations. Although both can cause diminished locking features, the first condition can be remedied, whereas incorrectly estimated shrinkage factors will result in permanently built tooling that has the wrong dimensions.

Occasionally, thermoforming processors must be able to make products containing large reversed draft angles that would not lend themselves to any type of forceful stripping or part removal. These angles cannot be accommodated by any of the common thermoforming molds. Huge undercuts combined with heavy-gauge plastic sheet can be thermoformed, but part removal will be impossible. To work around such an annoying situation, tooling modifications have to be implemented. In cases where one or two—but not many—reversed draft angle undercuts are produced in a given part, it is preferable for the undercut angles to be positioned in the same direction (front to back, left to right, etc.). For these difficult forming tasks, "breakaway" undercut inserts can be placed in the molds. These inserts become part of the mold for the forming and cooling cycle. However, during stripping and removal, those inserts are pulled away together with the formed part. They will thus not interfere with stripping and mold separation. The breakaway mold insert is then replaced in the main mold for use in the next cycle. Thermoforming can also be carried out using several identical pieces of breakaway mold inserts for the maintenance of production flow. Good candidates for this type of breakaway mold insert are such parts as shower and bathtub enclosures. The large soap-dish cavities formed in the sides of the enclosures require an extremely large reverse draft angle and could not be stripped from the mold in any other way.

For large quantities, or grossly oversized undercut shapes, molds can be made with splitting, tilting, or retractable features. Although such molds can be incorporated into the thermoforming process easily enough, their maintenance and airtight characteristics must be made to be repeatable. Accomplishing this will always be an impressive engineering feat.

For the sake of thermoforming expediency, screw caps have been formed using a continuous thread in the form of an undercut which is thermoformed into the caps. The thermoformed caps are removed with rotation (unscrewed) from the mold cavities with a rack-and-pinion device. This proves that there is almost no limit to uses of the thermoforming process. As unsolved process limitations are put to thermoformers, they will usually be able to find ways to overcome them, thanks to the versatility of the process.

D. Layer Reductions in Coextruded Barrier Sheets

The production of thermoformed containers made of coextruded materials has a unique limitation that it is important to understand. Coextruded sheet materials provide exceptional barrier qualities and considerable ability to eliminate chemical migration. The main reasons for using coextruded layers in a sheet for making containers is that some of the individual layers possess these unique barrier properties. When these layers are combined into a multilayered coextruded sheet (or even a laminated sheet), each layer brings its own barrier qualities to the combined sheet. In combination, the sheet layers complement each other and provide an exceptional packaging medium that can substitute for most known metal and glass packaging materials.

Such coextruded barrier material is often formed into containers using the thermoforming process. The thermoforming method is a natural way to produce packaging containers. The coextrusion method is the easiest one for converting the barrier layers into a combined sheet form; each layer thickness is easily controllable. For the best economy, each layer is extruded at the minimum thickness, where its barrier qualities will remain at satisfactory levels. Thermoforming practitioners must realize that the original thickness of a sheet will be reduced in the process of thermoforming. The prestretching and final forming can occasionally cause heavy reduction in the original gauges. This reduction of gauge can easily thin out a specific layer so that its original barrier qualities will be lost. This limitation must be known and planned for.

In addition to this natural overall thickness reduction, thermoforming has another limitation with respect to coextruded sheets. The natural gauge reduction is easy to compensate for with additional increases in layer thickness. However, each layer in the combined sheet composition is produced from different materials. The various layers are used for their different barrier qualities; they may not have similar or even close softening temperature points. In addition, they may also have differing flow characteristics and varying levels of submissiveness to stretching or tear resistance. When coextruded barrier sheets are thermoformed into products for the packaging of any sensitive preparation, such as food or drugs, the results of the thermoforming must undergo thorough testing and scrutiny. The true limitation in the thermoforming of such containers is found in the inability of the process to guarantee even reduction in the forming of the individual layers. Some barrier material layers may reach softening and flowing viscosities before other layer components do, thus creating inconsistent layering. Material may ooze out from between the firmer layers, causing a loss of barrier quality in those areas. Any container that has lost barrier tendencies even over only portions of its surface should be considered useless. It is important to know that thermoforming can cause such diminishing qualities in a barrier layer and that a greater thickness may be needed to compensate for the thinning. It is also important to subject the formed products to proper testing to ensure valid barrier properties. Measuring layer thicknesses in an already formed coextruded sheet or after thermoforming of the product is virtually impossible and cannot be depended upon.

4
Thermoforming Machines

I. INTRODUCTION

As stated earlier, the process of thermoforming easily lends itself to the use of repeated production cycles. Each time the same sequences are repeated, the same production cycle will be performed and identical products made; all conditions must be reproduced as well. When heat and forming force levels plus the timing sequences are duplicated identically, the resulting thermoformed products should be exact replicas of each other. To provide identical forming conditions for every cycle of the process, various types of thermoforming machines are employed. The many possible variations among thermoforming methods, product sizes, and volumes of product will demand an equal variety of thermoforming machinery. It is obvious that widely disparate products, such as a larger hot tub versus a smaller container, will be thermoformed best on thermoforming machines of different sizes and features. In fact, today's highly competitive business environment requires the utmost specialization of equipment, providing specific advantages for a particular thermoforming process to reduce costs and promote quality. Modern thermoforming means that specialized equipment is used for a specific purpose. To carry this specialized practice even further, most thermoformers set parameters on their fields of interest and capabilities. The segregation by a company of their equipment and their selective thermoforming procedure is usually predetermined by various material thickness ranges. Specific thermoforming machines are best suited to adapt specifically to thinner-gauge-material thermoforming, while others may be best adapted to heavier-gauge-sheet processes. In addition to gauge preference, thermoforming machines are divided into different categories by their production output capabilities. The thermoforming practitioner has to have some advance knowledge of the intended products in order to choose the proper equipment. Thermoforming manufacturing plants that try to cover more than one facet of production capability may find that several types of machinery are needed. Some of the equipment could be in the form of multipurpose machines, which are capable of dealing with a wide range of plastic sheet thicknesses. However, this all-around equipment usually lacks high-speed production capabilities; it performs best in small-volume productions. Such machines are equally adaptable for experimental, research, and development uses. Larger manufacturing concerns, even thermoplastic resin suppliers, are engaged in different levels of test runs on thermoforming equipment to obtain useful data for future planning. Of course, larger companies will extend their R&D work from smaller "lab" machines all the way to actual production-sized equipment; through the natural progression from small machines to larger ones, transferrable data are acquired.

To compile all the available thermoforming equipment into one functional list, the first criterion to be considered is whether the machinery is to run precut sheets or continuous rolls or webs. The basic thermoforming machine categories followed in this chapter are based on this factor.

Each type of equipment caters to a specific segment of the thermoforming industry with limited overlap. Each type of machinery can be adapted to run some products and sheet gauges more suitable to other machines, but not without capability curtailments and possible distress to the machine or product. It is best to know and recognize the limitations of the specific equipment before asking it to produce thermoforming in a way for which it was not designed.

The other basic criterion to follow in differentiating between equipment relates to its size. Thermoforming machinery is usually specified by its platen size, which is the determining factor as to what size molds can be fitted into the equipment and what size products, or in the case of smaller products, how many "ups" can be placed into the mold area. The platen dimensions are the second most important factor in categorizing thermoforming equipment. Specification of a thermoforming machine by platen dimensions must contain exact measures of width and length. The applicable platen square area is useful only when comparative listings are made and are not beneficial for any other reason. The actual platen dimensions will provide clues to mold fit or the number of "ups" that can be considered for placement within a specific area. These dimensions also determine the thermoplastic sheet size possibilities and trim-scrap factors, as well as influencing the entire economic outlook for the thermoforming procedure.

Thermoforming machinery can be found in every level of design, from the most simplistic and primitive installations to the utmost in sophisticated self-diagnostic machinery. All the thermoforming equipment will serve a specific purpose, yet all are aimed at producing a duplicate product. Machines that are made to produce just a few thermoformed parts do not need automation and will be priced accordingly. High-volume machinery carrying fully automated features, including self-diagnostics, can produce more than 100,000 thermoformed units per hour. There are dedicated thermoforming machines which are part of a complete production line system and are capable of functioning in-line with other equipment. These machines are capable of performing such tasks as extruding a sheet from the pellets of thermoplastics, carrying it through to the finished product stage, and even filling the containers. Equipment placed in a complete in-line system is performing a dedicated duty and usually accommodates continuous production of a single type of product. Complete in-line equipment may cost as much as $1 million. The higher the price, the more sophistication comes with the machinery, which can contain computerized microelectronics and the most recent "high-tech" technology.

Even with our modern technological sophistication, thermoforming can be carried out in a crude and simplistic way. Basic equipment can be put together in an almost "homemade" way and used for vacuum forming of simple items. Many of today's thermoforming practitioners have been lured to the thermoforming process by these simple forming procedures. In spite of modern sophistication, parts can still be made by such primitive methods and work as well as they did years ago.

To accomplish homemade vacuum forming, all one needs is a hose-type vacuum cleaner, a wire frame with a handle, a dozen or so clothespins, a thermoplastic sheet, a heat source (preferably radiant, e.g., a toaster oven), and a mold. The mold body can be made out of wood, plastic, or metal as long as it has a well-developed tapered design. The thermoplastic sheet is cut to the same size as the frame and held on it by the clothespins. The wire frame and clothespins act together as a clamp frame. With the frame held by its handle, the clamped plastic sheet is exposed to the heat source. As the thermoplastic heats up, it goes through its heat reactions. The mold is placed on a board or plate with the vacuum cleaner hose attached through a centrally located hole. The mold is positioned with a washer under each corner and is then sealed to the board with masking tape. When the thermoplastic sheets shows sings of softening, the vacuum cleaner is turned on and the frame is pulled over the mold. The vacuum force applied to the mold will pull and form the plastic against the mold surfaces. Holding everything in this position for several seconds will ensure good detail transfer and provide the necessary cooling. The vacuum cleaner is then turned off and violà: Your "first" thermoforming venture is completed.

How nice it would be if all thermoforming were this simple. In the remainder of this chapter we discuss all the various equipment in use today.
Thermoforming equipment is categorized in two major groups according to
thermoplastic sheet use: sheet-fed thermoforming equipment and web-fed thermoforming equipment.

II. SHEET-FED THERMOFORMING EQUIPMENT

This large group of machinery is designed to use only precut thermoplastic
sheet in the thermoforming process. Any of the customary thermoplastic sheet-
making manufacturers are capable of providing their product in cut-sheet form
upon request. The individual sheets can be produced by a batch process, or a
continuous sheet can be cut into individual sheets. The cutting and sheeting
of a continuous sheet is favored principally for the higher-gauge materials,
where winding on roll forms is either difficult or impossible. The lighter-gauge
materials can be made in sheet form as well, if desired, but are usually supplied
in roll form.

The sheeted thermoplastic is most often cut into a predetermined panel
size and the panels stacked. Each panel represents the material supply for one
cycle of thermoforming. The size of the panels can range from areas of a few
inches to many feet. Making larger panels depends on the thermoplastic sheet
producer's capabilities. Where a reduction in panel size causes virtually no
problem, larger sizes are limited by the size of the equipment.

It is imperative that thermoformers using precut sheets in their operations know the original machine direction (MD) of sheet manufacture. Most
extruded plastic sheets have a wood-grain-like flow pattern in the MD direction of the extrusion. Sheets cut from continuously produced webs can be cut
and stacked without an identifying direction. It is most important to know in
which direction the thermoforming processor needs the grain to face and to
specify this direction of cut to the sheet producer.

A 90° rotation in the grain alignment can create substantial changes in
both the thermoforming process and the final product. Such changes in grain
direction are often the cause of disappointing results, especially when the
change occurs in the progression from small experimental machines to full-
scale production machines. In cases where the sheet grain direction is incon-
sistently aligned, the resulting thermoforming will not duplicate earlier efforts.
Rectangular or elongated products thermoform much more easily and display
higher strength and less vulnerability to distortion when formed in alignment
with the extrusion directions of the sheet. This does not mean that thermoforming cannot be performed with misaligned sheets, but the quality will not
be up to par. Disappointing results that follow from directional inconsistencies
between sample products and actual production runs can be traced to the

common situation where due to economic pressures, the grain direction of the full-scale production is positioned at a 90° angle to the original experimental production. When the results are compared, they are often disappointing. Sales interests generated by small-scale test samples can turn into disillusionment when the final product is delivered. Proper respect and careful attention should always be given to the grain alignment of precut sheets. For sheet-fed machinery, the sheet direction makes no difference.

Sheet-fed machinery is available in every size from portable and tabletop models to "monsters." The smallest thermoforming machines have platens measuring only a few inches, while the largest have platens measuring 20 ft or more. The size of the equipment is closely related to the size of the finished product the thermoformer intends to make. The smaller machines usually come equipped with a fixed-size clamp frame since they use a predetermined sheet size. This is not necessarily the case with some medium-sized and large machines. A large number of these will have adjustable clamp frames. The basic large clamp frame either comes with adjustable sides that move inward to reduce the frame size or may have variously sized insert pieces that can be fitted into the clamp-frame structure to achieve the same effect. These adjustable features of the clamp frame are a great advantage when smaller parts are thermoformed. They allow the use of compatibly sized plastic sheets that eliminate the generation of excessive scrap.

Any of the sheet-fed thermoforming machines can be purchased with manual, semiautomatic, and fully automated control options. The price of the machinery will, of course, reflect the control option that is chosen, but all of the equipment is capable of giving satisfactory thermoforming results. However, with additional automation, less personal attention is required and better part uniformity is achieved, due to more precise control. For example, a sound or light signal generated by the equipment to a human operator may not produce the same response as that of an electromechanical signal given to an operating mechanism. The more automation that is incorporated in a machine, the better the uniformity and controls that will be attained. However, the cost of providing such automation often outweighs the benefits realized, and it will always be true that there is no substitute for a well-trained, conscientious thermoforming machine operator.

A. Stationary Thermoforming Machines

The stationary group of thermoforming machines is a very basic one. Thermoformed product production on this type of machine is very limited; only very simplistic, basic thermoforming can be carried out on them. Organized production output, which this machine is not really fit to handle, should be attempted only in the most underequipped situation. The real use for this equipment is in the production of one or only a few parts or extremely shallow depth-of-draw parts, such as magnetic door signs for automobiles. There are two versions of this machine: the stationary heater thermoformer and the stationary forming platen with movable heating machine.

1. Stationary Heater Thermoformer

The stationary heater thermoformer is the most primitive type of thermoforming machine. Almost everything done is accomplished in a stationary location, except for the clamp-frame mechanism. The clamp frame, with its captured sheet, can move a limited amount, but only in a vertical direction between the heater and the mold. The single heater unit is located above the clamp frame and sheet, which will be moved against a stationary mold, as shown in Figure 53. The stationary mold plate consists of a piece of sheet metal with a fine-grid-hole pattern that sits on and forms the upper surface of a vacuum box. The mold is placed on top of this plate and secured to it with double-faced tape or screws. With a female mold, any holes left uncovered by the mold are usually blocked off by masking with self-adhesive tape. This forces all the vacuum to be channeled through the mold. If the mold is male, no masking is necessary. The sheet is placed into the clamp frame and the heat source is turned on. As the sheet reaches the proper forming condition by going through its normal heating phase, as shown in Figure 36, it is lowered onto the mold. Upon correct contact, the vacuum forces are activated and the heater unit should be turned off. The vacuum force will complete the thermoforming and

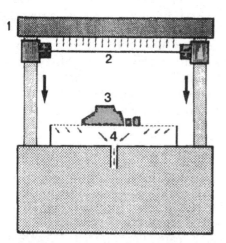

Figure 53   Stationary heater thermoformer: (1) heater unit; (2) thermoplastic sheet; (3) mold (male); (4) vacuum box.

the formed article will be cooled by the ambient area temperature and the cooler mold. The forming cycle is completed with the opening of the clamp frame; the thermoformed panel is then removed from the machine.

2. Stationary Forming Platen with Movable Heating Machine

The stationary forming platen with movable heating thermoforming machine also belongs to the category of simple, primitive equipment. The only innovation on this type of equipment is that the heater units are no longer stationary. The heater can be slid aside in a drawerlike fashion when the clamped sheet is ready to be formed. This shuttling heater can further improve the thermoforming process when built into a sandwich-type arrangement. A sandwich-type heater can heat the sheet simultaneously from both sides, giving it an advantage by substantially reducing the heating segments of the cycle. Some of these primitive types of thermoformers can also be made with swing-away heater units. In this case the heaters will pivot around a vertical shaft attached at one of the rear corners of the heater box. The pivoting action can provide a bonus in this heating method: As the heater is swung away from the forming area, it can be registered over the stack of precut sheets that are ready to be thermoformed. As the heater is positioned over this stack, it can provide preheating to the top sheet. To control the amount of preheating, depending on circumstances, the stack can be positioned to obtain the most ideal distance (height) from the heater units. With this heating method, the actual heating time can also be greatly reduced, thus giving better efficiency to the thermoforming.

B. Shuttle Thermoformers

The machines in this group are named "shuttle" thermoformers because their basic feature is that the clamped sheet is always shuttled in and out of a sheet-heating area. The shuttling clamp-frame feature with its clamped precut sheet provides better control of heating, the heating ability to promptly cut off, and absolutely no interference with the mold's cooling effect. There are some inherent disadvantages with this type of thermoforming machine, which will become more evident in further discussion. The manufacturers of thermoforming machinery generally recognize the shortcomings of the shuttle cut-sheet thermoforming equipment and have exerted substantial effort to improve on them by creating a number of variations. The many concepts that have been developed and are now offered for sale are aimed at improving the machine's basic energy and manufacturing efficiency. As will be seen, this has been accomplished only partially. There are four distinct variations of the basic shuttle former, which can still be considered the "workhorse" of the industry: basic single-station shuttle thermoformers, single-forming-station dual-oven shuttle thermoformers, duel-forming-station single-oven shuttle shuttle thermoformers, and dual-forming-station shuttle oven thermoformers.

1. Basic Single-Station Shuttle Thermoformers

The basic single-station shuttle thermoformer is an all-purpose thermoforming machine. This type of equipment can carry out all types of thermoforming methods and procedures, from testing to full-fledged production—a "workhorse" in the true sense. The basic single-station shuttle thermoformer can easily be modified to run and duplicate the thermoforming results of any other type of machine. It may not run at matching speeds or produce product at the same rate of speed as the other units, but it can be manipulated and adjusted to match the results of most thermoforming. Because of their ultimate adjustable features, these machines lend themselves to R&D or test equipment use.

The machines range from simple manually controlled or semiautomatic units to fully automated operations. The best come with fully automatic controls with manual override features. Since this thermoforming equipment is used in most fields of the thermoforming industry, the size range of these machines is equally broad. With the same features, machines range from tabletop models to very large equipment, some of which require 300 square feet or more of floor space for installation. To provide convenient tool setup height and mold parting-line height on some of the large equipment, the lower platen is made to retract into a pit placed deep in the concrete flooring.

All of the machines, regardless of size, provide common sequencing and simplicity of operation, and that is what makes them so popular. Figure 54 shows the basic single-station shuttle thermoforming machine. With the proper programming, any of the machines is capable of continuously duplicating the previously produced cycle and therefore making identical thermoformed products. The repeated steps will naturally depend on the ability of the thermoforming operator or the extent of the machine's automation. If all segments of a cycle are duplicated properly, the thermoforming results must match. Deviations in the thermoforming cycle will produce different results, whose effects depend on many factors and forming sensitivities. It is obvious that manually operated equipment is totally dependent on an operator's judgment and responses and cannot match the repeating ability of automatically controlled equipment. The resulting thermoforming and its shot-to-shot uniformity will be affected by the particular plastic's forming sensitivities together with the equipment and the operator's coordination.

This is as good a time as any to point out that when the thermoforming produces just a single satisfactory part and no more, the process need not be judged impossible. To continue to turn out satisfactory products, the only thing that must be done is to identify and repeat the thermoforming conditions of the satisfactory part. As soon as the correct matching conditions are provided

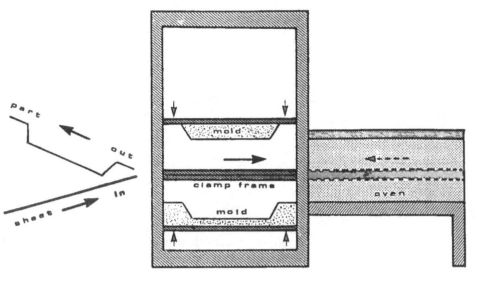

Figure 54   Basic single-station shuttle thermoformer.

in the forming, the outcome will be equally satisfactory and confidence in the thermoforming will be restored.

In cases where no satisfactory part is produced and all adjustments have failed to resolve the thermoforming problem, it is possible to conclude that the process is impossible to perform. Predicaments like this are often encountered with basic single-station shuttle thermoformers because this type of machine is likely to be used in product research and development.

Operation of the basic single-station shuttle thermoforming machine always begins with placement of the precut thermoplastic sheet. Placing the sheet into the clamp frame is usually performed by hand. To aid the operator and shorten the time required for loading, the stack of thermoplastic sheets must be placed conveniently close to the clamp frame and preferably at the same height.

Automation in sheet loading is also available. This particular aid, providing assistance in sheet placement, is most helpful when continuously rapid cycles are performed or when very large panels are involved. With this equipment, assistance in sheet placement is necessary primarily when placing extremely large and flexible panels. This task is generally handled by using a team of suction cups, a method borrowed directly from the cardboard and printing industries. The system consists of a "team" of suction cups mounted in an overhead position into a common frame that shuttles back and forth between the stack of sheets and the clamp frame. To pick up a thermoplastic sheet from the stack, the equipment shuttles over the top sheet. The team of suction cups is lowered to engage the plastic sheet. The vacuum is activated through the suction cups, which grab onto the sheet. To pull up and separate the top sheet from the stack, the procedure usually begins with the lifting of one or two suction cups at one corner of the panel. This is done to introduce air beneath the sheet for ease of separation. If an attempt is made to lift the sheet "square-ly" (all at once without the introduced air), the self-adhesion between the stacked sheets will overpower the efforts of the suction cups and will result in some pulling away from the suction cups. Just a single suction-cup disengage-ment can cause the loss of vacuum, causing all the other suction cups to fail and the entire system to malfunction. Any damage to or deterioration or crack-ing of the suction cups will also result in system malfunction. As mentioned, the top sheet is lifted and separated from the stack by the action of one or two of the suction cups, which initiate the lifting at one corner. This prelifting motion breaks the hold between the top sheet and the rest of the stack as air is displaced. When sheet pickup is complete and the suction cups have re-tracted into the frame, the entire frame system will shuttle over the clamp-frame mechanism. At the moment when perfect registration is reached, the suction cups are deactivated and the transported sheet is released. As soon as the suction cup transport is shuttled back, the clamp frame is ready to be closed onto the sheet. The sheet-loading mechanism can be adapted equally well to any other type of sheet-fed thermoforming equipment.

As mentioned, thermoforming machinery can be purchased and used with various levels of automation, from completely manual to fully automatic. The most automated equipment not only demands less work of the machine operator but will maintain accurate time control of all functions of the ther-moforming cycle, resulting in the utmost uniformity in the parts produced. However, the greater the automation, the more the thermoforming machine will require interacting and sophisticated controls, which will proportionally increase the cost of the equipment. Automation is most beneficial for long production runs where products are made in large quantities or where product duplication quality must be uniform. Again, we emphasize that automation and fully programmable controls should not be considered a suitable replace-ment for a dedicated operator.

With single-station shuttle thermoforming, a machine's cycle sequencing should always be made the same way, whether operated by manual, semiauto-matic, or fully automated controls (Figure 54). The clamp frame closes onto the sheet, capturing it by all four sides. If this action is made manually, the latching is done by hand or with an automatic clamp frame by a series of small pneumatic or hydraulic cylinders.

The clamp frame, together with the captured sheet, is indexed into the oven (heating area). For this function on manual machines, the clamp frame is pushed into the oven in a drawerlike fashion. On automatic equipment, the clamp frame travels in both directions (in and out of the oven), implemented by a long-stroke horizontally positioned cylinder. In the heating segment of the cycle, the thermoplastic sheet will be heated to an ideal softening level and will go through all the stages shown in Figure 36. In manual operation, the thermoforming operator will watch the progress of the sheet softening and when he determines the proper heat level, will pull the clamp frame out of the oven into the forming area. By activating a button or lever, the molds are closed onto the sheet, whereby manual activation of the other switches will initiate the forming forces (vacuum or air pressure). When the forming is completed, time is allowed for cooling the formed parts. This time allowance on manual machines is entirely up to the operator's judgment. After the cooling segment of the cycle, the operator manually activates the machine by a hand lever or switch to open the molds. At this point, the clamp frame is opened up by hand and the formed panel is removed. The equipment is now ready for the next cycle.

In a semiautomatic operation, some of the manually triggered functions are replaced by automatic switching mechanisms. For example, when the clamp frame is pulled back into the forming area and the mold closure activated, the forming forces will be triggered automatically. Naturally, any manually operated function can be substituted by automatic ones. With fully automatic thermoforming equipment, the operation of the machine is controlled by microswitches and timers or even a built-in microprocessor capable of consecutively triggering the entire operation automatically. The thermoforming operator's only duty with these machines is to push a button after sheet placement; the machine then takes over all functions. After the thermoplastic sheet is placed into the clamp frame, the cycle start button is pushed. This button activates the clamp-frame closing, which traps the sheet and holds it by all four sides. An improperly placed sheet could experience slight realignment by the clamp-frame mechanism when the closure is made. However, excessive misalignment cannot be remedied, and if it is not corrected, that panel will be lost for thermoforming during a subsequent step.

When the cycle start button is activated on the automatic thermoformer, the clamp frame closes onto the sheet and is automatically shuttled into the oven. At the point of full insertion, a microswitch is triggered to start the timer for heating. This timer is usually equipped with resettable adjustment that can be set to any desired time schedule. This schedule can be segmented into minutes and fractions of minutes. When using a timer, it might appear that heating of the sheet is regulated by time alone. However, further adjustment to the heating is achieved by controlling the output level of the heaters. The thermoformer requires both these controlling factors for ultimate heating control. As the timer depletes the preset time and reaches "zero," it will automatically signal the clamp-frame cylinder to retract from the oven. Again, after the full return to the forming area, a microswitch automatically activates the mold closure onto the heated sheet. As the mold closes, it activates another set of microswitches positioned by the thermoformer to activate the forming forces at the proper moment. The adjustability of these switches is very important for best coordination of the forming forces. By regulating the moment of activation through the mold travel, the operator determines when the forming force will be turned on and thus arranges the most advantageous timing for the particular forming. With this adjustment, the vacuum or air-pressure forming forces can either be activated sooner or delayed. All these options are within the adjustment capabilities of the switching device and can be made automatically.

At the point of full mold closure, another set of microswitches is triggered, starting the cooling cycle timer. This timer can be a separate electromechanical unit, or with a microprocessing system, will be the same digital readout timer. In either case, the timer for the cooling segment of the cycle is regulated by this timer, which is set to a predetermined schedule. As soon as the allocated cooling time is depleted and the timer reaches its preset cutoff point, the unit automatically activates the platens to open them together with the molds. It can also initiate a momentary blast of air through the mold to aid in part stripping. In addition, when the molds are opened and retracted, a switch signals the clamp frame to open automatically. The thermoformed panel is then removed and the thermoforming cycle completed. When a new thermoplastic sheet is placed into the clamp frame, the machine is ready to repeat the thermoforming cycle. At this point it is easy to recognize that the preciseness of the timing controls can pay off in the quality of the duplication of the thermoforming process. If and when all time segments of the cycle are exactly duplicated, the same forming conditions will exist and a perfectly identical product is assured shot after shot. In manually or semiautomatically operated machines, such uniformity cannot be guaranteed, even with the most highly skilled operator.

The manual "override" features of an automated machine make it the most adaptable to testing and R&D projects. With manual override controls, the ultimate forming conditions can be tuned and pinpointed much more quickly. As soon as the timing conditions are established, switching to acceptable settings is made faster and the operator's ability to respond to changing ambient plant conditions is increased and therefore becomes very practical.

The only two shortcomings of the single-station thermoforming machine can be found in its comparatively slow operating speeds and its inherent high energy consumption. The cycle speeds of this type of thermoformer, even in the most idealistic conditions, will be inhibited. The shuttle clamp frame and reciprocating mold movement operation can be no means by judged a perfect production plan. The thermoforming machine, with its systems of operation, is constantly held back from performing all other cycle segments while one is in progress. These segments, carried out as the machine moves from one phase to the next, together constitute an entire cycle. It is clear that to maximize the efficiency of any operation, only the excessive time element within a cycle can be reduced. However, there is a minimum time requirement for each segment of the cycle, and once this is reached, no further optimization of time can be realized. For example, any excessive heating of the plastic, although it may not prove harmful to the thermoforming, will demand additional time by itself as well as requiring an extension of the cooling-time segments. So it is easy to recognize why the thermoformer should watch the efficiency criteria closely.

The other shortcoming of the single-station shuttle thermoformer is related directly to its energy consumption. This type of thermoforming machine is often referred to as an "energy hog." This title is only partially accurate, and its implication can easily be discounted. The thermoforming machine ovens are normally turned on and set to a desired heat-producing level. As the machine is put through its paces, there is only a small portion of the cycle during which heating is needed. For the remainder of the cycle, the heaters are kept on. Most thermoforming machines come equipped with standard tubular heater elements. Switching off those heater elements will, in most instances, affect the heat output rate and is therefore detrimental to the forming process. The tubular heater element has a poor heat recovery time; it cannot react as quickly as the process demands. For this reason, these heater units are left on all throughout the production cycle. Taking into consideration machine size and number of machines placed side by side, this practice can account for 80% or more of a thermoforming plant's total electric power consumption.

The cost of such high energy consumption has been a concern of thermoformers from "day 1" and many complex solutions have been attempted. the first and most "ingenious" plan to resolve this drawback is to completely rework the concept of the machine. But before looking at innovations that do not actually offer satisfactory improvement, let us discuss the one proven method for reducing energy consumption, as concepts from this method should be recognized before examining the others. The engineering feat involved is discussed in Sections II.B.2 to II.B.4.

The most promising way to reduce the high rate of energy consumption is found in the use of quartz heater elements. As stated earlier, this type of heater element provides excellent heating-level recovery time when being turned off and on. This is due to the nonhindering effects of the quartz material on the infrared wavelengths. With this important benefit, if the ovens of the basic single-station shuttle thermoformer are equipped with quartz heater elements, it can make all the difference. As the heated thermoplastic sheet and the clamp frame leave the oven together, the whole oven can be turned off and kept off until just before the new sheet is ready to be heated. The oven switching element can be connected to the machine operation to allow sufficient time for the heaters to come back on just as the new heat cycle is ready to begin. With this type of heater and knowledge of how to use it, a lot of normally wasted energy can be saved and an energy-efficient machine can be reclaimed from a former "energy hog."

2. Single-Forming-Station Dual-Oven Shuttle Thermoformers

In the desire to make the shuttle thermoforming machine more cost-effective, several poor design implementations have been made to the basic machine. One of these miscalculated designs is the single-station, dual-oven shuttle thermoforming machine. As Figure 55 illustrates, the machine differs from the basic shuttle thermoformer discussed previously only in that it has a second oven. The clamp frame is constructed with two identical sections attached side by side. As it shuttles left or right into the respective sides of ovens, the other clamp-frame section will automatically shuttle to the center of the machine. The concept behind this dual-oven design is that one side of the clamp frame holds the heated sheet to be formed and cooled (after forming), and then the

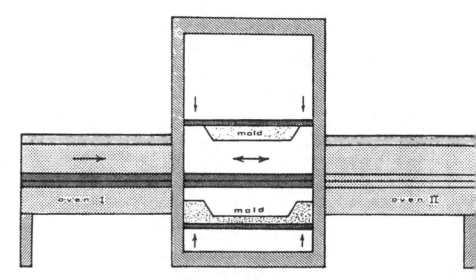

Figure 55   Single-forming-station dual-oven shuttle thermoformer.

formed product is removed and a new sheet loaded, while the sheet in the other side of the clamp frame gets heated.

It is easy to recognize that this design suffers from the definite misconception that it is saving energy. However, the only time it will prove useful is when the sheet-heating sequence can be accomplished in the same time span as the rest of the cycle. Even the slightest deviation can force a reduction in thermoforming quality in at least one side of the machine—not to mention the nerve-racking conditions forced on the machine operator, who would have to complete forming, cooling, stripping, formed part removal, and new sheet replacement into the clamp frame in the same time as the other side is rapidly heating. Even if the heat-producing levels of the ovens are cut down so that the problem of overheating no longer exists, the fact remains that there are two ovens constantly on instead of one. Successfully justifying both the economy and feasibility of this type of thermoforming machine is difficult indeed. The most careful considerations have to be made before its acceptance is possible.

3. Dual-Forming-Station Single-Oven Shuttle Thermoformers

The next controversial shuttle thermoforming machine is equipped with dual forming stations connected to a single common oven. The basic configuration of this machine is shown in Figure 56. The machine's twin-sided clamp frame is designed in much the same way as in the dual-oven shuttle thermoformer

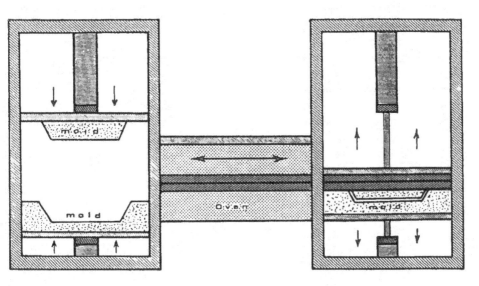

Figure 56 Dual-forming-station single-oven shuttle thermoformer.

machine. When one side of the clamp frame is in the corresponding forming station, the other side is automatically subjected to the sheet-heating procedures in the oven. The same criteria exist with this type of a machine as with the preceding one. The sheet must be heated in the same time span during which the other side of the clamp frame is subjected to the cycle of forming, cooling, stripping, part removal, and placement of the new sheet.

The same nerve-racking conditions as before will exist for the machine's operator since each side's functions must be accomplished within the same timetable. In addition to this, the dual-forming-station concept requires purchase of two sets of molds. Of course, the molds need not necessarily be of the same configuration, but a similarity in their shapes is very helpful in maintaining similar forming conditions. However, as pointed out earlier, it is rare that these same conditions will prevail; only a few such opportunities will present themselves. More than likely, the different molds will introduce enough deviation into the forming sequencing to render the use of this type of machine impractical. If the use of two identical molds is chosen, the initial cost of the second mold will outweigh the contemplated energy savings. To make matters worse for the worker, if only one person is loading and unloading, he must also "shuttle" himself between the two forming stations. When two persons are used, the added cost of the second worker will undermine any hopes of energy savings over the use of a single-station shuttle thermoformer.

4. Dual-Forming-Station Shuttle Oven Thermoformers

The dual-forming-station shuttle oven thermoformer is the third design attempt to improve on the basic shuttle former, and this one is as complicated as its name implies. The only merit in this plan is that when the machine is purchased, the purchaser will feel that he is getting two machines for the price of one. But when this concept is examined closely, it will be seen that the basis for this thermoforming machine lies in the placement of two single-station shuttle thermoformers side by side (Figure 57). The only innovation added here is that a single oven is made to shuttle between the two machines. With this oven arrangement, the benefit comes from the savings made by the oven shuttling over to the other side and performing another heating job instead of standing idle and consuming energy. At first, this shuttle concept sounds acceptable and most promising in the effort to save on energy. However, close examination will reveal that matching cycle spreads can be realized in only a few instances. Most thermoforming mold configurations and the production of the parts within them require nonuniform cycle times. Different thermoplastic sheet colors and gauges can further aggravate the timing intervals forced on side-to-side shuttling of the heater unit. It can also easily induce anxiety on the part of the machine operator. When identical molds, which will probably have the same cycle intervals, are used on both sides, the timing discrepancy

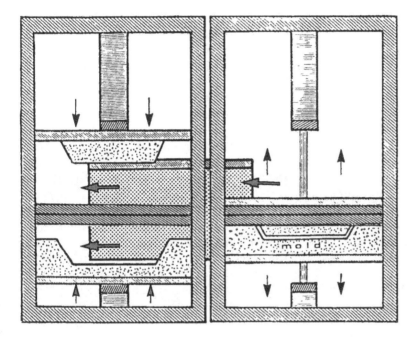

Figure 57   Dual-forming-station shuttle oven thermoformer.

can be reduced. However, the burden of the double tooling cost will probably outweigh the cost-saving benefits of the shuttle oven arrangement.

C. Rotary Thermoformers

In the pursuit of improved thermoforming speeds and production flow, the advantages of the rotational actions of equipment have always been well recognized. To improve on the shuttle actions, rotationally indexing thermoforming equipment, with its horizontal rotation from station to station, has long been a desirable form of thermoforming. The distinctly individual stations, which are located in a rotational format, have specifically assigned functions. With this concept, the thermoplastic sheet is carried through the various segments of thermoforming in a continuous cycle that creates good production flow. A complete rotation contains the full thermoforming cycle. To complete the cycle, the thermoforming machine will stop and index at three or more stations. Each station has a specific duty to perform and is built with that task in mind. A matching set of clamp frames is built for this machine, one for each station. The rotating frame structure carries the thermoplastic sheet through the entire thermoforming process, stopping at each station to perform a specific task. With this rotational format, the segments of the thermoforming cycle are performed simultaneously in each station and must be synchronized with one another. If one station's performance is out of step with the others, it will demand a sacrifice of time from the other stations. Minute timing delays will cause only a slight prolongation of the total cycle. However, longer delays can seriously affect the overall results of the thermoforming.

For the reasons cited above, a variety of multistationed machines are built and made available to the thermoformer: three-station rotary formers, shuttle oven rotary formers, four-station rotary formers with separate loading and unloading station, four-station rotary formers with preheat and final heat stations, four-station rotary formers with dual forming station, and four-station rotary formers with in-line trimming station. To determine which type of rotary former would be most advantageous to use and to pinpoint which segment of the thermoforming cycle will demand timing interference, it is a good idea to run simulated thermoforming tests in a single-station shuttle thermoforming machine. Of course, with some previous experience, testing may not be needed; the processor's preestimation and judgment can provide all the necessary guidance. As the various types of rotary formers are discussed and compared, it will become much more evident that one type of machine will compensate for the imperfections of the others, and it will be clear why their correct selection is so important for the most competitive production mode.

1. Three-Station Rotary Formers

The three-station rotary former is one of the most popular rotationally producing thermoformers, due to its ability to handle a large number of products. Its size can range from 12 in. by 12 in. to 10 ft by 21 ft platen sizes. As Figure 58 shows, the machine has three stations. Each station has a specific task to perform and the machine has three matching indexing clamp frames.

1. At the first station, the operator places the precut thermoplastic sheet into the clamp frame. The same situation exists at this stage of the thermoforming as with the shuttle thermoformer. To ease the sheet loading, it is always advisable to place the sheet stack as close as possible to the loading operation and preferably at the same height. Automated sheet loading is also available and can be successfully adapted to this particular thermoforming machine. Automated sheet loading is most advantageous with extra-large sheet panels and with the flimsy thinner-gauge sheets, whose handling usually requires several people since their placement demands an extra degree of care. The same equipment and procedures are usually used with rotary formers as with shuttle machines in this type of sheet handling. Sheet alignment is just as critical as with any other sheet-fed thermoformer. After sheet placement, the clamp-frame mechanism is closed. This can be done either manually or with

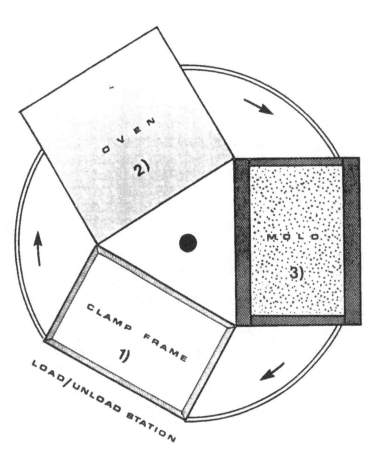

Figure 58   Three-station rotary thermoforming machine: (1) loading and unloading station; (2) heating station; (3) forming station.

automated pneumatic or hydraulic cylinders. When sheet capture is completed, the sheet will be ready to rotate into the next station.

2. In the second station, the sheet is registered into the heating oven. The machine usually has both top and bottom heaters, built into a so-called "sandwich heating" oven. With this setup, the sheet is subjected to one of the most efficient heating systems, one that will heat the thermoplastic sheet from both sides. The thermoplastic sheet in this heating station will go through the heating sequence shown in Figure 36, provided that neither the sheet panel size, thickness, nor manufacturing method necessitates the use of different steps from those in the normal sheet-heating process.

Rotary machines with manual control switches are just as common as semiautomatic and fully automatic models. When a manual machine is being used, a human operator must send an activation command for each cycle segment, so the operator's alertness and decisiveness are the key to the quality and uniformity of the thermoforming results. By contrast, automatic machine operations rely on equipment that is activated by a timer or electric-eye system. The operation of timers has been discussed in connection with shuttle thermoformers and they act much the same with rotary formers. The only difference is that instead of a shuttle movement between sequences, the machine makes a rotational movement to the next station.

The electric-eye system provides heat control not by elapsed time but through the development of sag alone, which occurs in the final stage. The electric-eye system uses an adjustable light beam that detects blockage by the developing sag. An adjustment in the electric-eye position can control the size of the sag and the actuation of the machine. As the sag extends beyond the normal horizontal position, it blocks the electric-eye beam. At the moment the electric-eye view is cut off, the machine is activated to advance the heated sheet out of the oven and into the forming station. A new sheet rotates into its place and the operation repeats.

3. While in the forming station, the heated sheet goes through the intended forming process according to the forming procedure selected and is formed and cooled into its final shape. When cooling is sufficiently complete and the formed part has been shrunk, the mold is opened. The elapsed time for forming and cooling is controlled either manually by the operator or, what is more likely, is tied into the rest of the rotational sequencing of the system. In the larger machines, the mold may require a pit in the floor into which it can retract.

After the mold has opened and retracted, the formed panel indexes into the first station for removal. As the formed panel is removed, a new thermoplastic sheet is put into its place and a new cycle begins. The formed panels are usually stacked and transferred for a post-trimming operation in which the edge trim is cut away from the actual thermoformed product.

These three-station rotary thermoformers truly offer an improvement over shuttle systems in production flow. In the rotational format, each station can handle simultaneously the three basic functions of the thermoforming procedure, from thermoplastic sheet loading to formed panel removal. With nominal-sized sheets, one person should be able to handle sheet loading and formed part removal. This person may also oversee all other functions of the equipment. While the formed part is being removed, initial quality control examination is also feasible, and if any adjustment in machine functioning is required, immediate action can be taken.

The only limitation to rotary thermoformers is their vulnerability to drafty plant conditions. It is easy to recognize that when in a rotary format the sheet is transferred from the oven to the forming station, it has to make its way exposed to the open air of the plant. At this transfer point, the carefully heated sheet can be subjected to an uncontrollable situation. The effects of draft are less critical with heavier-gauge thermoplastic sheet materials because their heat-retention capabilities are comparatively superior. However, the sheet material used on rotary formers can be as light as 20 mils in thickness. These thinner materials are vulnerable to cooling by air even without draft conditions. In rotary thermoformers, perfectly executed heating and sheet "pampering" can be spoiled at this stage. To minimize the lighter-gauge sheet heat loss sensitivity, partitions or curtains (of fireproof fiberglass) must be placed around the equipment to block any draft. Even better, the machine can be placed into a smaller plant room with evenly controlled ventilation.

Cooling fans should not be positioned so that they blow air directly or indirectly by deflection toward the heating oven or on the traveling heated sheet. Remember that occasional production of a poor-quality part probably indicates an unwanted and undetected air current problem. It has been proven that even an improperly directed comfort cooling system can play havoc with thermoforming. Each time the thermostat activates the system, the quality of thermoforming diminishes. The air outflow from the system must be directed elsewhere, not toward the thermoforming machines.

2. Shuttle Oven Rotary Formers

The shuttle oven rotary thermoforming machine is aimed at minimizing the cooling hazards of open-air travel of the heated sheet. This machine was developed in Great Britain and offered for sale worldwide. This design is based on the idea that the heating oven will accompany the indexing heated thermoplastic sheet all the way to the forming area. This allows very little chance for the preheated sheet to receive unwanted cooling. After accompanying the sheet to the forming area, the heater unit travels back to its original heat-station location and begins to heat the next clamped thermoplastic sheet. It should be emphasized that the added feature of the shuttle oven can only minimize the draft problem. The spacing between the clamp frame and the respective oven sides and the exposure to air at the forming station still create unnecessary opportunities for drafts to produce damaging effects. There is no better answer to the draft problem than eliminating it altogether.

3. Four-Station Rotary Formers with Separate
Loading and Unloading Station

The basic concept of three-station rotary thermoformers can be expanded into a more specialized machine by adding a fourth station. The purpose of the extra station is to extend the machine's capability by either improving production speeds or performing an additional task, which usually requires a secondary operation. A four-station rotary former is capable of performing all the tasks of the three-station rotary thermoformer and performing them equally well. However, the three-station rotary thermoformer is only capable of "run-of-the-mill" thermoforming tasks. As the functions of the four-station rotary former are explored, it will become more evident how the fourth station fits into the scheme of specialized thermoforming.

In the basic four-station rotary thermoformer, sheet loading and formed panel removal are accomplished at two separate stations. Figure 59 shows this concept by means of a top view of the machine. The use of two separate stations for loading and unloading is justified by the difficulties found in some sheet handling. Some materials are extremely flimsy or ultrasensitive to handling.

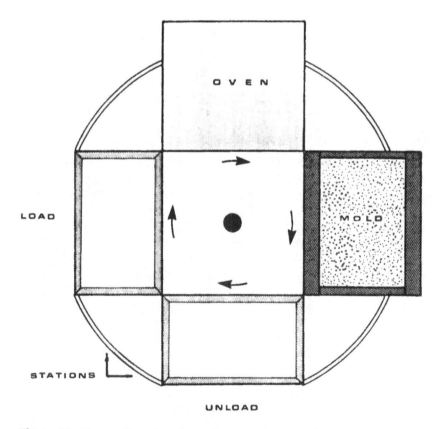

Figure 59   Four-station rotary thermoformer with separate load and unload stations.

Therefore, such material demands slow and careful denesting from the sheet stack and positioning into the clamp frame. The sheets may be so flimsy, in fact, that several people are needed to handle them. They may also have surface structures like that of woven cloth and thus cannot be handled with suction-cup lifters, or they may have a surface that is easily damaged, scratched, and so on. The difficulty encountered and the resulting slowdown in operation may not exist at the sheet-loading area alone. Unloading formed parts may prove to be just as time consuming. Using a four-station rotary former with separate loading and unloading stations will diminish such a bottleneck in the thermoforming operation. It is common practice to use at least two people with this machine, with one person handling only sheet loading and the other removing the formed panels. Additional personnel may be necessary for extra-difficult cases. However, in cases where the loading and unloading operations do not conflict with the timing of the other forming segments, the separate unloading station can be disregarded and both functions performed at the first station. In this way the operation can be handled by a single operator. The remaining functions of this machine operate in much the same way and have the same limitations as those of the three-station rotary former.

4. Four-Station Rotary Formers with Preheat
   and Final Heat Stations

This version of the four-station rotary former is designed to handle thermoplastic materials that due to their thickness or material makeup, suffer from slowness in their acceptance of heat. Thicker and denser materials heat up only on the surface, and heat penetration is much slower than with normal-gauge materials. Figure 60 shows the use of two heating stations, a preheat and a final heat station. Between the two stations, most hard-to-heat thermoplastic materials will receive the necessary exposure to bring them to a formable stage. Since this double-heating-station machine is used for extra-heavy-gauge material, the sheet's mass and heat-retention characteristics will prevent any draft-causing effect. When the thermoforming process calls for the use of lighter-gauge material, the preheating station can be turned off so that the only heating is performed by the second heat station. On some machines, the heaters in the first station are easily dismantled and the station is converted into a loading station. This provides an opportunity for changing the machine into a separate loading and unloading station machine. This feature makes this type of machine even more versatile.

5. Four-Station Rotary Formers with Dual Forming Station

The four-station rotary former with dual forming station is another ingenious thermoforming method for stretching the basic limitations of customary thermoforming. The employment of dual forming stations is not restricted to rotary

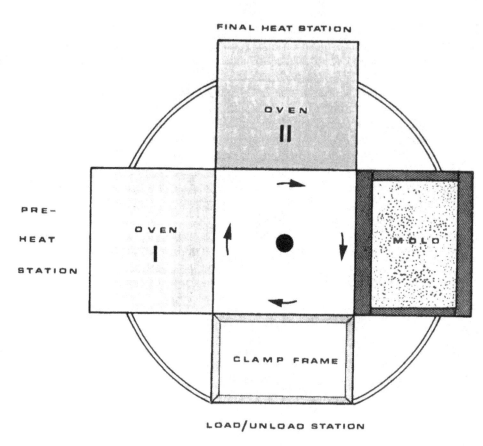

Figure 60   Four-station rotary former with preheat and final heat stations.

thermoformers. The same idea can be applied as easily to continuous, web-fed thermoforming. Figure 61 shows a four-station rotary thermoformer equipped with two separate forming stations. The concept behind this setup is that extremely deep-drawn shapes are difficult to form, especially if additional design intricacy is involved. In customary thermoforming, the mold usually performs two functions: to give shape to the forming plastic sheet, and to cool the formed part and set its permanent shape into place. In customary thermoforming, both tasks are handled simultaneously by the same mold. In the process, molds are usually colder than the forming sheet. As soon as the mold has given shape to the plastic, it will absorb its heat, setting it into a permanent form. This customary thermoforming arrangement will usually curtail deeper form-

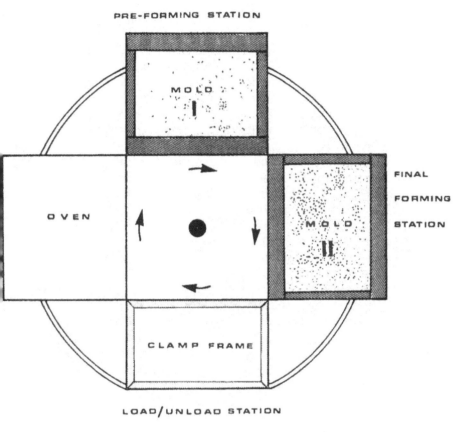

PRE-FORMING STATION

OVEN

FINAL
FORMING
STATION

MOLD I

MOLD II

CLAMP FRAME

LOAD/UNLOAD STATION

Figure 61   Four-station rotary former with dual forming station.

ing since the plastic tends to set up sooner due to the mold's cooling effects. This factor becomes even more obvious when extremely deep part forming is attempted. In trying to extend the depth-of-draw limitation and neutralize the mold's chilling effects, use of a dual-forming-station machine will provide the answer. The first forming station is actually preforming the heated thermoplastic sheet into an entirely different shape from that of the intended final shape. The preforming is usually done with a controlled-heat mold. The temperatures of the mold match those of the heated thermoplastic sheet. With this heated mold, the plastic sheet can be purposely deformed without losing its formability. The deformed shape chosen can be designed so that it will aid further shaping by the second mold. Material can be pulled and stretched away from a given area and redistributed to another, creating the most ideal reformable shape for the final forming.

The finished shape is made at the second, or final, forming station. Here the previously preformed and stretched sheet is formed and cooled into its finished shape. It should be obvious that with this thermoforming machine, the particular arrangement by the processor will stretch the depth-of-draw limitation. This is because the final shape will be produced not out of a flat heated sheet but from a preshaped form. The difference between the two mold shapes can be so elaborate that no relationship between them can be recognized. The success of this method depends on the use of two separate molding stations. However, the real key to this project's success lies with the processor's ability to hold and maintain correct mold temperatures at the first preforming station. The slightest deviation from the actual sheet temperatures can deter the achievement of this thermoforming goal.

One additional benefit to be gained from use of the dual forming method is outstanding material distribution. With this system, material mass can be relocated out of one specific area and redistributed to another, as desired. Basically, the preforming mold is specifically designed and shaped to deform the sheet into the most advantageous shape by which the best forming results can be obtained. Little guidance can be given as to which would be the most advantageous premold shape. Thermoforming practitioners usually depend on their best calculated guess and proceed from there. They often rework the shape or even make new forming shapes, according to the results of the last forming, before total satisfaction is achieved. Temporary mold materials allow mold experimentation before arriving at the final shape. Certainly, thermoformers who are repeatedly involved with such preforming techniques can arrive at a final preforming shape sooner than can those who rarely perform such tasks. Dual forming methods are not necessary for most thermoforming and should be considered only in special circumstances where unusual depth-of-draw limitations or material distributions might complicate the process.

6. Four-Station Rotary Formers with In-Line Trimming Station

Nearly all thermoformed products require some type of trimming after the forming cycle is complete. In the trimming operation, the excess material which was needed for the clamping and bordering of the actual product is trimmed away. That is, these excess areas, although necessary for the forming procedure, are not usually part of the final product. When using sheet-fed thermoforming machines, trimming is usually done in a secondary operation. The formed panel is removed from the thermoformer and transferred to separate equipment where trimming will take place. The separate trimming process does not have to be run at the same speeds as the thermoforming and even if it runs slightly slower, no major difficulty will arise. The only time posttrimming will cause economical disadvantages is when it will require additional person-nel to run the process. To remedy this situation, a four-station rotary former with a built-in trimming station can be employed. The basic principle of this machine's function is the same as for the three-station rotary former, with the addition of a fourth station for the trimming process. Figure 62 is a top view of this machine showing its stations and functions. The key to this machine's success lies in its ability to provide trimming within the same timetable during which the other stations are performing their assigned tasks. In cases where the trimming operation would take longer than the rest of the forming cycle, the entire process would have to be slowed down. However, slowing the overall process and favoring the trimming station may not prove to be economically

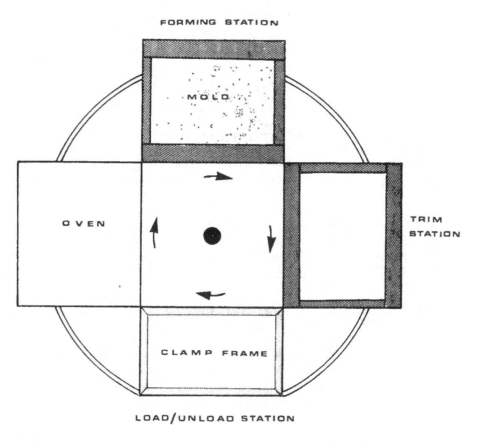

Figure 62   Four-station rotary thermoformer with in-line trim station.

feasible. These specific time criteria must be examined and measured beforehand, or serious economic consequences could result.

The versatility of a four-station rotary former provides an opportunity for switching off or dismantling station functions or even substituting other task-performing equipment. The same opportunities for substitution exist with the trimming station. The station can be made idle by not performing a trimming task at all, or if desired, can be substituted for by different types of trimming equipment.

The trimming apparatus installed in the fourth station can range from the most common type of punch press to an elaborate computer-guided laser system. Naturally, the installation of this trimming station will be made according to a customized program, which will be closely tied to a specific thermoforming project. The desired trimming apparatus may be installed in a rotary thermoforming machine by the thermoforming equipment manufacturer on a prearranged basis or can be added on by the local work force. In most instances the trimming equipment is purchased as standard equipment from a specialized trimming equipment builder. This equipment is built for all types of utilization as well as for different industries, not just for thermoforming. For example, ordinary punch-press equipment with its customary flywheel assembly, used for metal punching, can be adapted and installed into the fourth station of a rotary thermoformer.

A high-velocity water-jet cutter can also be installed into the trimming station. The cutting pattern can be guided by a mechanical track system or a computerized-memory tape-guided servomechanism. The high-velocity water-jet trimming system can just as easily be installed in a separate posttrimming machine. However, incorporating it into the rotary thermoforming machine format will provide an ideal production flow. The same criteria also exist for the use of laser-beam-generating cutting systems. Both high-velocity water-jet and laser-beam cutters do a fine job of cutting within rotationally in-line operations. The only critical factor to be examined is the speed of cutting, which must not jeopardize the heating, forming, and cooling segments of the thermoforming cycle by being too slow. It is a well-known fact that high-velocity water-jet and laser cutters must be slowed down as thermoplastic sheets of higher gauge thickness are introduced. This factor is not limited to the original material thickness of the thermoformed part but includes those areas where the cutting angle is not square to the cutting surface and needs to cut through a sloping wall, which at those angles will have to penetrate a much greater thickness. In these cases, the cutting speed must be further reduced to accomplish full cutting and can and will add cycle-time extensions to the overall thermoforming process. Proper planning and thorough study of the final thermoforming shape should provide a clue to the feasibility of using such thermoforming equipment with a built-in trimming station.

D. In-Line Sheet-Fed Thermoforming Machines

To make the thermoforming of cut sheets even more productive than can be obtained using rotary formers and to add the benefits of multiple-zone stops in the oven, the in-line sheet-fed thermoformer was designed. This type of thermoforming equipment, with its continuous advancement, closely resembles web-fed thermoforming equipment. The only distinguishing feature is that this machine still uses precut thermoplastic sheets and the sheet-carrying chains contain traveling clamp-bar units to capture the individual sheets. The chains are looped to return the same way as they came, indexing forward with each advancement to return and pick up the next sheet. The chain loop may contain a large number of clamp bars, which are engineered to pick up individual thermoplastic sheets from a denestable stack. The denesting of individual sheets is usually accomplished by pushing them out of a stack through properly controlled stop blocks or sheet gaps. These allow only one sheet to leave the stack at a time. The sheet feeding of the thermoformers from the stack is not limited to the "bottom feeders." It can also be accomplished with "top feeders" or even with a suction-cup feed mechanism. However, bottom denesting feeders are the most popular units in use today.

Once the individual sheet is placed into the sheet-carrying mechanism, the clamps close on it and advance into the oven as the next sheet unit gets fed into the system. The individual sheet will probably stop at least twice before advancing into the forming area. Naturally, with each advance, a new sheet is fed into the machine, creating a continuous process. The real advantage of the in-line sheet-fed machine over the rotary former comes from its full protection of the sheet from draft. The indexed clamped sheet travels through the heating unit as if it were traveling through a tunnel with station stops. It then indexes into the forming area without ever going through open air. The forming is made in the customary manner using moving platens on which the molds are mounted. After the forming and cooling segments of the cycle are completed, the mold opens and the formed part indexes forward, leaving the sheet transport and clamp frame behind. The clamping mechanism loops beneath itself and returns the way it came.

The sheet transport/clamp mechanism is the key element of this machine. The sides of the sheet are captured by the pins of the pin-chain system, just as on web-fed machines. The front and back edges are captured by clamp bars attached across the two chains. In this way the chain with the clamp bars can make its normal loopback for return travel. The closing and opening of the clamp bars and the pressure lock maintained between their top and bottom members is controlled by the chain-rail system through a variety of methods. The formed panels are usually ejected from the machine as the loading clamp bar opens and the chain loops beneath for the return trip. The formed panels can be made to drop on top of one another to create a stack that can be moved to the trimming operation. Depending on the type of cutting and trimming tools utilized, the actual trimming may be done right next to the unloading of the formed parts. If fabrication, gluing, painting, or laminating have to be made prior to trimming, the formed panels should first be moved to the fabricating area.

III. WEB-FED THERMOFORMING EQUIPMENT

To further improve the productivity of the thermoforming process and to eliminate completely the constant sequencing of loading and unloading procedures, web-fed thermoforming equipment was developed. In this type of thermoforming production equipment, a continuous web of thermoplastic is used in place of individual precut sheets. The web form of thermoplastic is a natural extension of the sheet-producing operation since the sheets are made that way. In a few instances, the web currently being produced is fed directly into the thermoforming operation. In most cases, however, the web is wound in roll form for later use. Materials that lend themselves to coiling and roll making are the thinner-gauge solid materials and most medium- to heavy-gauge foam materials. The production of rolled material stock can range in gauge from fractions of a mil up to 200 mils in solid thermoplastic sheets, and foamed thermoplastic can easily be managed in roll form as much as 1 in. in thickness. The web can be made to any width and a single strand can be slit into many narrower strands to provide the desired and most economical web width for the specific process. Roll diameter and roll weight are two optional specifications which may be required to match the size-handling limitations of the thermoforming equipment. Rolls may have 5- to 6-ft diameters and weigh as much as 600 to 1000 lb. Large rolls can offer economic advantages since small rolls will demand more frequent roll replacement. Large rolls, due to their size and weight, require the use of specialized roll-handling and supportive equipment. Rolls weighing more than that which can comfortably be handled by the operators require roll-lifting equipment, preferably with some type of powered arrangement. Roll-lifting devices are generally not offered as standard components of thermoforming equipment, and their selection and installation must be chosen separately from the thermoforming machinery. Roll-loading equipment may consist of power lifting machines that operate as independently mobile power lifts or may entail built-in overhead-crane-type hoists. In any roll-handling operation, the roll core size must match all the operating and handling equipment size capabilities. The roll's core size can range from 2 in. up to 10 in. The core size must be included in the specifications for the sheet order. Smaller cores allow more sheet to be wound per roll but can sometimes interfere with the unwinding of certain sheet materials. Some thermoplastic sheets wound onto rolls with a small core size tend to retain permanently the coiled shape near the core of the roll. The set-in curvature will create resistance when unwinding. When this condition is present, the web will split and break before it can be straightened out. Repeated breaks in the web as the roll diameter is reduced by unwinding the material near the center of the roll will create constant interruptions in operation. Splits and breaks in the continuity of the web not only interfere with the material indexing flow through the thermoforming machine but will demand constant rethreading of the sheet onto the chain mechanism. If this problem is encountered with regularity, the remainder of the roll should be replaced with a new roll and the troublesome material sent for reprocessing. Naturally, when this situation is encountered, the scrap factor calculation will rise way above acceptable levels. One good example of this problem is the stiffness that plagues some overaged polystyrene foams during unwinding. In this case the remedy would be to use cores no smaller than 8 in.

In spite of some of the difficulties described here, web-fed thermoforming equipment is more appropriate than any other type of thermoforming machinery for running high-speed, high-volume production. With the web of plastic being fed through the continuous thermoforming operation, the material can be indexed into all the phases of thermoforming without stopping. Only when the material runs out or the equipment fails will the production flow be interrupted.

A. Trim-in-Line Machines

Thermoforming procedures performed in a continuous stream from a plastic sheet web provide many more advantageous manufacturing opportunities than are available using the sheet-fed method. The web of sheet material not only simplifies the sheet-feeding procedures but provides a carrier for the remainder of the forming sequences. The web edges, which fall outside the actual heating area, can be adapted to guide and carry the entire web throughout the thermoforming process. Using this method permits the continuous web to be fed and indexed into the thermoforming machine and subjected to thermoforming shot after shot, creating thermoformed articles next to one another in web form. Each article is produced identically to the others and all are equally spaced from each other. The use of the web form of thermoplastic sheet not only ensures product uniformity but also offers greater production speed. In addition to the benefits of thermoforming speed, formed articles in web form lend themselves well to automated trimming. In-line trimming procedures not only offer a faster one-shot operation but may also provide better accuracy of the trim line in relation to the formed part. Continuous web-fed thermoforming is a natural candidate for in-line trimming procedures, for which there are two distinct methods. With the first method, trimming is done with a cutting implement right in the molds and is therefore called "form-and-trim-in-place thermoforming." With the second method, the trimming is produced immediately after forming in a second station within the same machine and is therefore called "form-and-trim-in-line thermoforming." Both in-line procedures have a number of variations and come with different implementation processes and characteristics.

1. Form-and-Trim-in-Place Machines

If the procedure of forming can be followed immediately by trimming right in the molds, without moving the formed article, the article cannot fall out of registration. A part molded and trimmed with combination tooling will display and hold excellent uniformity in relation to its body shape and trim line. Some thermoformed products demand flawless uniformity and allow no discrepancy whatsoever. Such products are generally containers that require accurate mating with their respective lids. Of course, any container that looks better without the slightest off-trim is a good candidate for the form-and-trim-in-place thermoforming method. With this technique, articles display uniform flange dimensions throughout their entire circumference. There are two different procedures for achieving the form-and-trim-in-place technique, each performed with its own setup, and they are not intended to produce identical articles. The two techniques should not be considered as interchangeable and should never be used for that purpose. Each type of machine, fabricated specifically for one or another of the two form-and-trim-in-place operations has a different goal and uses its own setup for entirely distinct reasons.

    a. Heat-form-and-trim-in-place machines: This type of thermoforming machinery is locked into a special thermoforming method and cannot be used for any other. The thermoforming machinery is designed to produce and follow the forming sequences of trapped sheet forming, which was described in Section II of Chapter 3 and shown in Figure 40. The machine itself operates through movement of the upper platen and the thermoplastic sheet is contact heated at the same location as that where forming and trimming are carried out. This type of thermoforming machine may be listed in manufacturers' catalogs as "contact heat pressure formers" or "trim-in-place forming presses/cut-in-place thermoformers." Either way, the name "heat-form-and-trim-in-place thermoforming" describes this equipment well. These thermoforming machines carry the trade names Thermtrol and Kirkhof. Because this type of machine has experienced declining sales volumes, this make of machines manufacture was discontinued. At that time, the trend was moving to favor of noncontact radiant heating types of machinery. However, these old equipment remained in active service and earned their keep by repeatedly producing vitally important thermoformed products for the industry. It is also somewhat ironic, that due to diminished interest among new machine buyers, this type of machinery was no longer available as new machines. In case someone offering this type of machinery for sale on the used machine market, even today, immediately snatched up, and it is usually demand a good sale price as well. One Eastern Canadian machinery manufacturer has maintained interest to supply such type of equipment and currently two U.S. manufacturers put out notifications that they are considering to reintroduce and produce this type of a Form-and-Trim-in-Place machinery. This new contemplated machinery, will have some modifications from the original concepts due to the improved features we have established since the old equipment was manufactured. The equipment's functions and basic concept are shown in Figure 63.

The platen sizes of these machines range from as small as 12 in. by 12 in. to as large as 46 in. by 42 in. The actual molding area is somewhat smaller than these overall measurements, of course, to allow extra space around the edges for proper heat control and heat transfer into the thermoplastic sheet.

Tooling for the machine can be made from a single one-up format or multi-up format which can be accommodated within the overall platen dimensions. The bottom platen always remains stationary and is fixed to the frame of the machine. The stationary platen consists of two bulky steel plates affixed one on top of the other. The bottom plate, the actual heating plate, has elec-

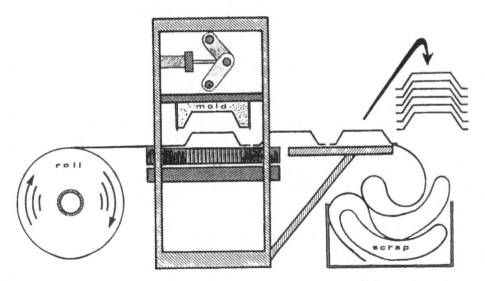

Figure 63   Heat-form-and-trim-in-place thermoforming machine.

trical cartridge heaters embedded in it together with a built-in thermostat for proper heat control. This plate transfers its heat into the upper plate, which is made of hardened tool steel and has a grid of holes drilled into it. The holes are approximately 0.0135 in. in diameter and are placed at about 1-in. increments over the entire plate surface. All the holes are connected to a central plumbing system that will permit introduction of air-pressure or vacuum forces. The combined plates are heated to precisely the temperatures required for a particular thermoforming process. The mass of the plates offers excellent temperature fluctuation control and makes this heating system virtually failsafe. The heating system is also completely draft-proof. The hardened upper plate is also used as a cutting surface against the trim knives and its grid of holes provides the introduction of forming force—air pressure, vacuum, or both—for the thermoforming. To avoid unnecessary loss of forming force, which can occur outside the molding areas or between the mold cavities, a thin-gauge buffer plate is sandwiched between the two steel plates. This buffer plate is patterned after and duplicates the mold arrangement configuration and will block those holes that have fallen outside the individual mold cavities within the configuration, thereby channeling the forming force to the mold cavities themselves.

The top platen, the only moving part of the platen system, holds the mold, which has a cutting knife attached. The molding technique follows the forming sequence shown in Figure 40. The top platen movements are usually generated with pneumatic or hydraulic forces and are often coupled by a toggle joint action. There is a two-stage movement of the top platen:

1. Mold closure onto the thermoplastic sheet with the knife edges, creating a seal
2. The ramming action of the platen for trimming the formed parts

These movements are usually accomplished with a two-stage cylinder or can be performed with a double-cylinder system. With the toggle setup, extended travel of the mold closure can be supplied simply by the toggle action and partial pressurization of the cylinder. This creates enough pressure against the thermoplastic sheet to seal it against the cutting plate. Further pressurization of the cylinder will "lock" the toggle in place and create the necessary platen force for cutting of the sheet. With this motion the formed part will actually be trimmed by the customary cutting knives discussed in Sections VI.D and VI.E of Chapter 2. Notching the knife's cutting edge creates tablike cutting interruptions that allow the formed part and scrap skeleton to advance out of the molding area while preventing the formed part from dropping onto or coming in contact with the heated platen, which, due to its heat, could rapidly distort the formed and trimmed article flanges or lips.

Sheet feed and sheet advance can be accomplished by a simple sheet pulling device and are always performed after the forming sequence and adjacent to the forming area. Sheet movement is by way of a set of traveling grippers or by "pull rollers." The grippers latch onto the edges of the web and pull it forward within a preregulated distance. A mechanism is then activated to disengage the grippers along both sides of the web and the grippers then travel back to their original position for the next advance.

In the second method of sheet advance, counterrotating pull rollers nip the continuous scrap skeleton for the duration of web travel. The pull rollers, also called "nip rollers" because of their nipping action, have a timed rotation for accurate web advance. With measured revolutions and the on/off controls of the nip roller drives, constant web advance can be matched to the reciprocating-mold-movement timing and mold dimensions.

A third and somewhat newer version of sheet advance is accomplished by use of a pin-chain sheet transport system. The pins of the two parallel running chain loops penetrate and securely grab each side of the web and carry it through the entire form-and-trim-in-place thermoforming procedure.

All three sheet advance systems are equally satisfactory and can provide uniform sheet movement for this type of thermoforming operation.

When the formed part advances out of the molding area together with its scrap skeleton, the part is separated from the scrap by hand-operated or automatic equipment and is nested into a stack of formed parts. The collection and nesting of parts are not at all difficult, because the size range of this type of product is generally small, rarely longer than 18 to 20 in. Stacking by hand is usually simple for this type of production. When the formed part comes off the line it is grabbed by hand next stop after exiting the mold. The tabs break away with the slightest grip, releasing the part. After separation from the scrap skeleton, the parts are easily nested into a stack.

With an automated stacking procedure, the separation of formed parts from the scrap skeleton is made simple by threading the scrap skeleton web sharply over an idle roller and turning its direction of travel from horizontal to vertical. The sharply angular turn will tear the tabs and separate the part, as shown in Figure 64. The freed parts can be handled and organized for stacking in many ways, such as by a moving-belt system, a downward-sliding chute, or even a mechanically operated stack collector. An automatic part-stacking system is not usually a standard component of thermoforming machinery because it is closely related to the shape and size of the molds, and this dictates a range of more sizes and specializations than the equipment manufacturer usually cares to handle. Since this specialized equipment is always coupled with the mold characteristics and dimensions, it is either built together with the molds or custom built by the thermoforming practitioner.

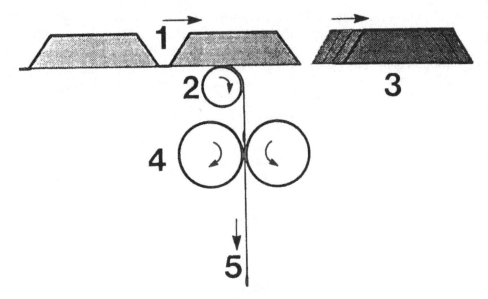

Figure 64 Formed article and scrap separation after thermoforming: (1) formed articles in web; (2) stripper roll; (3) stripped parts free to move for stacking; (4) nip driver rollers; (5) web of scrap skeleton.

The benefits of this thermoforming machinery lie in its simplified operation, easy cycle controls, and minimal susceptibility to outside interference. In addition to these benefits, this type of thermoforming machinery uses comparatively inexpensive molds. With its single moving platen, the equipment proves to be a very reliable production machine. With proper maintenance and ideal tooling setups, a long and satisfying thermoforming production life can be realized.

The only shortcomings of this type of thermoforming machinery lie in its method of thermoforming, not in the equipment itself. As pointed out in Section II.A of Chapter 3 and in Figure 40, contact heating cannot provide heat rapidly to the thermoplastic sheet, especially when heat is applied to only one side of the sheet. With a contact heating system, due to the slower heat transfer, a longer cycle time is required and a greater-than-25-mil sheet thickness should not be considered. However, to obtain more reasonable cycle speeds, the installation of sheet preheaters can offer improvements that will extend sheet usage capabilities up to 35-mil thicknesses. The sheet preheating units use a bank of radiant heater elements that utilize infrared energy instead of contact heat.

The preheating elements can be purchased from the equipment supplier or can be put together from available components and installed by plant personnel.

The other limitation of this thermoforming machine is its restriction to the use of female molds only. This limitation curtails the depth-of-draw ratios that can be used, which ideally should not be greater than 1:1. Thermoforming produced on this type of machinery will yield poor material distribution with excessive thinning of the walls near the 1:1 ratio limits of the material. Due to the fact that the wall thickness of the products produced by this method tend to get thinner, product makers generally will not thermoform products deeper than 2 in.

Thermoformed products produced on heat-form-and-trim-in-place machines always display clues that identify their manufacturing method as well as the type of equipment used. One of these clues comes from sheet material thinning, which has been described above. The other clue can be found in the flange or lip formation of the thermoformed article. Since the mold has its own cutting knife edge and trimming is made against the heated plate, the lip of the resulting part can be produced in only one way and always displays a slightly curved tip near the cut. If an absolutely straight, flat flange formation is desired, this type of machinery should not be used.

b. **Form-and-trim-in-place machines.** These machines also fall into the category of specialized equipment. Since they are made for a specific type of thermoforming, they are tied to a specific thermoforming method and cannot be used for anything else. Their name gives the best clue to their functions. The thermoforming method used with this machinery was discussed in Section II.C of Chapter 3 and shown in Figure 42.

All thermoforming functions performed on this equipment begin with a roll of thermoplastic sheet. The roll is placed on a stand from which it is unwound and fed into a transporting chain-rail system. The unwinding of the roll is usually implemented with power apparatus that matches the reciprocating mold movement and the stop-and-go sheet indexing of the thermoforming machine. The unwound sheet is carried by the transport chains through the heating ovens, where it will be advanced and indexed two or three times before moving into the molding area. As the sheet indexes through the oven, various zone-heating operations can be accomplished, according to the predetermined needs of the material to be heated. At the first heat zone, the sheet is preheated, receiving maximum heat "bombardments" from the various types of radiant heat elements that are used. The ovens can be equipped with any of the radiant heat-emitting elements mentioned earlier, depending on the type of thermoforming or operator preference. When the ovens are fitted with top and bottom heater units, it is best to provide two-sided, sandwiched heating to obtain the fastest heat cycles. Zone heating is also most advantageous because in heat zone-to-heat zone indexing, the thermoplastic sheet can be exposed to the best heating conditions for the type of plastic material being used. Zone-heating systems allow the sheet to be gradually penetrated by heat—matching its maximum heat acceptability levels—instead of being overly bombarded with heat, which would damage the sheet surface without truly heating the inner core of the sheet.

The same oven criteria prevail in all remaining web-fed thermoforming machinery. Heating tunnel ovens, with their specific number of heat zones and sheet stops, perform under the same rules. Some thermoforming machinery in use today have an undersized oven with one, or at most two, sheet stops. This equipment was originally built for light-gauge thermoplastic forming alone. To improve such short heat conditions and make this equipment more useful for heavier-gauge sheet materials, the oven tunnels must be extended to at least three stopping stations.

The appearance of a finished product's gauge, in relation to the gauge of the material from which the finished thermoformed article was formed, can be misleading. Although displaying a rather thin-gauge wall thickness, the product has actually been formed out of material that may have had an original sheet thickness of 200 mils or more. Materials such as this require slow, gradual heating for best heat penetration without surface scorching. For this reason, a longer heat tunnel oven with at least three stations is the best heating system.

After exiting the oven tunnel, the heated sheet is immediately indexed into the forming area, where thermoforming and trimming take place. It is also very important that there not be more space than is absolutely necessary for clearance between the oven tunnel and the mold. Any excess space left at this location will allow some sheet cooling and make the thermoforming vulnerable as well. One thermoforming procedure using the method of forming and trimming in place with a special mold arrangement was shown in Figure 50. The formed and trimmed articles are ejected by mechanical means or by a blast of air coming from the back side of the mold. There are special air-jet nozzles built for the latter purpose, which move the individual parts and blast them out, after mold opening, into specially built magazine and stacking racks. Like roll-handling equipment, these racks are also not a standard part of thermoforming machinery because their specialized nature ties them to a specific type of mold. Each rack is custom built to match the article and mold configurations.

It is unfortunate for the thermoforming equipment purchaser that these specialized stacking racks are neither part of the thermoforming machinery nor part of the tooling order package. They must be purchased separately and are often produced to exacting specifications to meet the specific customization process of the thermoformer.

As the parts are trimmed out of the thermoplastic web, the leftover scrap skeleton remains in web form and is either wound up or threaded into a scrap bin. The thermoforming equipment usually has a scrap winder, which is an orderly way to collect the remaining scrap skeleton. The wound scrap can be pulled off when the scrap winder reaches its full capacity or between roll changes. If the scrap skeleton is randomly fed into a scrap bin, it will rapidly fill the bin, and constant compacting will be required to maintain full load capacity. Both the wound-up scrap bundles and filled scrap bin are usually transported away from the thermoformer and handled according to the company's specific scrap-handling procedures.

For the most ideal scrap-handling system, the exiting scrap skeleton can be reduced to plastic chips, or in the case of foams, to plastic "fluffs." As the thermoplastic scrap skeleton is reduced by granulating equipment placed in the path of the scrap web, the smaller particles can be blown through an overhead pipe system or conveyed by an underground belt. The reduced scrap can be collected and stored in a centralized area for resale, recycling, or perhaps a not-so-beneficial method of final disposal. The implication of various methods of scrap use and their financial benefits are discussed in Chapter 6.

2. Form-and-Trim-in-Line Thermoforming Machines

This particular group of web-fed thermoforming machines, with their form-and-trim-in-line thermoforming methods, has enjoyed many years of useful service. The application of these machines to various thermoforming activities has ranged from R&D programs to full-fledged production. Using the same thermoforming principles, the equipment can range in size from a small 12 in. by 12 in. platen to as large as a 30 in. by 30 in. platen. This machine can be differentiated from the others by the fact that it retains the advancing sheet in web form with a transport-chain system that carries the sheet straight through the entire operation into the trimming phase. Actually, the transport-chain system handles all carrying and indexing functions throughout the entire thermoforming process. The thermoplastic web is fed into the machine, and the pins of the pin-chain sheet transport system penetrate the sheet's edges on both sides and carry the sheet into the oven area for heating.

In its travel the sheet will stop and index through two or three heating zones before entering the molding area. The sheet is usually heated by radiant infrared heater elements, which come as standard equipment and usually consist of tubular heating elements. However, any of the radiant heating elements (Section III.C of Chapter 2) can be substituted for the tubular elements if a specific condition or advantage is sought. The two or three heat stops that the sheet makes should be sufficient for proper heating. Besides the advantage of heating efficiency, other heater functions are necessary with any of the web-fed machines. It is necessary that the oven system be free to move away from the

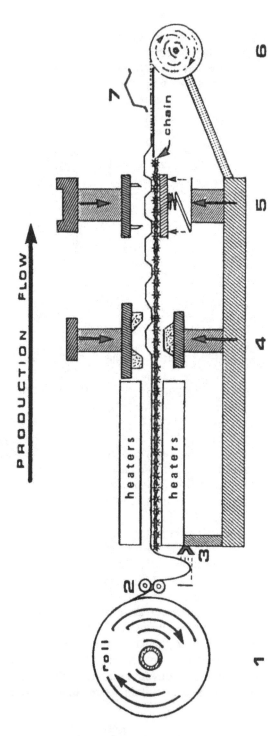

Figure 65 Form-and-trim-in-line thermoforming machine: (1) roll stock; (2) nip rollers; (3) electric eye; (4) forming station; (5) trimming station; (6) scrap wind; (7) part stacking.

heated sheet in the event of an emergency or for normal production shutdown time. It is important to be able to disengage the heating phase from the total operation because due to the continuity of the sheet, there is always a section of the web undergoing heat exposure. At normal or emergency shutdown, the ovens must be made to move aside or apart or at least be manually openable in a clamshell fashion. If such provisions are not made, the heating elements will retain enough heat to cause serious damage to the exposed portion of the sheet, even if the electric power has been disconnected. In addition, the continuity of the sheet can be cut in the back end of the thermoforming machine when a problem develops. The remaining sheet is then manually indexed out of the equipment to clear the heat tunnel. The operation sequencing and the specific task-performing locations and concepts are shown in Figure 65.

The trim-in-line thermoforming machine begins its operation with the roller-stock unwinding station. With this type of equipment, the same roll-handling implementations and core-matching criteria exist as with previous roll-fed thermoforming equipment. One of the innovations incorporated into this equipment is the placement of two in-line roll stands at the back of the machine. With these, the machine has a primary roll to feed the production and a secondary roll as a stand-by for uninterrupted feeding as the first roll runs out. Actual feeding can be achieved with nip rollers or the use of a pivoting dancing roller (Figure 66). Both will be synchronized to match the stop-and-go sheet advance. The reason for this prepulling of the rolls is to create a "slack loop" in the web to remove the strain on the material caused by the chain-sheet transport system. Otherwise, the roll's bulk and dead weight will not accom-

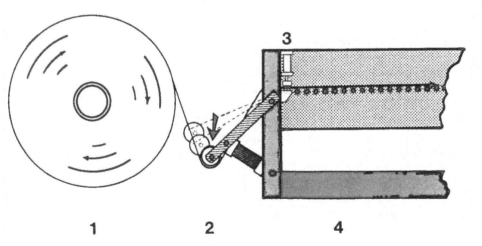

Figure 66 Unwind stand with dancer roll bar: (1) roll stock; (2) dancer roll apparatus; (3) air-cylinder break; (4) back end of thermoforming machine.

modate rapid acceleration and deceleration of the sheet indexing movements and might easily be damaged. The slack loop formation and the on/off controls of the drive mechanism are generally triggered by an electric-eye beam sensor. When the beam is blocked by the preadvanced slack loop, the drive system is turned off. As soon as the loop of material is used up and the beam view is opened, the feed drive is turned on again to produce a new slack loop in the web. The sheet web is picked up and carried into the thermoforming process by the pin-chain sheet transport system. The pins on the chain linkage are spaced continuously in $1\frac{1}{4}$-in. increments and penetrate the sheet just like spikes, providing positive web holding. The pin-chain transport system is set up so that each side of the machine has a loop of chain traveling along its entire length, forward along the top of the loop and returning beneath it. The two loops are supported by sprockets and driven by a sprocketed connecting shaft powered by a rack-and-pinion cylinder system. To provide accuracy of indexing to the drive and chain advance, an air-driven clutch is usually incorporated. The duty of the clutch is precise engagement and disengagement with an electronically measured distance of chain travel. The sheet-advancing drive cylinder can be powered by pneumatic or hydraulic forces.

The thermoplastic web is carried through the entire thermoforming process (including the trimming operation) by the same sheet chain-transport system. The initial stop for the web is in the sheet-heating ovens, where it will usually stop in two or three different heating zones. As the web is carried through the heating procedure, the sheet transport chain is also exposed to the heat source. Most web-fed thermoforming equipment that provides heating must be equipped with some method of chain shielding and cooling, which becomes part of the chain-rail sheet transport system. Newer systems are usually equipped with a plumbing network that carries cooling water through the system's entire length and provide shielding and sufficient contact cooling to the moving pin-chain loop.

After proper heating, the web will index into the molding area and undergo thermoforming. Shaping is made with the mold in either a one-up or a multi-up format. Any thermoforming method can be used for shaping and can be selected from the thermoforming techniques discussed previously. The proper technique is the one that will suit the forming procedure most ideally and at the same time, benefit the specific thermoformed part. After forming is completed by the pin-chain indexing mechanism, the part and its surrounding web material move out of the forming area. During this advance, it will move two or three indexing spaces forward and end up in the trimming area. With this type of thermoforming machine, the entire panel will undergo simultaneous trimming. The mold's format (its number of ups) will be matched to the trim tooling configuration. The trimming station has its own moving platens, independent of the forming platens, which can usually be considered as a distinct press arrangement. The trim station's press appointments are generally made much stronger than those in the forming station because trimming usually requires higher levels of power and thus creates greater stress on the platens and on the frame. For example, with pressure-forming procedures, the forming station's platen forces will be satisfied with a 7-ton pressing capability, while the trimming station can expect 50 or more tons of cutting pressure. The beefed-up frame and platen setup is needed to overcome any mechanical deflection caused by the exerting forces. The in-line trimming thermoformers are probably driven and actuated by pneumatic forces in the sheet advance and the forming-and-trimming platen closing movement. However, pneumatically operated platens can generate only a limited pressure force, which is underpowered and not enough for a normal trimming operation. Therefore, a hydraulically developed boost is used for the actual cutting since fluid pressure can develop the necessary force. Larger thermoforming equipment is operated fully by hydraulic forces, which are run from a self-contained unit.

The formed articles are trimmed out of the web using steel-rule dies; these will cut against a hardened steel cutting surface plate. The criteria for successful cutting of the plastic will arise from the prevailing rules of steel-rule die cutting, including the specific type chosen and the perils involved. Figures 31 and 32 should be consulted together with the accompanying text. The power demand for cutting is usually generated by the length (linear inches) of the cut. Multi-up molds, with their many individual trims or long trim lines with intricate patterns, can easily add up to lengthy cutting dimensions that require higher cutting power. To achieve the utmost in power cutting, pneumatic power is used only to actuate the travel of the (top) platen. When the toggles are made to lock, the hydraulic forces are introduced from underneath through the bottom platen. This platen, with its hardened cutting plate, travels only a short distance, but with the hydraulic force, its brief period of travel is loaded with power. The hydraulic forces can be gauged and set to any desired pressure reading for the purpose of directing excess hydraulic fluid to bypass the main line and thus prevent a pressure overload. There are additional precautions to be incorporated into the procedures to limit cutting-edge travel and prevent the blade from jamming against the hardened cutting plate. Such jamming can shorten the cutting-life expectancy or damage outright both sides of the cutting equipment (Figure 33). It is at the trimming station of this thermoforming machine that both its advantages and disadvantages can be experienced.

The express purpose of this thermoforming machine, with its in-line trimming format, is to provide a single machine responsible for all phases of thermoforming. A single line can produce thermoformed articles from a roll-stock thermoplastic sheet. This concept appears sound until the machine indexing tolerances receive higher levels of scrutiny. uniform and registered cuts are hard to produce with this type of setup. This is the machine's main failing. The discrepancies in the trim location versus the desired trim line are in the very nature of this equipment. This condition has three independent causes, which can interact. The first cause is found in the slight discrepancy caused by the advance mechanism. The cylinder-driven rack-and-pinion drive system may have stroke variations that can be traced back to the stroke-limiting factors, such as dirt or water in the air supply. Overpowered cylinder actuation does not have proper deceleration and can create bouncing or "creep-backs" on the full strokes. Cylinder stroke variations can be corrected by cleaning the air system or adjusting the pressure, as required. The air-operated clutch can also contribute to the trim variations. In this case, however, the problem cannot be corrected. The two opposing clutch faces come with wedge-shaped teeth. The tooth pattern starts from the center and fans out toward the circumference of the clutch, increasing in size as it goes. The actual width of the teeth is the guiding factor to over- or under-indexing of the machine. As the two faces of the clutch engage, the tapered tooth patterns are situated either one tooth over or one tooth under at each engaged indexing point. Due to the tapered shape of the teeth, the clutch face will slip into place with completed engagement. If the clutch tooth patterns have $\frac{1}{32}$-in.-wide teeth, the registration will be made with $\pm\frac{1}{32}$ in., which could mean that the trim variation from one article to the other can have as much variation as $\frac{1}{16}$ in. When a thermoformed product is designed with a $\frac{1}{16}$-in. flange, it is easy to understand how this equipment can cause complete flange wipeout by cutting the flange off on one side while leaving a double flange width on the opposing side.

The second major cause of misindexing of formed parts in the trim station involves pin penetration. Although the pin penetration and sheet-holding abilities are satisfactory for sheet advancement, they may not be accurate enough for close-toleranced indexing. The holes made by the pin can easily get enlarged by either mechanical pull forces or applied heat. Either way, the fixed position of the pin-chain travel and its correct locating factors will be lost and no controllable situation will be possible.

The third and least problematic cause of misindexing in the trim station comes from the slight variations in the shrinkage rate of the plastic material. This shrinkage problem is minute and noticeable only when extreme tolerances are expected. It is also true that if any of the problems mentioned previously should exist, they will overpower this one and make it insignificant. This machine has linear adjustment provisions to remedy shrinkage-related misalignments as well as to make correctly aligned initial setups. This distance between the forming and trimming stations is adjustable and the whole trim station can be pulled forward or backward according to the particular alignment requirements.

This type of form-and-in-line-trimming thermoforming machine is often overrated in its ability to hold to small tolerance levels during trimming. Equipment suppliers list specifications as small as ±0.025 in. However, actual obtainable tolerances are far greater than those generally listed and will become even worse as the machine ages. Knowledge of this limitation is helpful in preventing unrealistic expectations as to meeting close-toleranced product criteria with this machine. Nonetheless, parts that do not have flange-size-variation sensitivities can be made effortlessly on this type of thermoforming machine. The production output rates can be substantial, with an average of 6-second cycles and a maximum of 3-second cycles. The rapid cycle-time repetition requires matching speeds of sheet advancement and trimming together with mold resident times. It is during the mold resident time that forming and cooling take place. With such a high rate of thermoforming speed, the cooling segment of the cycle will be in jeopardy. The rapid cycles, together with repeated contact between the heated thermoplastic sheet material and the molds, will raise the mold temperatures. The 6-second or less cycle time does not permit the heat to escape into the machine framing but rather, causes it to accumulate in the mold. If no steps are taken to cool the molds with rapid cycle action, the necessary cooling will not take place, and soft and deformed articles will come out of the molds. Such high rates of production always demand auxiliary mold cooling. This can be achieved in many ways, but the most efficient is with built-in cooling channels placed into the mold base. Through these cooling channels, a cooling fluid—usually water—is circulated. This fluid comes from either the plant's centralized cooling-water source or an independent mold temperature control unit.

The temperature control unit is a separate and self-contained system which is not a part of the thermoforming machine, although it can be purchased from the same supplier. The temperature control unit has only one duty, to remove any excess heat from the mold and maintain the mold temperature at a preset level. The temperature control unit contains a reservoir of fluid that is pumped and circulated through its internal cooling coil system and through two flexible hoses. These hoses are equipped with "quick-disconnect" fittings attached to the mold's plumbing system. The fluid could be either chemically treated water or a glycol formulation which, at constant circulation, will remove the excess heat buildup from the molds. Use of a closed-loop system of temperature control can provide full-scale temperature ranges for the mold. Cooling channel formation in the molds is discussed in Chapter 5.

Despite the less advantageous trim accuracies with this type of form-and-trim-in-line thermoformer, the machines often find great popularity among thermoforming producers. As long as unrealistic expectations of trim accuracy are not being placed on the machine, the production result of this thermoforming equipment can be most satisfactory. The two or three size variations of this type of equipment offered by several machine manufacturing concerns can serve the needs of a large range of thermoforming production requirements.

B. In-Line Posttrimming Thermoforming Machines

Continuous web-fed thermoforming machines placed in-line with a separate
but synchronized trim press would qualify as in-line posttrimming equipment.
The benefits of using this type of in-line thermoforming system can be found
in the high production rate. Thermoforming machines in this system can be
run at speeds exceeding 30 cycles per minute. In addition to its higher thermoforming speed, this machine's platens are comparatively larger than those of
the in-line machines previously discussed. Machines in this category can have
platens ranging from 25 in. by 25 in. to as large as 56 in. by 90 in.
   The basic thermoforming concept here follows the forming techniques
discussed for form-and-trim-in-line machines (Section III.A). However, part
trimming is accomplished farther downstream in separate equipment placed
in-line with the thermoforming machine.

1. Basic Forming and In-Line Posttrimming
   Thermoforming Machines

The thermoformer in this in-line setup is a separate machine that uses standard
roll-stock thermoplastic sheet material or can accept a directly extruded thermoplastic web. The web is fed into the thermoforming machine, which converts
the flat sheet stock into a series of thermoformed products that remain in the
web until subjected to postoperated trimming. The roll-stock handling and
feed system also follows the aforementioned criteria and most common practices. Since this type of machine has a larger mold format, the use of greater
web widths, and higher production speeds, larger-diameter rolls are commonly
used. With larger rolls of thermoplastic material, special equipment and precautions must be followed for proper roll handling and management. Unwinding and feeding the roll into the thermoformer must be made either with any
of the counterrotating nip rollers described earlier or by the older method of
cam-action dancer roll bars. These bars are made to cover the greatest width
of roll stock that the thermoformer can accept. The two ends of the dancer roll
bar are attached to a set of pivoting cam arms secured to the machine frame.
The pivoting cam action is produced by an air cylinder, which produces the
up-and-down movement of the dancer roll. To make this pivoting system work
better, the web is clamped at the entrance of the machine. The clamping action
is used only to prevent pullback from the part of the sheet already advanced
into the machine. This clamping action, when actuated, will only permit the
unwinding of the roll stock by the pivoting action of the dancer roll bar (Figure
66). As the machine demands more material, both the clamping mechanism
and the pivoting arm cylinders are deactivated, allowing them to pull up the
slack loop in the sheet. After sheet advance, both the clamp and the pivoting

arms will be reactivated, producing another web slack loop and unwinding another portion of the roll stock.

a. The thermoformer: The introduction of the web to the transport-chain system is made with an adjustable sheet guide apparatus that properly aligns the web for pin-chain pickup. If the web is not fed into the machine parallel with the chain rails, the angle error will accumulate with the continuing indexing. This will result in a runoff from the pin chains on one side and a jam-up on the other. Improper sheet feeding can cause constant transport-chain failure and "rail jams." The sheet guiding apparatus is the most impor-tant feature of the back end of this type of thermoforming machine. The sheet guiding apparatus can be made with simple slotted guide bars, or in a more sophisticated setup, using a self-aligning micrometer adjustment.

The thermoplastic web is carried through the oven in this type of ther-moforming machine the same way as before and preferably, with at least three stopping stations. As before, the oven consists of a top and bottom heater unit to provide more efficient sandwich heating. The heater elements can be made with any of the radiant-energy-generating heater units; however, as an industry standard, tubular heater elements are generally used. In a more delicate situ-ation, ceramic heater elements are used. Combination heater units are also offered where the top oven has tubular heater elements and the bottom oven is equipped with ceramic elements. However, this combination oven setup cannot provide suitable programmed heating, only a pretense of such. Not all of the thermoplastics or design criteria demand the elaborate heating program that ceramic heaters can provide; therefore, ordinary tubular heaters will pro-vide satisfactory sheet heating in most cases. The sides of the ovens in this thermoformer are usually well protected from uncontrolled drafts. But a rather large opening remains at the point of web entry and the molding end of the machine. This is where most of this equipment is vulnerable to draft, especially when placed in the path of a draft produced by the plant layout.

The ovens in this thermoforming machine also come with various safety features. For ideal oven safety, the ovens should be engineered so that the oven heating surfaces can readily be moved away from dangerously overheated thermoplastic. Some ovens are built in a C shape, with the radiant heater ele-ments placed both above and underneath the sheet and interconnected by supports along one side. With this design the entire oven system is capable of shuttling away from the web. There are other oven designs that offer similar features, such as clamshell-type openings or center-split clamshell heater ov-ens. In this latest type of oven design, the oven splits in the middle parallel with the machine so that the top opens up like a hatch and the bottom drops downward (Figure 67). For the utmost in oven safety, especially when foamed thermoplastic materials with their greater flammability are used, it is best to

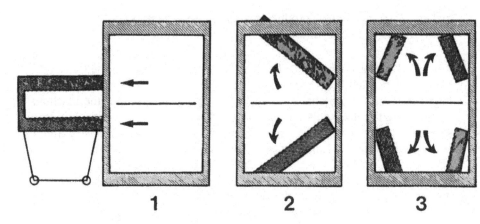

1   2   3

Figure 67   Heating discontinuation oven safety features: (1) C-type roll-a-side oven; (2) clamshell oven; (3) double-hatch oven.

have a self-contained fire extinguisher system plumbed into the oven of the thermoformer in addition to the retracting oven feature. The fire extinguisher system should be activated by a conveniently placed single button and there should be spray nozzle heads in the oven. Oven fires from burning plastics can be put out with several of the popularly used gas- or liquid-filled fire extinguishers that contain CO_2 or the more sophisticated Halon formulation. It is also advisable that such fire-extinguishing efforts should be designed so that they are functional even when electrical failure is encountered. There are some equipment manufacturers who offer a built-in fire extinguisher system that not only reduces the fire hazards but can minimize costly cleanups after a resulting oven fire. (Remember that a glowing-red heater element can reach 600°F or higher; this is far above the ignition temperature of any of the thermoplastic materials. When these materials burn, they tend to produce a large amount of black smoke and carbon deposits.) One other type of oven safety feature is still in use, although no longer offered on new machines. This is the use of air jets manually operated during an oven fire. The concept behind such air jets is to provide immediate cooling down for the overheated plastic sheet as well as the ovens. It is also expected that the force of air from the jets will blow out the flames. Such systems are not at all effective, as they may produce a far worse situation than if the fire were left alone. The airstream may feed more oxygen into the fire and make the flames blow like a torch. It is best to dismantle such air-jet fire safety systems and replace them with CO_2 or Halon fire extinguisher systems.

The heated thermoplastic web will index from the oven into the forming area by way of the same pin chain. The forming area is where the heated sheet is converted into the final shape by the forming and cooling segments of the cycle. It is here that the moving platens, with the mold halves, are set in (see Figure 70). There are many factors that regulate just the platen functions, including the wide range of sizes and different mechanical means of operation. The goal with these machine platens is to achieve unrestricted movement and sequencing together with sufficient power and speed. In addition to those criteria, the platens must mate parallel to each other and be capable of holding this arrangement under the most demanding conditions throughout a long series of runs. Since the platens are mostly large, parallel platen conditions will require quality and strength in the platen supports. There are several engineering designs for platen supports, all of which aim at resolving platen deflection and platen guide bearing or bushing wear while reducing equipment component cost. The platen supports can be engineered with two large round steel posts placed in the middle of both sides of the platen or with four smaller posts placed at each corner of the platen. Naturally, each type comes with respectively sized brass alloy bushings or sliding ball bearings. There are thermoforming machines that are engineered with rectangular or square machined jibs in place of the round posts for platen guidance. The choice of one particular system over the other must be decided by their comparative cost and size or thermoforming criteria. One system may initially cost less but require more maintenance. A cheaper system may also not meet the higher quality demands of a specific production scheme.

Since the platen movement is made in an up-and-down direction, most of the motion is a sliding motion made on the guide posts or jibs. Despite provisions incorporated into the system for ease of sliding by the use of bushings and bearing units, the sliding surfaces must be lubricated regularly and subjected to periodic maintenance service. On some thermoforming equipment, lubrication is by manual greasing. The more sophisticated machinery is equipped with a central lubricating system that will deliver lubrication to all moving friction points.

It is ironic to see overanxious maintenance personnel place so much lubricating grease into the system that the excess bulges out and clings to the surrounding equipment components. The grease globs can fall on top of the indexing sheet at any time and when the mold closes, the grease will be sucked into the mold's inner chambers and contaminate it for many forming cycles to come. Such careless lubrication has caused enough contamination of molds and thermoformed articles to cause production stoppages. Thermoformer practitioners should put a stop to this poor lubrication practice and enforce cleanup programs as well.

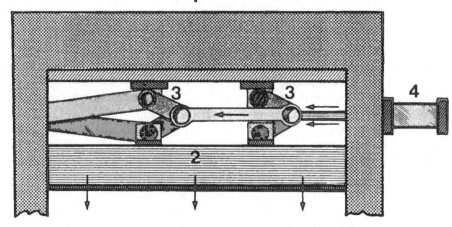

Figure 68   Toggle-operated platen: (1) upper machine frame; (2) upper moving platen; (3) toggles; (4) activating cylinder.

The up-and-down movement of platens is usually provided by pneumatic or hydraulic forces. These forces are introduced through cylinders of different diameters with respect to the size of the machinery. To obtain more precise movement and a positive locking feature at the end of each platen stroke, a toggle joint mechanism is incorporated between the actuating cylinders and the platen body (Figure 68). This mechanism works with a knee-joint-like action; the thermoforming machine frame and platens are tied together with individual steel bars jointed together with a large pin. At the joint, the two bars can act like a knee, bending together or completely straightening out. When this straight-line formation is obtained, the toggles are locked in full stroke position, and any end-to-end pressure exerted on them will meet with the full structural strength of the combined units. When toggles are not locked into a straight formation, the exertion by the platen is provided only by the mechanical strength of the driving cylinder and its driving force.

Platens operated by direct actuation of cylinders connected to the same pneumatic or hydraulic forces cannot provide positive travel against each other and one will probably override the force of the other. This overriding of each other's travel is caused by unequal reaction of the cylinders to the air pressures introduced into them; this condition will not be diminished even when the cylinder size and exerting pressure forces are substantially increased. This means that identically sized cylinders connected to opposing platens and actuated against each other will engage in an unpredictable way to shifting one way or the other from the original meeting point. Placement of positive stop blocks or a stroke-limiting device can be used to stop territory takeover and cylinder overrides. However, the most positive stroke limiting and locking of the platen position is provided by the use of toggle joints. Because of the movement of the toggle joint, a smaller-stroked cylinder can provide a longer travel distance to the platen than the method without a toggle (Figure 68).

The platen of this group of thermoforming machine can be driven with mechanical forces other than pneumatic and hydraulic. Mechanically driven platens are generally driven by electric motors that operate mostly on dc power. The motor rotational movements are transferred via a gearbox to a cam shaft. This cam shaft may have one or a series of cams, which will control the entire machine's functions. A conjugated cam system combined with a top and bottom toggle setup can also drive the platens. The lobe of the cams will cause the platens to close or open, and by various adjustments in the cam position on the shaft, will provide the changes in the actuating sequencing. Cam-operated thermoforming machines are classified as mechanical machines, and their advantages are obvious when their power consumption and thermoforming speeds are compared with those of the customary pneumatically or hydraulically operated machines. Mechanical thermoformers can operate one-third

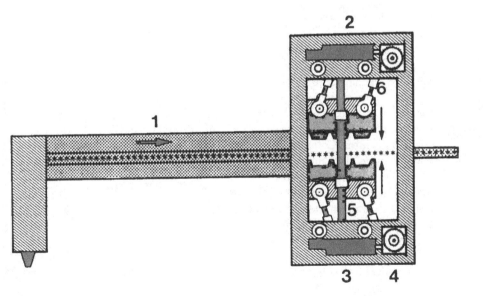

Figure 69 Cam-type operated thermoforming machine platens: (1) oven tunnel; (2) frame of the forming station; (3) dc motor drive; (4) gearbox; (5) platen guide post; (6) one of eight moving cam arms.

faster and with fewer breakdown incidents than conventional machines. Their operation is mostly vibration-free and does not require fine tuning of the flow control valves to obtain smooth-running production.

There is one more type of mechanical thermoforming machine that is unique. Its individuality comes from the fact that its entire platen system is driven by nothing more than a series of cam arms. Four separate cam arms are connected to the four corners of each platen, ensuring parallel and equal drive to the entire platen plate. The four independent cam arms are tied together with other cams and the whole assembly is tied to a common shaft driven by a permanent-magnet dc motor (Figure 69). The top and bottom platens can have their own independent dc motors, providing individual actuation possibilities. The platens can also be coupled together with only one dc motor, which will drive both platens. The cam arms have length adjustments to provide the necessary stroke-height adjustment for each platen travel. When this machine is in operation with all those interconnected cams operating in union, it is easy to imagine that someone has brought back to life the old steam locomotive principle applied in a modern way. Using this principle makes this type of thermoformer run not just well but smoothly. In spite of the older engineering principles, this thermoforming equipment and its modernized engineering version help to reduce energy consumption in a very timely way.

All of these types of thermoforming machines, whether actuated by pneumatic or hydraulic forces or by a mechanical setup, share the same basic thermoforming scheme. Any of the thermoforming methods discussed previously, with their various tooling implementations, can be incorporated into the manufacturing plan of this type of thermoforming machine. The only limitation this machine places on thermoforming projects is its size limitation; it cannot accept thermoforming tasks larger than its platen sizes, nor can it form products deeper than its actual top or bottom platen stroke. The mold weight each platen can handle is also a limiting factor and should be specified by the machine manufacturer. Complicated multifunction molds, with their large number of "ups," can weigh as much as 700 to 1000 lb. This weight can easily overburden the platen-lifting capabilities of this thermoforming machine. In addition to these common types of limitations, there are specific limitations to be found between the different makes and models offered in the competitive thermoforming machine market. Unfortunately, it takes time and effort to produce a comparison of the various features. What is normally offered as a standard component on one piece of equipment may be an optional item or not be offered at all by other suppliers. It is the responsibility of the thermoforming equipment purchaser to be selective and to choose the equipment features according to specific thermoforming criteria. Although competing equipment suppliers have an essentially matching lineup of machinery for every conceivable thermoforming task, it is a well-known fact that there are specific model choices for the discriminating thermoforming practitioner. The various types of sheet indexing implementations—chain-rail systems and even the various types of machine frames—can produce a difference of opinion among the prospective users.

With this particular group of thermoforming machines, the web will index out of the forming area after the forming is completed and with one more indexing, will completely leave the thermoforming machine. The formed sheet web is disengaged when the pin chains turn downward for the return loop, thus pulling the pins out of the web. The web, with the thermoformed articles, will advance index after index out of the thermoformer and will be threaded some distance away into a separate but synchronized in-line trim press (Figure 70). With this "downstream" postoperative trimming, thermoformer indexing preciseness could be less critical because it will not influence the finished article trim accuracies. Any inaccuracy in the thermoforming indexing will cause deviations only in the spacing between each formed panel, not between the individual rows within a panel. In the case of underindexing, minute deviations may not present a problem. However, with exaggerated underindexing, the feed fingers may not be able to slip behind the formed parts or the "feed buttons" and can slide off when actuated. In addition, if the advancing stroke is made too short, it is possible that the mold will trample on the back portions of the previously formed panel.

Where overindexing is experienced, the resulting deviation can add excess material to each shot and therefore increase the overall material consumption. When the web is overindexed, the feed fingers will engage higher and earlier than the catchable feed surface (thermoformed part side or the feed button). When activated, the finger will lower itself until contact is made; the full registration stroke will follow. This causes problems only when the feed fingers are accelerated too rapidly and the thermoformed material is much too soft to withstand the impact of the feed fingers and yields to destruction or tearing.

Minute indexing variations in either direction should not affect the actual trimming result in the in-line postoperative trimming method. In the posttrimming system, the formed parts are made to register either individually or by rows for accurate trim location. Despite the reduced sensitivity to indexing variation in the posttrimming operation, some of these thermoforming machines are equipped with superior index alignment implementations, such as electrohydraulic pulse motors, digitally controlled units or dc motor-driven indexing mechanisms, and solid-state logic indexing systems. All of these systems are aimed at providing the most accurate controls for the sheet-advancing system for the particular brand of thermoforming machine in use. The selection of systems should be influenced by the specific thermoforming tasks and the product criteria, which could favor one type of setup over another. Typi-

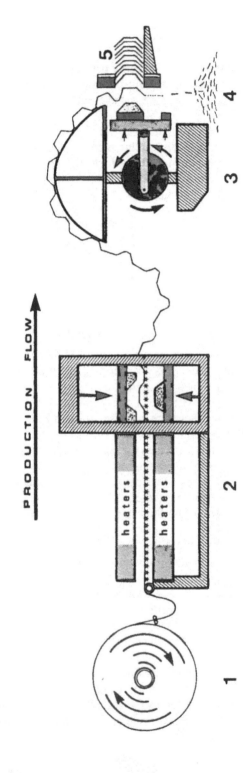

Figure 70 Forming and in-line posttrimming thermoforming machines: (1) roll stock; (2) thermoforming machine; (3) trim press; (4) scrap chips; (5) stacked thermoformed parts.

cally, the most ideal thermoforming machine concepts are not to be found on a single machine from a single source. Rather, some features will be available from one supplier, while other features will be found on competing equipment.

b. The trim press: To trim the thermoformed articles out of the formed web, trim-press equipment is called into play. The separate machinery is placed in-line with the thermoformer to perform the task of trimming in a synchronized manner. The released web, with its formed articles, will continue moving forward. When enough formed web accumulates between the machines, the web is threaded over a canopy rack. The web should be fed into the trim press from the thermoformer with just enough slack so that the web will not drag on the floor. The operator must provide plenty of excess to account for the running-speed discrepancy between the two machines. The slack between the machines will aid in their timing synchronization and will help maintain the web material registration. The use of the overhead canopy rack also permits a directional movement change to the web from horizontal to vertical. As Figure 70 shows, the horizontally released thermoformed web is threaded over the canopy above the trim press. This guides and turns the web over its large-radiused surface so that the web enters the trim area vertically. This vertically fed trimming operation has the advantage of creating horizontally stacked parts, which are ideal for handling and packaging. Such vertically fed web positioning, with the horizontally moving punch and die set, is quite common and can be considered a traditional setup. By this method, the thermoformed articles are not only trimmed out of the web, but can be stacked and even counted automatically.

There are also some vertically moving trim operations which manage the trimming out of a horizontally indexing web. With the vertically moving trim press, the trimmed articles are either dropped downward (usually in random fashion into a collection box) or can be trimmed upwardly into stacks. Both vertical trim operations can be made to work, but accomplishing accurate trimming registration proves to be more difficult. On the other hand, the slack and the greater distance between the thermoformer and the trim press in the horizontal trimming setup, cleverly provided by the canopy, offer greater opportunity for ease of alignment. The vertical feeding of the web also allows faster cycling of the trim press, with the result that the sheet indexing and its acceleration and deceleration can be managed much more easily. The vertical sheet-feeding arrangement allows the sheet to realign itself after it is initially located in the trim station. Any overly pushed—and overly traveled—web might be pulled back by the entire weight of its large slack hanging over the canopy.

The side-to-side web guidance into the trim station begins with the canopy guide rails. The entire canopy is made in such a way that it will align and direct the web with its frame structure. The sides of the curved canopy structure are often interconnected, preferably with roller-type installations. The rollers not only make sheet advancement easier by reducing the web's drag, but also do not harm the surface, which is preferable for softer thermoplastic sheets or easily damageable foamed materials.

The final locating and indexing of the web with its thermoformed articles is made with the "feed cage" system, so called because of its close resemblance to a cage side. The vertically positioned guide bars align the web with the trim tool. The guide bars are positioned directly above the trim tool and are made so as not to interfere with its movements. As the parts slide through the guide bars the bars slip into the "guide path" that is created by the space between the formed articles. To avoid surface damage to the article's sides, specifically created guide buttons—"slots" if they are elongated—are formed into the scrap area between the parts to create the guide path for the guide bars.

For horizontal registration (i.e., across the web), a separate group of moving feed fingers are used. This feed finger actuation is tied back to the same rotating shaft that moves the trim tool in and out of the trimming position. The feed finger actuation and its timing are activated by a cam wheel with predetermined lobes. The cam rotation is picked up by a cam follower that has the necessary arm adjustment capabilities for regulating the stroke distances. A shorter distance of web advancement allows the trim press to make more strokes per minute while longer advances will slow down the trimming cycles. For example, a trim press running with a $4^1/_2$-in. advanced distance may run 100 to 120 cycles per minute, while $13^1/_2$-in. advancement settings can accomplish only 60 cycles per minute. Of course, there are trim-press machines that can offer speeds of up to 200 cycles per minute using the minimum advancing strokes. However, their speed must be matched by the thermoforming machine output rate to live up to their full potential.

Regardless of speed, the feed fingers must move and operate the web advance while the trim press, with the trim tool, is in an open position. As soon as the trim platen moves into the trimming mode, the feed fingers must have accomplished the web advance and be fully retracted in order to clear the trim tool outer edges and avoid smashing their tips. Although it is possible for the feed fingers to accomplish web advance by pushing against the formed parts themselves, this is not always ideal. Although this provides a larger area to push against than a guide button or slot would, applying a pushing force against the sidewalls of a formed part may not always prove successful. Formed parts may display irregular sidewall designs or substantial material thinning which cannot accommodate a pushing force and will distort under the pressure and alter the stroke distance. Any distortion will result in a misalignment of trim in the MD (front-to-back direction). Therefore, if the sides of the formed article cannot structurally support the web advancement, it is best that the pushing forces be applied to a specially formed guide/feed button. This button is designed specifically for part advance and is usually formed into the web in the scrap area between the formed articles. Due to their specific purpose, the feed buttons are always shaped in the most advantageous way: not so large that the material weakness prevails but at the same time with enough detail for the feed finger to engage on firmly. The feed buttons will remain on the scrap skeleton after trimming and will be disposed of according to the reclaim procedure being used.

The feed finger tips, which make the contact with the web, are usually made of synthetic material (Teflon, nylon, etc.) and shaped for best damage-free operation. As the feed fingers retract from the web and the trimming procedure is completed, the fingers make contact with the next row of formed parts. As soon as the trim platen, with the trim tooling, is opened, the feed fingers push the next row into the trimming area (Figure 71). With each stroke of the trim press, the parts are trimmed and the web advanced with each trim tool retraction.

As the punch enters the trim die, it cuts through the material and pushes the formed article into the die throat. At this point the remaining scrap skeleton surrounds the punch. At the moment the punch retracts, it carries the entire scrap skeleton with it and retains it frictionally. For this reason, a stripping implementation is added to the punch to pull off the scrap skeleton. In considering the various options for stripping, the simplest and most inexpensive methods are also usually the least reliable and their life expectancy is short. Using compressible foam (rubber or synthetic) materials or steel springs can produce the necessary stripping action when fastened all around the punch perimeter. Upon the trim tool engagement, the foam material is compressed under the scrap skeleton. When the trim tool is opened, the expanding foam will strip the scrap skeleton off the punch body. The other major method of stripping is more precise. This method is performed with mechanical strippers that are direct extensions of the guide bars and become actual parts of the feed cage system. In this case the entire cage produces a short swinging action by pivoting at its top and moving the most at the bottom, next to the trim dies. This swinging action becomes an oscillating motion when synchronized with the trim platen movement. The stripper frame attached to the oscillating feed cage will move enough to strip the scrap skeleton off the retracting punch body and free and cut web (scrap skeleton) for the next advance. The trim tooling on both the punch and die sides is built such that the stripper frame can clear the space between them without damage.

With a multi-up mold arrangement in the thermoforming machine, the mold format may contain several rows of formed parts across its width. This is especially common with this type of thermoforming machine, with its larger platen format and higher speed and volume production capabilities. Since the mold contains several rows of individual mold cavities, the trim press will feed

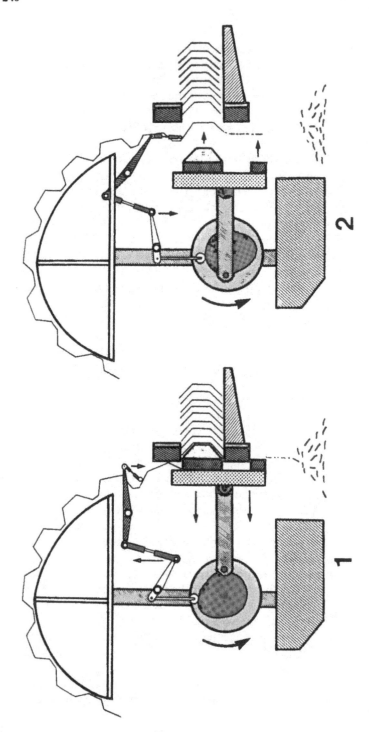

Figure 71 Trimming and web feeding synchronization: (1) trim punch in, feed fingers feed up; (2) trim punch out, feed fingers feed down.

and register the web with its formed parts one row at a time. As trim registration is made, it will force the trim press to run a faster cycle than that of the thermoforming machine. For example, where the thermoforming mold contains four rows of mold cavities, the trim press, since the trimming is made row by row, has to run four cycles to the one of the thermoformer.

When short products are being produced, the trim press can be set up with trim tooling capable of handling double-row trimming. In this arrangement both the trim tooling and the receiving table must be able to accommodate two rows of products, one on top of the other. There are only two criteria to be considered with double-row trimming methods. The first is that the combined length of the two formed articles and the space between them cannot be greater than the maximum advancing stroke capability of the trim press. The second is that the product coming out at the receiving end must be capable of being handled by packaging, loading, and all other secondary operations. If these downstream procedures cannot keep up with the highly productive double-row trimming, this type of tooling and trimming arrangement should not be considered.

Trimming with the postoperative trim press is always made in-line with the thermoformer. However, it is not always necessary that the trim press work in the same direction as the thermoforming production flow. However, this is a popular practice, as shown in Figure 70. On the other hand, the trim press can be turned completely around and made to trim backward in a reversed production flow. This backward directional trimming is chosen either because of limited plant floor space or the desire to use a single operator for both machine operation and product packing. The only criterion for the backward trimming procedure is that the thermoformed product web be compatible with the trim press.

For the most ideal trimming setup, the web of thermoformed articles is generally positioned such that it is approached by the punch on its female side. Trimming a three-dimensional article out of the web from its female side makes both the registration and distortion-free trimming easier to produce. It is obvious that if the thermoformed article were approached on its male side instead, the pressure of the trim punches might cause distortion and outward flaring, resulting in undersized and irregular trim lines after cutting. In cases where the thermoformed article is designed with a downward-turned flange, the flange curve itself will dictate which is the female side for trimming and consequently from which side the punch should approach the part (Figure 72). In this case the trimming is actually made from what would otherwise be considered the male side. Such a part will therefore require a hollow punch to clear the formed article body while it is trimmed. There are a few cases when trimming has to be approached from the male side of the article to capture additional advantages or cater to a specific condition. However, these cases are few indeed, and the justification for such a trimming setup, with the re-

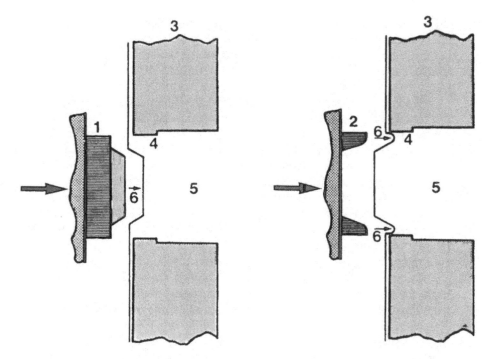

Figure 72   Trimming approach from the female side: (1) trim punch with "pilot"; (2) hollow trim punch; (3) die; (4) die lip; (5) die throat; (6) female-side-approach trimming.

spective flipped mold positioning in the thermoforming machine, must be carefully considered.

   To obtain access to the female side of the formed article on the web, the trim press can be made operational from either direction. To encompass such directional changes for any of the above-mentioned reasons, the trim press must be turned around and its overhead canopy must be positioned for proper web slack looping. The mold placement in the thermoforming machine must be switched to match the desired production scheme. Figure 73 shows the reversed trimming arrangement with the switched mold position. When a comparison is made with Figure 70, the difference in setup can easily be seen.

   The trim-press functions can be expanded to include automatic stacking of the parts together with counting the number of strokes to keep track of the quantity of parts produced. The stacking is easily performed, especially with the horizontally running trim platen. As the punch penetrates the die, the

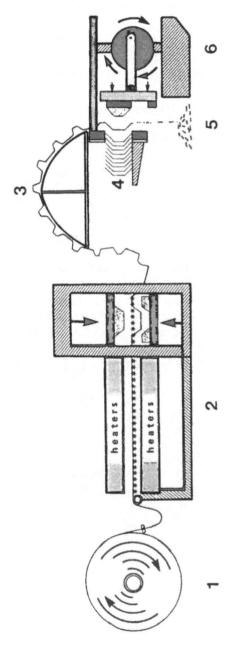

Figure 73 Reversed trim-press arrangement: (1) roll stock; (2) thermoforming machine; (3) overhead canopy; (4) packing table and stacked parts; (5) scrap chips; (6) trim press.

shearing plastic part is pushed past the die lip and carried through the die throat, which is open in the back. A table is bolted onto the back end of the die, its surface level to match the bottom of the die throat. The table will receive the trimmed and nested articles as they index out of the die throat, pushed with each subsequent trimming. The punch side of the trim tool can contain one or both of two mechanical aids that will, in effect, add even more precision to the trimming operations. One of these implementations is use of a contoured punch tip for "piloting" for improved part registration. As this contoured punch contacts the formed article in the web, some limited realignment can be achieved to correct any error made by the feed guide and feed finger registration. The second implementation is a device that can be incorporated within the punches to provide push-out for the stacks of trimmed parts at a predetermined count. Activation of the device is tied into the electromechanical counter. This setup can provide a single extended trim stroke at the predetermined count sequence in order to clear the die throat and create a correctly measured stack of parts for packaging.

The trim-press machinery is always powered by electrically driven motors. The motor is used to drive a large flywheel or can be converted directly to cam action. In either case, a rotational motion is converted into a reciprocating motion that matches the overall production sequencing. The key to achieving effortless trimming is to engineer into the system a smooth movement that cuts cleanly through the plastic without hesitation. In addition to attaining a sufficient power level for smooth operation, the trim press must synchronize with the thermoformer. It must have the means for speed adjustments, which not only can be made to match its own speed to the thermoformer's but if desired in the interim, can be sped up or slowed down to allow it to catch up or simply to be readjusted. Such speed adjustment is always easier to make at the trim press and should not be made at the thermoformer since many more criteria would be affected if the thermoformer speed were altered.

One of the most important factors involved with the trim press is that of the utmost accuracies required in alignment of the punch and die mating. In the trim press, where the punches and dies have zero or minimal clearance allowances, this mating from shot to shot must be made repeatedly and in an absolutely meticulous manner. The slightest deviation in the mating between the punch and die will result in damage or even complete tooling wipeout. Just one misaligned engagement can cause damage serious enough to halt production. With these extreme alignment sensitivities, the trim press must have excellent provisions for maintaining its platen movement accuracies. Centralized and automated lubrication will minimize any short-term-caused wear as well as its resulting problem. The trim press must also have high-quality tool steel guide posts and guide shafts of substantial size so that their deflection would not be a cause of misalignment. Bushings and bearings must meet the same high-quality standard. The heavy weight of most common punch tooling also creates some burden on the equipment's moving parts, which can cause poor alignment between the punch and dies. The two main culprits of weight-related misalignment come from normal wear and eventual seating of the moving equipment components. There are two engineering options that help to overcome such misalignment predicaments. The first method is less costly and simpler; however, it is less reliable. The alignment is ensured with the help of tapered "alignment pins," which are installed on the outside frame of the actual trimming tool. Alignment pins, with their conical shape and hardened tool steel construction body, are made to register into a matching female body cavity. Their full engagement is made prior to the actual punch and die engagement, correcting any discrepancies of alignment. Naturally, with this method, only slight discrepancies are neutralized, and even then this method cannot be used for extremely long production runs. Eventually, the wear and deterioration can also destroy such realignment features. For extended service life, the second method is much better for avoiding misalignment between the punch and dies. This method is preferred by most tool and die makers. The method consists of securing the punch and dies to their own independent "die set." The die set is made with its own die shoe body, which contains two or four die pins in addition to those of the trim-press platen. Using a self-contained die set system will guarantee utmost alignment with any trim press, even if it is old or worn, and will not permit trim tool misalignment or wipeout. Inherently, this type of independent die set system, applied across the board to all trim tooling, can raise the overall tooling cost to staggeringly high levels.

In addition to their standard operational functions, the different types and makes of trim-press units usually come with variously designed operational controls, web guides, overhead stationary, or even rocking web canopies, feed cages, and so on. Some trim presses are available with specially constructed catwalks built for the operator's convenience in threading the leading edge of the web into the feed system of the trim press. The packing tables that receive the rows of trimmed products can also be customized and are not necessarily part of the trim-press package. Packing tables may be designed for specific products and specialized operations that are not standard for the thermoforming industry. For example, whereas flat-sided products can use a standard flat table surface, curved or round products may demand matching table surface contours for ideal stack segregation.

Some packing tables are built with a special adaptation in mind and are outfitted with such gadgets as static eliminators, ionized air blowers, and plastic dust and angel-hair particle removal collectors. There are operations that incorporate the packaging of the thermoformed goods with the packing table, where the outcoming rows of products are put into bags and corrugated cartons. The combined output rates of the thermoformer and trim press are usually great enough to warrant the installation of automated packaging equipment, set to handle all the packaging needs of the thermoformed product. Such implementation may contain automatic bag openers and sealers, box-making setup and gluing machines, and so on.

The last supplemental component attached to the posttrimming thermoforming system is the scrap granulator. This reduces the scrap skeleton to a much more easily handled form, which is especially important with high-volume output machines. Since the scrap skeleton is generated at such a high rate, neither the scrap bin nor the scrap winder can accommodate it. Some trim tooling comes with a scrap shearing knife made as part of the whole trim tooling and bolted underneath the punch and die portions. The purpose of this knife is to engage with the opposing die edge on each trimming cycle so that it will shear across the remains of the scrap skeleton hanging out of the trimming area. This knife chops up the scrap skeleton, and even if this should not provide substantially reduced scrap, further handling can certainly be made easier. The chopped-up scrap skeleton remains can be fed into boxes or any under-the-floor scrap-collecting system, such as exist in some thermoforming manufacturing plants. These systems consist of interconnected trench channels built within the plant flooring. The thermoforming lines are positioned such that the trim press is near or directly above the trench. A large portion of the trench is covered and protected with weight-bearing lids, except where openings are needed for dropping in the chopped-up scrap materials. The opening of the trench is placed at the trim press where the scrap will fall. The trench base is lined with a moving-belt system to carry the scrap skeleton remains into a centralized area, where it is collected for further processing: complete reclamation or recycling. The trench system not only requires a major and costly installation, but eliminates further plant layout modifications. It also works only with thermoforming machine lines which dump compatible scrap materials that can be mixed by the collecting system. If different (either chemically or by color) materials are mixed together, their reprocessibility and usefulness are reduced substantially.

Thermoforming plants that engage in relatively high volumes of production of different materials may use an independent granulating machine for each thermoforming line. The granulators are engineered in a low profile so that they can be fitted under the trim press just beneath the exiting scrap skeleton's path. As the scrap skeleton is fed into the granulators, the machines will chop them up into small chips of plastic. With the aid of an in-line blower, the chips can be blown through a larger-diameter pipe system to a further location. The pipe systems are coupled with the blowers to carry the reduced particles of the thermoplastic chips or fluff into a centralized reprocessing area designed for storing, handling, and even recycling of the scrap. The incoming pipes can be coupled together or segregated for the individual systems that match the incoming material. A free flow of the reduced plastic scrap through the pipes requires not only the blowers but also some negative pressurization of the storage silos. The venting, or depressurization, of storage silos must be made through proper dust collectors, which will prevent the finely powdered thermoplastic from escaping. For further precautions and for proper scrap storage management, both the thermoplastic material maker and storage equipment supplier should be consulted.

Although the granulators have independently operating functions from the trim press, they are electrically connected to it and their on/off controls are placed in the same control panel. The connections are often made such that a single button can activate or deactivate both pieces of equipment. Granulators may be purchased together with the trim press, although they may not be manufactured by the same company. Some granulating equipment builders specialize in this type of machinery, and equipment can be purchased directly from them. Again, it is the final responsibility of any thermoforming practitioner to select the most ideal granulators and attachments to fit and handle the specific trim press. The decision should be made with the particular thermoplastic material, plant layout, and scrap management factors in mind.

The in-line production system does not always end with the part-receiving table and the scrap granulator. In many instances, additional equipment is used for further work on the outcoming trimmed thermoformed products, all of which may be placed totally in-line. Many thermoformed parts are subjected to additional manufacturing steps to be made more complete and more valuable. Further downstream equipment can perform such operations as "lip rolling" to the flanges, spin welding, or applying multicolor prints to the product exteriors. The trimmed and stacked products are fed into the specific machines and automatically denested, processed, and nested back into stacks. All this is done in total synchronization with the production line system.

2. In-Line-Printing-and-Forming Posttrimming
   Thermoforming Machines

When printed images are required on the thermoforming articles for decorating or labeling purposes, the printing can be applied after the article is formed and trimmed. For postoperative printing the individual thermoformed part must be denested, printed, and dried before renesting can be attempted. The only disadvantage to this method is the extra time and effort needed to accomplish all those steps.

To gain a more precise production flow, the printing can be applied to roll stock. There are definite advantages to this method. First, this type of printing can be done on any customary printing equipment and system developed for continuous flat stock printing. Second, the drying of the printing ink can be managed due to the extra spacing between the printing machine and the thermoformer. In addition, the final drying can be accomplished by exposure to the heat of the long thermoforming oven. Most solvent- or water-based inks will be almost dry by the time the advancing web reaches the final heat station, and completely dry when the molding area is reached.

The web is printed after the roll is unwound and fed into the printing press. The printing press is a separate piece of equipment placed in-line between the unwind stand and the thermoformer. The three independent pieces of equipment, although placed in-line, are separated by a variable distance, in which slack loops are produced in the web. The slack loops are necessary to compensate for the different equipment speeds and movement modes; the printer provides continuous movement while the thermoformer runs in a reciprocating fashion.

The printing can be patterned in two distinguishable ways on the thermoformed article. The first type of printing is placed in a random fashion, while the second method is placed on a "registered form." The random print placement is actually created with numerous print images placed in repeated intervals in an orderly manner. However, when such a repeatedly printed sheet is subjected to the thermoforming procedure, the printed images will end up located on the thermoformed articles in a disorderly fashion. Out of the numerous print images, at least one or two patterns will be located to provide distinct legibility. The rest of the images will fall either somewhere else on the thermoformed article with various degrees of distortion or on the scrap area, where it may be partially or completely cut away in the trimming. No identical print image location can be found from one article to the next. In the thermoforming process, where the randomly located print images are used and a large number of images are trimmed away, the printing ink color will contaminate the recycled thermoplastic scrap. This is because to achieve satisfactory print legibility, bright, contrasting colored inks are used. Such intense color differences, even in a minute quantity, will cause discoloration when the scrap is recycled. For example, white thermoplastic having red or black print images cannot be expected to remain white after recycling. With random printing, color contamination effects often influence the color choice for the sheet material and the printing ink color. It is best to choose colors that will blend well together in the reprocessing of scrap without sacrificing print legibility. An orange-yellow thermoplastic sheet with red print or an aqua sheet with blue or yellow print will do just that. Where there is no choice but to use sheet materials and printing inks that alter the recycled material color, such materials can still be blended back with darker colors or sold to reprocessors, who may recycle them into a new, darker material. However, this is not the most cost-effective way to handle scrap.

Registered print images on thermoformed articles are much more difficult to achieve with the preprinting system. The printing itself is not at all difficult or different from random printing and the same printing procedures are used. The real task comes after printing, when the printed image must be made to locate and register exactly on a thermoformed article. The goal of this printing method is to locate the preprinted and heated sheet so that the print image will fall exactly where it is intended. With this setup, the thermoformer indexing capabilities are very critical and the machine should have additional registration and advancing mechanisms over those of the ordinary thermoformer. In this case the thermoformer has to have an electric-eye index registration sensor that will read a printed mark on the sheet advance and stop the indexing the moment it reaches the predetermined position. On the printing surface of the web, an additional marker is placed beside the actual print image, usually well outside the actual forming area. This mark is generally a dot or a sharply defined cross. The thermoformer indexing drive coupled with the electric-eye system will produce well-accelerated indexing with substantial speed reduction at the deceleration. During the last portions of the indexing motion, the forward sheet movement is slowed to a gentle creeping motion; as soon as the marker crosses under the electric eye, the web movement is stopped. This will ensure registration of the print images in the machine direction (MD). However, if any overadvance is experienced, the web cannot be drawn back, nor can the advancing mechanism be moved in the reverse direction. The registration will therefore be out of control.

Side-to-side registration is managed by the sheet guides at the web entrance to the thermoforming machine and maintained further by the sheet-transporting pin-chain system. It should be recognized that none of these types of registration systems can provide absolute registration alignment. In choosing registration-type printing, both the print image and its location on the thermoformed articles must be made with some error allowance, which could hide or at least make the registration discrepancies less noticeable.

With the use of preprinted web materials in the registered setup, all the printed areas will be located on the thermoformed product, causing no color contamination of the scrap skeleton and thus no reclamation sensitivity. However, any rejected production results should be managed according to the randomly located printed scrap procedures. If the rejects are segregated from the main stream of the bulk of the scrap skeleton and separately reprocessed, none of the scrap will be contaminated.

The use of a preprinted web in the thermoforming process will, one way or another, bring with it more chances of causing or finding errors. With this additional factor, the overall finished product and reject rate ratios will also be changed. The result will be higher rejection rates and a higher cost of production. Allowances in cost estimating should also be elevated to match production results when preprinting is incorporated into the thermoforming procedure.

C. Dedicated Thermoforming Machines

Most thermoforming machines are intended for general-purpose thermoforming, which is categorized by the specific type and size of machine. The purchaser who obtains the thermoforming machinery for a special manufacturing purpose never knows what other jobs may have to be produced on the equipment in the future. For this reason, most equipment is made with a wide range of capabilities. For example, all web-fed thermoforming machines are equipped with a full range of sheet advances. At the same time, their platen movement will dictate how deep an article can be formed on the particular machine. Any decrease in depth from the maximum number specified will be made within the scope of the machine. These general-purpose thermoforming machines may offer differing levels of adjustment from one manufacturer to another. Determining which criteria limits are more or less important to a specific thermoformer will make one supplier a better choice than another. It is well accepted that in today's highly competitive thermoforming industry, certain thermoforming machines may fit a particular thermoforming method or thermoplastic material and therefore equipment is used exclusively in a specialized way. Since particular models or brands are found to be more adaptable to a specific type of thermoforming, their specialization for running a single type of product makes them "dedicated" to that purpose. Three types of dedicated web-fed thermoforming systems come under this category: assigned, in-line extrusion, and complete pellet-to-product thermoforming systems.

1. Assigned Thermoforming Systems

Thermoforming machines in the "assigned" category can be any of the thermoforming machines discussed previously, including both sheet- and web-fed equipment. Any time a particular thermoforming process, with its specified forming method, is assigned to a thermoforming machine on a long-term basis, the production line becomes dedicated. Some thermoforming manufacturers assign a single article to each machine, and that machine may run continuous production of the particular item three shifts per day, six or even seven days a week. Although the specific machine has a wide range of manufacturing capabilities, the constant production of one type of article does not require that it use all of its available manufacturing modes and adjustment ranges. In this situation, the thermoforming machine is usually set up with an assigned mold, a specific advance stroke and platen strokes, and even a fixed trim criterion. The only adjustment called for after the setup is made is in sheet heating or cycle speed, to compensate for the different ambient and sheet temperatures or the thermoplastic sheet variations. Such an assigned thermoforming setup is most common in large thermoforming plants, where in their multi-line plant layout, each machine is used to manufacture a specific product. Scheduled production runs and sufficiently large warehousing facilities usually allow enough downtime for a preventive maintenance program and even occasional repairs.

Since the assigned thermoforming machine has a full range of adjustment capabilities, it can be pulled out of an assigned service and be refitted with other tooling and production modes. When changes are made from one production service to another, similarities in tooling and production modes can greatly reduce changeover times. At the same time, major tooling changes with entirely different production modes can be a time-consuming task. If this is encountered often, certain adaptations to the tooling and installations to the thermoformer platens can minimize the time involved in a tooling change. Production scheduling personnel are usually very selective as to which equipment in a multimachine plant receives assignment changes and which machine is more receptive for tooling or even production mode changes. It is often found that identical makes and brands of machines, even when placed side by side, do not function in the same way. This temperamental variation in equipment is even more evident with thermoforming machines than with most other types of machinery. Two identical pieces of thermoformer equipment will probably have to have different settings when the same tooling is used and the same parts produced. This variation between "identical" equipment becomes evident when the machines are placed in close proximity and subjected to the same tool setups. The differences assumed by the machines can be so severe that a thermoforming machine would simply not be capable of running one type of tooling and production mode (almost as if "refusing" it) while running perfectly well with other tooling. It is no wonder that each piece of equipment is often viewed as having its own personality.

2. In-Line Extrusion and Thermoforming Lines

In this group of thermoforming systems, the thermoforming machine, instead of receiving a thermoplastic web unwound from roll stock, is placed in-line with an extruder. The extruder converts the pelletized thermoplastic resin into a continuous web, which is fed directly through the sizing roller stack and into the thermoformer machine. These in-line setups offer definite advantages not only for a better in-line production arrangement but also for reducing energy consumption. The extruded web retains large portions of its extruded heat going through the sizing roller stack and entering the thermoformer. The thermoformer heating ovens need only heat the web a small amount or not at all in order to perform the thermoforming. This small amount of heating is dependent on the thermoplastic material and also on the type of web-handling equipment used. There are different makes of systems that incorporate either a slight- or nonreheating approach, and their process systems are covered by patented features. Each production system, although somewhat varied in its relationship to thermoforming, falls under the same thermoforming rules and thermoforming criteria as always.

The key element in the extruder-thermoformer in-line production setup is that the two, or in the case of posttrimming machines, three pieces of equipment must run simultaneously in a synchronized manner. Any breakdown or out-of-specification production will affect the equipment further downstream and can force a shutdown of the whole production line. It is also obvious that to achieve economically feasible production results, the equipment that is tied together must have a longer production time and not be subjected to frequent tooling or material, material thickness, or web-width changes. Any changeover to other tooling might demand numerous alterations in equipment setup and result in many interacting production mode changes. A change in the web width in an extruder is a major task, while a similar change in the thermoformer can usually be accomplished quite easily.

Most web-fed thermoforming machines come with adjustable chain-rail systems that allow the use of a narrower web width than the maximum widths specified. The minimum usable web is usually defined by the chain-carrying sprocketed shafts that tie the left and right chain loops together. The maximum and minimum acceptable web widths are usually a part of the specific thermoforming machine's specifications. The ability to adjust the transport-chain system to narrower-width webs is a very important element of web-fed thermoforming machines. This system allows setting up of the thermoforming machine with smaller tooling that does not require a maximum web width and therefore offers material savings by reducing the size of the edge trim scrap. It is, of course, unreasonable to expect to use large thermoforming equipment to produce a small article running in a one-up mold configuration. Web widths in a specific thermoforming machine can be reduced only within a certain limited range.

The chain-rail adjustment system can be adapted to purposes other than just making the thermoformer accept different web widths. The scheme is also used in a special and somewhat out-of-the-ordinary way which relates to the fact that some thermoplastic sheet materials tend to sag excessively. Some, if not all, of the sag can be taken up by adjusting the chain rails so that they no longer run parallel to each other but open up to run wider apart at the molding end of the machine. With the outwardly moving chain-rail system, the resulting sag in the heat tunnel will be taken up by the time the web indexes into the molding area. Of course, only a limited amount of tapered adjustment can be obtained safely without causing abnormal wear to the chain links and sprockets.

Another use is found in the reverse adjustment, when the chain rails are brought to travel inwardly toward the molding area. In this situation only oriented (biaxially prestretched) thermoplastic web material is used. The inwardly moving chain rails will allow the sheet material to give up some of its orientation and as a result, increase the original sheet thickness. For example, a 0.015-in. oriented polystyrene sheet fed into the chain rails can display a 0.020-in. thickness at the forming area. A partial loss of its orientation should be expected and a display of higher brittleness can also be anticipated. Running the chain-rail system out of parallel is very rare. However, knowledge of the trick can come in handy on the few occasions when it is needed.

3. Complete Pellet-to-Product Thermoforming Systems

For years it has been a dream to have a manufacturing setup that will begin the production of articles just as in all other plastic processes, with thermoplastic pellets, and continue nonstop to produce the finished product. The scrap in this process is automatically recycled and nothing leaves the plant except the final packaged thermoformed products. Today, this dream is a reality; thermoforming manufacturing lines like this are currently offered by several companies. The complete line can be built by a single manufacturing concern or can be put together from different sources. It is of utmost importance that when these machines are placed in-line with one another, they run completely in unison without production flow interruption. The main benefit of a complete process system is the minimum labor required to oversee and run the machinery. The entire equipment line can be run by one person and can be coupled with a computerized monitoring system that can self-monitor many aspects of the entire line. Use of pelletized thermoplastic raw material permits a larger quantity of shipments, with on-site storage possibilities that allow the user to purchase raw material at the lowest available price. In addition to these benefits, the in-line system, with extruder-thermoformer machines, can reduce the power consumption of the line by 30% or more. The reduction in electric power usage is substantial, especially with extensive production runs. Most of the power savings come from the fact that the extruded web, as it is made, can retain enough of its original heat, and therefore only a minute quantity, if any, of reheating is required for the thermoforming process. The normal type of roller stack which receives the molten hot web from the extruder does not have to cool the web completely, only its surface. The roller stack basically conditions the web by regulating its heat level after extrusion, thus producing the most advantageous results for downstream thermoforming. The amount of cooling, gauge sizing, and true conditioning that a thermoplastic web receives are entirely dependent on the thermoplastic material makeup. Some thermoplastic materials require only a small amount of conditioning, whereas others demand a far greater heat reduction. Each type of thermoplastic material is governed by different conditioning rules that must be satisfied to obtain the best thermoforming results. When immediate thermoforming is performed in a newly made web, the roller stack is built into the back end of the thermoforming machine, leaving room for only a very short oven tunnel between the

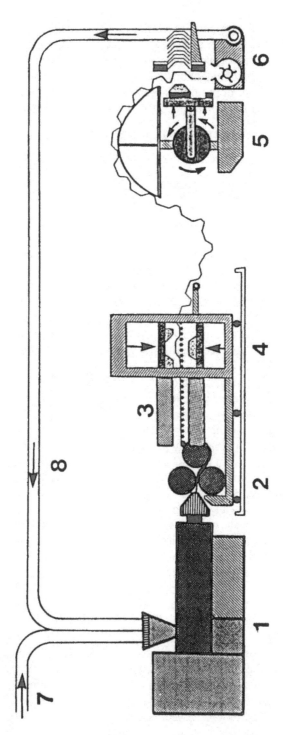

Figure 74 Pellet-to-product thermoforming systems: (1) extruder; (2) conditioning rollers; (3) short heating tunnel; (4) forming station; (5) trim press; (6) scrap granulator and blower; (7) virgin resin intake; (8) scrap return line.

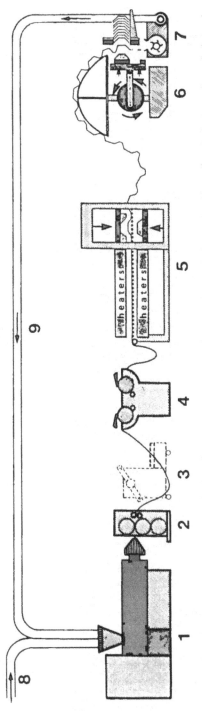

Figure 75 Pellet-to-product with preprinting thermoforming system: (1) extruder; (2) roller stack; (3) optional winder and unwind stand; (4) printer; (5) thermoforming machine; (6) trim press; (7) scrap granulator and blower; (8) virgin resin intake; (9) scrap return line.

roller stack and the forming area. This short oven tunnel may have one or two sheet indexing stops for elevating the heat of the thermoplastic web if it should be needed. If no additional heating is required, the heat tunnel temperature can be lowered to a point where its only purpose would be to act as a heat protector. Figure 74 shows the pellet-to-product thermoforming equipment setup.

In cases where a printer is incorporated into the complete pellet-to-product thermoforming system, the roller stack must almost completely cool down the extruded web. A well-cooled and firm surface of the plastic web will accept printing impressions much better if it has no undue distortion or surface alteration. The fluid coolant circulated through the rolls of the roller stack allows all levels of temperature control and with it, the proper degrees of cooling effects on the outcoming thermoplastic web. The actual temperature reduction by the rollers is governed by three factors: roller temperature, which is regulated by the circulated fluids; roller turning speed, which is tied to the extruder output rate; and the diameter of the rollers, which controls the length of time during which surface contact is made. Since these three factors can be adjusted only within a given range, a thermoplastic material or tooling change may force a change in the roller-stack specifications as well. A complete pellet-to-product thermoforming system combined with a preprinting press is illustrated in Figure 75.

A complete production-line system such as this, even with its many economic advantages, can still have a few shortcomings. One of the major obstacles to this thermoforming system is its price, which has kept many prospective buyers away from this type of setup. A complete pellet-to-product system for most larger thermoformed product-size capabilities can cost a substantial amount; when multiline productions are considered, the heavy financial investment will be duplicated by each line. In addition to the high cost, the complicated connections between the individual components of the synchronized system do not readily allow frequent tooling changeover. Each changeover, if it is not within the same specification criteria (product size, depth of draw, number of ups, web material width and thickness), will demand major alterations to equipment functions. Depending on how elaborate a changeover is made, such modification in the running conditions of the equipment may require several days of setup work time. The retuning and final adjustments are usually made with the subsequent equipment as it is turned on and as the machines come up to running conditions. In the interim, their production results will be strictly out of specification. These enormous changeover difficulties strongly suggest that this type of equipment should be used with one tooling and for an extended period.

The last disappointment encountered in the use of the complete thermoforming process line is the way in which the breakdown of one equipment component will force the entire system to be shut down. Although today's engineers are capable of creating equipment that does not often break down or require long downtime for preventive maintenance, the line will, from time to time, experience production stoppages. Today's production equipment is also equipped with electronic microprocessor line-condition monitors with self-diagnostic features that may not prevent a final breakdown but can forewarn the operator that it is coming. An electronic monitoring system can reduce the chances of further expensive damage or even possible destruction of an entire line component. It can minimize the chances of unexpected breakdown, which always comes at the most inopportune time.

Some production-line users have been sufficiently irritated with component machinery breakdowns that they have installed optional roll winder and unwind stands between the extruder and the thermoforming machine. This equipment is placed on steel tracks and can be rolled into location if one of the two main pieces of equipment is forced to stop. With this special setup, if the thermoformer becomes nonfunctional, the extruder can still produce the extruded web. The web is simply rolled up and placed into storage for use later. If the extruder should go on the blink, the unwind stand is pushed into the line of operation and the stored roll stock is used up. Naturally, this is a poor way to maintain production and product supply, but it does work when needed.

There is one more variation of this pellet-to-product thermoforming system—complete turnkey operation offered by one thermoforming machine manufacturer. This unique type of pellet-to-product system is aimed at running mostly shallow thermoformed articles, such as lids for various containers. The equipment is kept to the minimum floor space requirements and its production flow, although remaining in-line, changes direction 90° at the extruder die. The most unique aspect of this system is its use of an annular blown film-type die that extrudes a tubular-shaped film. The cylindrical film is laminated and squeezed together by two chilled nip rollers, creating a single flat, narrow web approximately 5 in. wide. Since the tubular extrudate is squeezed together while hot, it creates a permanent bond between the two layers. The squeezed-down extruded web receives an ideal amount of molecular orientation when a bubble is blown into the film between the annular die and the nip rollers. The larger the bubble, the greater the molecular orientation in the web will be. The flatly squeezed web will retain enough of its extruded heat to be subjectable to the thermoforming process. In this particular pellet-to-product thermoforming machine, the molding procedure is accomplished by the rotary mold arrangement described earlier (Section III.E of Chapter 3). Although this pellet-to-product arrangement is designed for lid making, there is a larger version of the machine that can produce $1^{1}/_{2}$-in.-deep parts without the aid of a plug assist.

It is incomprehensible that such a unique and ingenious machine design has experienced declining interest before it has reached its true potential. Although this specific machinery design may be experiencing some decline, not proving to be popular at this time, its engineering concepts may be revitalized and readapted at some point in the future.

Pellet-to-product thermoforming systems have several patented components or segments. Each patented feature is unique to the particular machine supplier and is offered only by that supplier. It is advisable to investigate these features and fully understand their specific benefits and shortcomings. Each particular thermoformed product, with its own thermoplastic material and required output volume, creates a specific situation for which one type of equipment is more applicable than another. There are several makes of equipment for sale today with tempting advertisements covering all their unique features. But nothing can beat the actual "tried-and-tested" production results.

D. Customized Thermoforming Machines

There are thermoforming situations where the thermoforming procedures are unique or a single type of product is being made with no plans to change the production or the specific product. For these permanent situations, thermoforming equipment that provides all-purpose functions may not be necessary or may even be impractical. The customization of thermoforming equipment, whether it is ordered that way or specially altered at the plant, creates a "committed" piece of equipment. What this means is that the piece of machinery can no longer be used for anything else but that which it is now intended to make. It is simple to understand that sheet-fed thermoforming equipment is committed to the use of sheeted thermoplastic and cannot accept web thermoplastic, and vice versa. In the same way, specific thermoforming equipment can be specialized even further. For example, an all-purpose thermoforming machine with a 0- to 4-in. platen movement, when used for nothing other than shallow $^3/_4$- to 1-in. depth-of-product thermoforming, can be refitted with short-stroke cylinders. This will eliminate unnecessary large and costly cylinders that in the long run can jeopardize all other functions of the machine. In this example the nonessential volume of the cylinders can act as moisture condensers, trapping unknown quantities of water, and causing corrosion buildup on the unused cylinder walls and even decreasing the cylinder stroke travel. On the other hand, the shorter-stroked cylinders will commit the reworked thermoforming machine to shallow product making only. This is not a sacrifice since the production plant is only engaged in specific shallow-drawn product making. In fact, in such a case it is highly desirable to alter the all-purpose thermoforming machines to their committed thermoforming task operation.

Thermoforming equipment often receives major modifications that can be so extensive as to employ cutting and welding in the change. It is also a well-known fact that the thermoforming industry's more specialized manufacturers often modify equipment for their own specific use. Such modifications are kept secret and not even revealed to the original equipment builder, to keep such customization ideas from being leaked to the competition. Plant-visiting policies often restrict outside people such as vendors and sale personnel from getting near such equipment. When the modified equipment is sold on the used machinery market, it is converted back to its original shape or condition when possible. If it is not economically feasible to do so, the equipment is cut up and sold as scrap iron. The customization of a thermoforming machine can be so elaborate and extensive that it can hardly be recognized or traced back to the original manufacturer. Major equipment modifications can confuse the issue of machine liability. Equipment modifications do not weaken the liability of the original equipment manufacturer. Equipment liability decisions by the courts have not favored the original equipment builders, even when machines have been altered. This trend will probably continue unless new legislation is passed.

One of the newest specialized customizations of an in-line thermoforming machine is found when two separate forming stations are installed in-line, right after one another. The purpose of duplication in the molding stations is for very same reasons as discussed in Section II.C.5, Rotary Thermoformers: Four-Station Rotary Formers with Dual Forming Stations, in Chapter 5 (see Figure 61). When forming a flat shaped sheet in the first molding station with a temperature controlled but elevated temperature mold, a preformed shape will lend itself better for the final shaping by the second mold. The second mold will have an advantage in reshaping an already prestretched material which is no longer in a flat form, but in a preformed shape. The second forming station will finalize the shape to be made and at the same time provide the necessary cooling of the plastic part. This two-stage forming done one after the other, can and will allow maximum depth-of-draw ratios with the utmost material distribution that cannot be accomplished by the customary single station molding. There can be arguments raised for or against the use of this type of double forming stations; however, the results, if it is produced right, cannot be debated.

Additional setup would also apply to this customizing method, when two rotary thermoforming machines are set up in such a particular manner, that the two machines are overlapping each other at their forming stations. It could be also expressed in such a way, that the two machines have a common molding station. The machines arranged in such a manner that rotational movement overlaps at the forming station, as the rings of the Olympic Circles overlap each other. Naturally one machine is perched up higher while the other machine rotates at a lower level, so their clamp-frame rotation can not interfere with one another. Both machines are loaded with the thermoplastic sheet and index into their respective ovens. The sheet loading is repeated as it would be with any ordinary rotary thermoforming machines. When the sheets are ready for forming they will index into the common forming area. At that time, the platens are activated and the top mold will drop to the top of the sheet line while the bottom mold lifts to the bottom sheet. As the two molds close, it can squeeze the two sheets together and it will produce a twin-sheet forming procedure. See Section III.C.4, Special Purpose Molds: Twin-Sheet-Forming Molds, in Chapter 5. Naturally some modifications of the clamp frames opening has to be made for easing of the finished part removal.

When two machines are overlapped with this type of machinery setup, one perplexity is found. If the machines require a pit to be cut into the flooring for their platen retraction, and if this setup is no longer desired as a permanent installation, such pits have to be filled. Such difficulties may not be easily overcome and repeated locating and relocating machinery for this type of forming procedure may not be wise.

E. Form-Fill-and-Seal Equipment

This group of machinery does not really qualify as full-fledged thermoforming equipment. It is more precisely categorized as "packaging" equipment. However, in the course of their production runs, these machines incorporate the thermoforming of the package cavities and therefore deserve inclusion in a comprehensive discussion of thermoforming equipment. The making of packaged goods usually starts with a roll of thermoplastic web fed into form-fill-and-seal equipment. The infeeding thermoplastic web is subjected to heating and then forming in a continuous way. The segment of the web thermoforming is done in the same manner as with any web-fed thermoforming, and its outcome is governed by the same rules. The same variables in the use of thermoplastic materials and their reactions to the heating and forming procedures will have comparable satisfactory or unsatisfactory results. The depth-of-draw ratio limitations and the use of plug-assisted thermoforming will offer the same complexities as will any of the ordinary thermoforming procedures. So the thermoforming aspects of this packaging machine are basically the same.

The purpose and benefit of using form-fill-and-seal equipment are established by the fact that a single piece of equipment provides all task-performing responsibilities: creating the containers, filling them, sealing them with a cover film, and cutting them into individual unit packages. This proves to be a definite advantage over the usual group of multicomponent equipment lines, which perform most of the tasks through separate functions one after the other. Many packaging contractors are now using this single-line packaging arrangement with satisfaction. Most form-fill-and-seal equipment makers are well versed not only in the thermoforming segment of their lines but all other downstream packaging functions as well. Their equipment has been tried and proven throughout the industry.

Two unique features common to all form-fill-and-seal equipment make them somewhat different from ordinary thermoforming machines. The first is the narrower web widths that are generally used. Form-fill-and-seal equipment is not designed to work with the extra-wide webs of thermoplastic materials. The narrower widths allow fewer cavities across the web, thus avoiding a numerically impossible product-filling or product-loading predicament. Too many cavities at once can create havoc with most filling operations. The second unique feature is that after filling, the entire web is heat sealed with a film cover. This sealing procedure is made immediately after the forming and filling are completed, and all of this is done in web form. Package separation and cutting are done after sealing the cover film, with ample time for cooling and setting the heat seals. Some of the equipment not only handles all the form-fill-sealing functions but even ejects and stacks the products in an organized manner after cutting.

The actual forming mold cavities and the product-transporting receptacles can be made as separate equipment components. However, the most common practice is to use a series of traveling-belt-mounted cavities that serve a dual purpose. The cavities are first used as a mold into which the heated thermoplastic sheet is drawn. As they advance further downstream, the cavities are filled with the intended content, which can be liquid or solid. The content can run the gamut of product forms, from food preparations to medical or hardware items. Naturally, each product-filling operation will have to be approached according to the individual physical form or characteristic of the product. Liquids are usually injected into the formed cavities in a premeasurable volume and with great care to avoid spillage. Any material spilled onto the flange area can contaminate and inhibit proper heat sealing between the base receptacle and the cover film. The cover film used to seal the thermoformed units can be made of thermoplastic sheet material, foil, or even paper. These cover materials, each with an entirely different material makeup, have a coating of heat-activated sealing layer on the container side. This special coating layer will aid in adhesion of the film to the container flanges and at the same time, offer ease of opening and peel-off when needed. The form-fill-and-seal production line can easily be adapted to use preprinted cover stock and to make registered indexing over the formed and filled cavities, thus providing accurate and correct labeling. Since most of the finished product is made in one production line at the same time as it is filled, the date coding of such packages can be routinely applied at the same time. Full-function packaging equipment like this can be placed into "clean room" or even sterile environments. This makes it ideal for food or medical packaging uses.

The cavity configurations that can be produced on this type of specialized equipment have been greatly improved. Flat and shallow package configurations are still the most widely used packaging shapes. However, deep-drawn single- or multicavity-configured containers are easily produced today with excellent wall thickness distributions. Containers up to 6 in. deep can be produced successfully without major difficulties arising on the rest of the production line.

This type of production line can also be fitted with computerized diagnostic systems to troubleshoot processing failures and forewarn the operator of imminent breakdowns. The diagnostic systems can also be tied into the manufacturer's main diagnostic computer system so that most process analysis is via telephone.

One of the debatable points about the use of form-fill-and-seal equipment versus the conventional separate thermoforming, filling, and sealing machine setup has to do with its combined output rates. It is true that the forming and trimming speed of separate machine lines cannot be challenged by combination equipment. The web width alone can greatly reduce productivity, which is even further diminished by the subsequent product-content-filling and cover-film-sealing procedures. (The product filling is actually the main cause of the lower production results.) But conventional thermoformer machines are only equipped to produce the container. For the filling and sealing steps, additional equipment must be added, and such tasks may, in fact, prove to be more of an effort and more time consuming than the use of combination equipment. Each separate handling of formed parts can damage a large number of them, rendering more product waste. Precise justification of form-fill-and-seal equipment versus conventional machinery is entirely up to the packaging contractor. A thoroughly calculated production estimate can favor one type of machine setup over the other when comparing the same packaging, but in a different packaging situation, the conclusion may be different.

IV. CORRUGATED PLASTIC TUBING AND PIPE MACHINES

Until now, no one has considered classifying plastic tubing and pipe manufacturing methods combined together or under the thermoforming process. But I choose to include this manufacturing method to be listed in this section, under the machine heading. Considering how this specific equipment is used to make corrugated plastic tubing or pipes, we can recognize that this process is almost identical to thermoforming (see Figure 76).

The only prevailing difference is that the thermoforming process always starts with a flat sheet or web formed of configured plastic. The corrugated

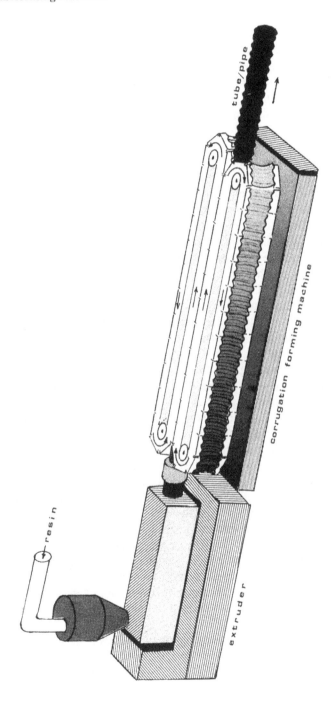

Figure 76 Corrugated tubing or pipe-making machinery.

plastic tubing or pipe start with a tubular shaped extrusion instead. This tubular shaped plastic extrusion can be made by any of the popular extrusion setups and can be produced by a variety of thermoplastic resins and preferred dimensions and wall thicknesses. Among the many choices of materials, polyethylene is the most popular material that is generally used in this type of product. This is mostly due to availability and reasonable pricing, plus it will usually have the best performance levels for the particular product applications. Polyethylene materials have a wide range of color and density, and have a rather flexible to a most rigid characteristic matrix.

Usually the specified polyethylene material is extruded in a tubular shape through an annular die. The needed diameter and its wall thickness are controlled to a predetermined specification. Some of the tubing can be made as small as .25 in.–3 in. in diameter, while the heavier pipes may be made from 3 in. to as big as 72 in. in diameter. The tubing or pipe extrusion is made as other products of this type are made and the same rules and techniques govern the procedure as the smooth-walled tubing or pipe manufacturing is done. The method of corrugation of the tubing and pipes is comparable to thermoforming. As the extrusion is made, the tubular shaped plastic is still hot and formable; instead of allowing it to cool, it is reformed to a corrugated configuration. Naturally, everything is done in a continuous in-line production. The hot plastic tubing or pipe is fed right from the extrusion die into a moving mold structure that contains the corrugated configuration profile.

The mold is made up of a series of mating mold halves that wraps around the newly extruded tubing or pipe. The molds are made up in a series of configurations, traveling with the extrusion. As one set of molds comes in contact with the extruded material, a new set of molds will move into its place. These moving mold structures are fixed into continuous loops, which all travel at the same speed as each other and the plastic tubing or pipes. Each time a mold comes into contact with the plastic the other half of the mold mates with it, forming a tunnel-like formation and as it is done consecutively, a long, lengthy mold tunnel is created. Naturally, at the end of the moving mold tunnel the concluded mold halves are separated and they disconnect from each other and move into the return travel mode to be engaged again at the beginning of the mold loop. This way a perpetual mold formation is accomplished as long as the two loops of mold halves and the extruded material is traveling at the same speed, and the same direction. The two opposing mold halves as they come together, become the tunnel-like formation contained on the inside of the corrugated design pattern.

These design patterns may have tightly or sparsely made concentric configurations or a continuous threadlike design pattern. The corrugation pattern could be rounded or made with a sharp 90° angle or any angle of transition from one corrugation to the next corrugation. It even can have a combination

of rounded and sharp transition angles from one corrugation to the next corrugation. The design, can even be made irregular, making a corrugated pattern, then leaving a portion of the plastic uncorrugated, smooth as it was extruded or in some cases an extra larger bulge can be formed into the plastic where the product design is required.

The corrugation or the given pattern is formed into the plastic tubing or pipe using a vacuum that is applied through the molds via vacuum holes drilled into them. The vacuum holes are connected to the vacuum system via specially designed plumbing. The connection should maintain good access to the vacuum. Mold fixtures that do not permit any vacuum leakage to the atmosphere and do not interfere with the travel of the molds are necessary. This connection should be made air tight and they should allow the molds to slide by, without interruption to the vacuum. As the vacuum is introduced to the molds, the hot and soft plastic tubing or pipe will be pulled (sucked) against the mold surfaces that has the corrugated configurations. Enough air will be present in the tubular center space to permit this forming to take place. As the forming is accomplished and the mold continues to move down stream of the molding tunnel, the vacuum is disengaged. Cooling of the plastic will be initiated and the plastic tubing or pipe will be set into its permanent shape. As all this is accomplished in the continuous travel in the molding tunnel, at end of the line the molds are separated and pivot into the return mode. The two mold halves loop, act very much like two caterpillar chain tracks, pressed at the molding tunnel area against each other. On some installations the two opposing mold chain loops are positioned on top of each other, or the other system may have two horizontal tracks facing each other. Both system accomplish the same purpose, as long as the mold's unification is made perfectly, and the mold halves are mated uniformly. This is a continuous process and can accommodate an unlimited length of tubing or pipe. But for ease of handling, the length is predetermined and the tubing or pipe is usually cut into a manageable length. Its manageable length depends on the application and the end use. Corrugated tubing is found mainly in vacuum cleaner hoses or swimming pool vacuum hoses. There are other uses for this type of tubing which would be too numerous to mention. Anywhere, where greater flexibility is needed, this type of corrugated tubing can be more useful than straight-walled tubing.

Corrugated plastic pipe has found a great niche in drainage or irrigation pipe usage as well, where the corrugated feature offers greater strength than a straight-walled configuration. Larger dimension pipes are made as large as 3–6 ft in diameter with a sufficient wall thickness configuration. When buried in the soil, it must be able to support a considerable weight of soil above, without collapsing. Being made of a plastic material, there is little deterioration due to corrosive conditions. It is not too rigid, like metal, ceramic or other existing materials; therefore it is favored in many soil or waste drainage installa-

Chapter 4

tions. A lighter wall-thickness version of this pipe can be used for transferring water or other liquids at low pressure from one place to the other, or even can be used for air conditioning ducts.

Forming corrugations into the plastic tubing or pipes follows the same criteria as the thermoforming process. The size and the number of vacuum holes is just as critical in this process as it is in thermoforming. The oversized holes will show up on the plastic tubing or pipe surfaces as small buttonlike formations. The level of vacuum force is equally important; if there is a leak or not enough vacuum, the corrugations will not form in full detail. In this process the problem or difficulty is greater because the molds are constantly traveling and the chances are more likely to develop an unwanted vacuum leakage through the long path of the mold tunnel.

The task of forming the corrugated pattern into the plastic tubing or pipe configuration is not limited to vacuum forming. It can be accomplished by using pressurized air as well. The only criteria in using pressurized air is to accomplish a good airtight seal. Air pressure introduced at the extrusion through the middle of the extrusion die must be captured by sealing off the previously extruded end of the corrugated tubing or pipe. It is much easier with the smaller sized tubing style. As the smaller tubing is made the beginning of the tubing portion is sealed off promptly and that will block the escape of pressurized air. Then the corrugated tubing is simply wound up in a coil arrangement. When the roll is full a cut is made and the new roll will start with a "pinch-off" seal to continue the process of winding.

The convenience of using pressurized air is currently not available for corrugated pipe making; therefore most manufacturers rely on the vacuum forming method. Unless someone is capable of accomplishing an economically feasible air pressure seal, pipes will continue to be made by the existing vacuum process method.

5
Molds for Thermoforming

I. THE BASIC CONCEPT OF MOLDS

To produce thermoformed articles, some type of tooling implementation is required, regardless of article size or quantity. To carry out most production tasks, complete tooling is needed. Complete tooling can have a wide range of components, including special sheet feed and holding units, trim dies, and related stacking and part-receiving magazines. The two major components of the tooling are the mold and the trim die. In some instances the tooling consists of just a mold. However, it is rare that the thermoformed article does not undergo some type of trimming or finishing after the forming cycles. Most sheet-fed thermoforming operations rely on a secondary trimming method, and such trimming procedures can be produced by any of the tools and techniques described earlier. In some instances the two major components, molding and trimming, are combined into a single piece of tooling that will provide both operating functions. This particular forming and trimming procedure was discussed in Section II.C, Chapter 3.

When these main components of the tooling (mold and trim die) are ordered for a thermoforming job, they are usually engineered and built together as a complete tooling package by the same tool and die shop. In this case, the toolmaker takes full responsibility for the entire tooling; this will include the mechanical fit of components into the specified equipment and their interrelated measurement and function. If the two components are purchased separately, their compatibility becomes the responsibility of the thermoforming practitioner. Tooling obtained either way must function well together to achieve satisfactory thermoforming results.

The mold can easily be considered the heart or main component not only of the tooling but of the entire thermoforming process. The mold forms a heated thermoplastic sheet into a product shape, converting a two-dimensional sheet into three-dimensional formed articles. The mold is the instrument through which the formed plastic will attain its shape. In this shaping of the thermoplastic sheet, the mold must perform well in order to transfer its details as fully as possible and allow the perfectly detailed article to be removed without damage. The mold itself can spell the difference between perfection or imperfection in its own detail transfer and thus between satisfactory results or poor-quality products.

A. Tooling-Up from Idea Stage to Finished Mold

Most product ideas, whatever the source, often begin as humble sketches or drawings. From the sketches, refinements and more details will be developed that will characterize the concept or the shape of the basic thought. Often, when the initial sketches show enough promise, fully developed drawings and possible blueprint designs are produced to provide different viewing angles of the product idea. After the drawings are produced, it is easy to convert them to actual physical shape. Bona fide pattern makers or people who are familiar with pattern-making procedures can easily read the drawings and develop a three-dimensional structure, following the shapes and measurements described. Building a pattern usually follows customary pattern-making procedures and materials. The three-dimensional physical shape is often not made of a single piece or block of pattern material (block of wood) but is assembled from segments or components. In the pattern-making procedure, the maker often dismantles the original form into its conceptual components or segments, which can be produced more easily on pattern-making equipment than a one-piece configuration. The pieces are assembled and fitted with glue or other fastening methods to create the final mold configuration. Pattern-making skills are not extremely difficult to learn. However, some woodworking knowledge is helpful, combined with some understanding of the various configurations and their development. Some people have exceptional talent at visualizing shapes and configurations easily; and can, in no time at all, convert a blueprint into physical form. Pattern shops with highly trained, skillful pattern makers can handle any type of pattern-making endeavor. Smaller, in-house pattern facilities that serve only a single thermoforming plant are limited to the specific types of patterns for which they are set up. Such limited, specialized skills can easily be learned and practiced by internally trained personnel within the shop parameters.

Prior to the initial pattern-making procedures, the pattern maker has to analyze the job as to which form is easiest to create. He must first decide whether it is easier to begin pattern production with a male or a female mold. The intended final configuration of the mold will help determine whether a male or a female mold will be more functional for the thermoforming process. The choice of the best form is usually guided by the forming technique and the quality of material distribution chosen. To make such a decision, the various molding methods and their descriptions in Chapter 3 should be consulted. In addition to the final mold shape choice, the prominent surface or sides of the resulting forming have to be determined. The correct selection is most important because it will decide which side of the formed article has to display the best details and which side will be allowed to carry the expected "detail loss" (also covered in Section IV.A of Chapter 3).

When both determinations have been made for the final mold configurations, the next step is to designate the way the pattern should be made for the particular mold configuration. Although the pattern choice for a mold is set, it is possible that a reversed-shape pattern will be easier to create. For example, when the aim is to produce a female mold with straight cylindrical or tapered cylindrical cavities, instead of trying to carve them accurately out of a block of pattern-making material such as wood, it is easier simply to turn the block on a lathe into a cylindrical shape and then split it into halves, which are then mounted onto a board backing. After gluing and pattern preparation, a reversed pattern can be made from the original pattern work, to create a pattern with the initially intended female configuration.

When dimensional criteria dictate precise measurement of the plastic article, the article must be made on accurately sized molds. This is where the perfect details of the forming results will come into play. To obtain the most accurate details and dimensions on at least one side of the plastic, mold surface contact becomes most important. It is at these contact points along the surface that the plastic will pick up its details, adopting the intended shape. It is of the utmost importance that this criterion be followed throughout the mold-making procedure, all the way back to the original pattern. Allowance for material thickness and correct surface contact must be planned for the final version of the mold.

B. Shrinkage Allowances

Besides correct placement of material thickness in the pattern and tool making, one additional criterion has to be satisfied. To arrive at the intended thermoformed product size, allowance for the natural phenomenon of shrinkage must be incorporated into the pattern and mold sizes. Thermoplastic materials expand and shrink when subjected to temperature changes and will therefore experience shrinkage after cooling. For the control of size reduction after forming, the molds are purposely built oversized to compensate for shrinkage. Once the maker is armed with the correct shrinkage rate for the particular plastic, building oversized molds will eliminate most surprises from resulting shrinkage. This is particularly important when producing mating components that must fit together correctly; for example, containers and their lids must have a snug, snap-action fit to serve their purpose.

Each particular type of plastic displays its own shrinkage rates, which should be specified by the specific resin manufacturer or supplier. The shrinkage rate and its range depend largely on the type of resin and its molecular weight or blend. For general purposes, a wide range of shrinkage rate can be used. However, for critical dimension controls, those numbers should be narrowed down to the actual shrinkage values. Table 1 provides the wide shrinkage ranges of some of the most popular thermoplastic materials used in thermoforming.

The shrinkage rates for a critically important fit should be specified and when material changes are anticipated, the material shrinkage levels must be matched between the old and new resin supplies. Discrepancies among shrinkage rates are one of the main causes of improper fitting results between previously well-fitted parts and put a damper on quick, hastily made material substitution. It should be noted that some thermoplastic materials can have such a great shrinkage change that it may affect the product fill capabilities or container capacity of a thermoformed receptacle. Such variations in shrinkage must be investigated before full-scale production is attempted.

In the pattern-making procedure, the pattern itself usually represents an intermediate form of tooling where the pattern should be converted into a more permanent type of tooling. Patterns are generally made of wood, such as gelutong, mahogany, basswood, or sugar pine. These high-quality, uniform-

Table 1 Thermoplastic Shrinkage Rates

Material	Shrinkage range (in.)
ABS	0.004–0.009
Polyester	0.015–0.025
Polyethylene	0.015–0.050
Polypropylene	0.010–0.025
Polystyrene	0.002–0.006
PVC	0.001–0.006

grained woods are ideal to work with for shaping and piecing together a pattern or mold form. Other woods are just as usable, but due to their grain irregularities, the maker must be extremely selective as to what portions of the lumber are usable. Using a choice wood quality usually provides some saving in the labor phase of pattern-making projects. Local wood material availability is also a guiding factor as to the selection of one type of wood over another.

The wood pattern by itself can be used as a temporary or sample-making mold. Using wood as a mold material can only be considered temporary. The wood itself does not keep its original dimensions accurately. Ambient humidity conditions may decrease or increase the size of the wood. Since a pattern or wood mold is made up from several pieces of different grain-line direction, each piece may acquire a change in its own dimension, causing "stepdown" or overhangs among the pieced-together components. For that reason, wood molds should be considered only interim structures. It is also true that wood materials provide poor heat transfer qualities. In repeated thermoforming use, each surface contact between the heated thermoplastic sheet will eventually raise the wood mold's temperature to the same heat level as that of the plastic sheet itself. Since the mold's retained heat does not allow sufficient cooling for the plastic to set, the mold will not render any useful quality for further thermoforming past a certain point. At this point, the operation has to be shut down until the heat in the wood mold dissipates. This poor heat transfer quality will never allow reasonably fast cooling times, thus causing long periods of waiting.

When wood patterns are made for temporary molding purposes only, they are made larger than actual size to compensate for the plastic's shrinkage. When the patterns are used for the purpose of casting to produce the permanent metal molds, additional shrinkage factors must be incorporated. This additional shrinkage factor is built into the patterns to account for the shrinkage that takes place when the molten metal cools down. Pattern makers often use "shrink rules" for this purpose. These special rulers resemble standard measuring rulers except that a shrinkage factor is built into their calibration. For example, with styrene materials together with aluminum tool casting, a $^7/_{32}$-in. shrink ruler is used. This ruler has $^7/_{32}$ in. of shrinkage per foot incorporated into its calibration. The $^7/_{32}$-in. shrinkage designation combines a $^1/_{16}$-in. allowance for the styrene shrinkage and a $^5/_{32}$-in. allowance for the aluminum castings shrinkage. With this total shrinkage allowance, the actual pattern is built to oversized with proportional size increases that have been distributed over all of its dimensions.

When the proper amount of shrinkage is not included into the mold, the outcome of the molding will be a proportionally undersized part. This is true especially when copying an existing item where the newly created mold shape is duplicated directly from an already thermoformed article.

C. Basic Mold Constructions

One of the most beneficial reasons for choosing the thermoforming process over other manufacturing methods is to make useful plastic products with substantially lower mold costs. The mold can be made much more easily and from less costly materials. Both of these factors can be so overwhelmingly advantageous over the tooling costs of other manufacturing methods that this alone often determines the choice of the thermoforming process over other process techniques. The reduction in tooling cost, however, is not the only reason for this choice; the higher output production rates create equally advantageous benefits.

1. Mold Materials

Making molds for the thermoforming process can be accomplished using a wide range of materials. The first thing the user has to establish is the size of the production run. The thermoforming process can be applied feasibly to a short production run and a "temporary" mold can be used for the job. Long production runs that produce a high volume of products need a more durable and "permanent" type of mold. For either purpose, different mold materials should be used, and with them, matching related tooling cost can be obtained. The less costly tooling materials are easier to work with and assemble; however, their life expectancy and dimensional stability make them useful only for temporary or short-term molds. On the other hand, specially treated metallic molds can have substantial "permanence" and years of production longevity unless premature damage or lack of product sales force them to be shelved.

As mentioned earlier, wood is a popular tooling material for thermoforming. Most patterns and temporary molds are made of wood; they can be shaped and worked by all common woodworking tools. Complicated shapes can be put together in sections or segments rather than as a single piece, making them easier to build. The pieces can be glued, nailed, and screwed to one another, creating the final structure.

When the woodworking procedures are finished, a sealing compound can be used to coat and protect their outer surfaces. Vacuum holes and air channels can be built into the wood structures together with metallic threaded mounting inserts. The use of threaded metal inserts is a good idea because frequent screw removal will wear and damage mounting holes used repeatedly. In addition to all the aforementioned advantages, wood molds will usually be subjected to frequent modifications. The ability to make alterations, such as adding or removing surface patterns, makes wood the ideal material for experimental mold making.

One of the drawbacks to wood molds is their reaction to humidity, which causes dimensional changes which affect the highly accurate and sensitive mold dimensions. Another, more serious inadequacy is that the wood itself will retain the heat placed into it by thermoforming. A wood mold quickly becomes unusable due to its noncooling tendency. This will force the thermoforming operations to shut down until the wood mold can cool down. Since wood is an insulating material, cooling may take several hours and may stop the operation entirely for the day. This heat buildup in a wood mold is obviously the direct result of the molding frequency. With fewer molding cycles, the heat buildup will not be as severe or start as soon.

To improve on wood mold qualities and make molds with improved durability, the wood patterns can be converted into plaster or epoxy form. These higher-grade materials provide better dimensional permanency and maintain their quality over a longer period of use. The mold's life expectancy is expanded together with the length of operation. The higher densities of plaster and epoxy make the molds stronger and give them more heat absorbency. The plaster mold is usually made of "hydrocal," a cementlike grade of plaster compound. The white powdery hydrocal is mixed with water to create a slurry, which displays a good liquefied, pourable viscosity. The slurry is poured over the wood pattern, which is prepared with a retaining frame surrounding the pattern body. Prior to the casting, all exposed surfaces of the pattern and the accompanying frames must be coated with some type of release agent. Wax or lacquer mixed with oil compounds or silicon greases provide excellent release after the hydrocal has set. To avoid a large amount of air entrapment or air bubble collection, the slurry viscosity is substantially thinned. After pouring the mixture, the entire setup is subjected to vigorous shaking on a vibrator table to force the bubbles out of the slurry. It might also be placed in a vacuum chamber to undergo air bubble evacuation. With either method, the trapped and encapsulated air is reduced, making the poured casting internally more solid and with fewer surface imperfections that must be patched up later. After the hydrocal has set, the pattern and the hydrocal are separated. The hydrocal will display a reversed impression of the original pattern. If duplication of the original pattern is desired, a repeated hydrocal casting is made. Two things must be remembered when using hydrocal. One is that a moisture-resistant release agent must be used to avoid adhesion of the casting to the original pattern or previously made pattern. The surface bonding can be so severe without the proper release agent that any attempt at separation of the two sides will result in the destruction of both. The second factor to remember is that hydrocal molds must be completely dry before thermoforming is attempted. The heat generated by the process converts the retained moisture into steam within the castings and will develop enough internal pressure to cause cracking or even rupture of the molds. After complete drying, the mold can be sealed with a sealant compound and, if necessary, have pieces inserted or other materials glued to it. A broken or chipped hydrocal mold can easily be mended or patched. The size of repairable damage depends on the repairman's skills and the thermoforming criteria encountered. The repair work may take more effort than that needed to make another hydrocal casting.

There are toolmakers and pattern shops that prefer to work with epoxy casting materials. In this type of casting, the water-based castings of the hydrocal method are replaced by a non-water-based two-part epoxy chemical substance. Since water is not used to create a pourable media, the wood pattern will not experience the pattern reaction to moisture that it does with hydrocal. Epoxy will also not produce grain raising and dimensional moisture-related changes. On the other hand, it may take longer to set up and will generate a much higher level of heat in the setting phase. The higher level of heat generation can be just as destructive or damaging to the pattern as moisture. If low-melting-point components such as wax filets are used in the original patterns, they will be melted away, creating an alteration in the final configuration. All these factors must be worked out prior to casting. Without the use of a proper release agent compound, the epoxy casting material can also lock against the casting surfaces, and removal will inevitably result in disaster.

Epoxy molds offer much improved surface details, dimensional stability, and structural strength. Their density and durability make them ideal materials for making low- to medium-production-volume molds. Their dimensional stability is excellent and their heat absorbencies are not poor, but they are, of course, no match for metallic molds. The heat-transferring quality of the basic epoxy materials can be substantially improved by blending in small metallic particles. Aluminum powder, pellets, shots, and even needles can be blended into the epoxy mass. Their purpose is to improve the epoxy's heat-transferring qualities and at the same time, boost its strength. The closely packed metallic particles tend to allow better heat channeling between the embedded pieces. Further improvements can be obtained by embedding cooling lines of metal tubing. Mold coolant can be circulated through the tubing to enhance greatly the material's cooling ability.

The two-part epoxy materials are usually purchased in premeasured cans. When the two compounds are mixed together, they react to form the epoxy compound. Each epoxy material has a different "pot life," which is the amount of time the material takes to set up after mixing. Shorter pot life makes the casting harden more quickly but allows less time for the mixing and blending of the metal fillers. A longer pot life is more advantageous for getting rid of the entrapped air bubbles that have worked themselves into the mixture. Working with epoxy castings does require some skill to achieve choice mold replicas; such skills are improved when working often with this material. Epoxy suppliers are usually very helpful and will come to the aid of the unexperienced practitioner. Epoxy molds, with their durability, can be considered as production-type molds, ideally suited to short to medium-sized production runs. A well-constructed epoxy mold can be expected to perform well for up to several million production units without major breakdown or failure.

The last group of mold materials are the metallic molds. Metal molds can be produced out of a wide range of available basic metals or their alloys. There are two basic methods for shaping metals into a mold configuration. Metals can be heated, liquefied, and then poured into castings, or they can be mechanically cut, milled, and turned into the desired shape on regular metalworking equipment. Metal tools have substantial permanency and therefore qualify as "permanent" molds. With additional surface treating they can be made virtually indestructible. Of course, like anything else, if they are subjected to abnormal use or abusive practices, they can be destroyed. Metal molds offer the most outstanding heat channeling and heat transfer qualities. This makes them ideal as cooling surfaces for the forming of hot thermoplastics. The most popular metal used in the mold making for thermoforming is aluminum. This metal is ideal because it has specifications that fit the thermoforming process best. Aluminum is lightweight and easy to work with on any common toolmaking equipment. It has one of the best heat transfer qualities among the inexpensive materials. On top of this, aluminum does not corrode easily like steel does. Aluminum's only shortcoming is that it is soft and easily damaged by nicking, scratching, or banging. Otherwise, it is the most ideal mold material in use today and for that reason is the most commonly used.

There are other metals used for mold production that are not as popular as aluminum but are demanded by specific conditions or situations. Brass or titanium alloys provide a much stronger mold composition, making them far more damage resistant than aluminum. They are also less vulnerable to corrosive environments, and tooling made out of them will probably outlast the life expectancy of most product designs. Tooling made of steel or brass alloys can be chromed or nickel plated if the product criteria demand it. They can also be Teflon coated for helping in the release of the formed articles. The possible variations of metal mold composition can be as complex and wide ranging as the thermoforming process itself. Many different materials based on different concepts and intended purposes can be used. Innovation in moldmaking technology is a perpetual process. In time, more thermoforming ideas will be developed together with many variations of tool-material combinations.

2. Mold-Making Methods

With all the mold materials available, the mold maker has several options. First, a decision has to be made as to whether a male or female mold is the correct choice for the final mold configuration. Second, plans should be made as to how best to achieve the specific mold features and design as well as the mold arrangements. When all this information has been collected, the third critical factor to be determined is how many thermoformed articles will be needed. Having answered all three basic questions, the tooling-up procedure can begin. The number of articles needed will dictate whether the mold is to be used for temporary, short-term, or long-term production. For small-volume production, molds made of wood, plaster, or epoxy would be the best choice. With at least some basic pattern-making skills, simple-configuration molds can easily be constructed and if necessary, converted from male to female, or vice versa. Sufficient sidewall angles and large radii on the mating surfaces can minimize the release problems when casting is produced. It is not possible to create negative or reversed draft angles or undercuts in any of the castings, and attempts to do so will lock the casting and the original pattern together, prohibiting their separation. All epoxy and most plaster casts will hold their original sizes after casting; however, there are some unique expanding plaster compounds which, after casting and cast removal, tend to grow in size. Such materials may be capable of expanding after casting to gain some of the shrinkage allowances required in plastic production. However, the mold maker should not place too high an expectation on the outcome, and the expanding plaster should be used with some caution. The expansion this material experiences is not in true proportion and some areas may gain satisfactory amounts of shrinkage compensation, whereas others will not, resulting in some design distortion.

As stated earlier, short-run production molds can be made of wood, plaster (hydrocal), or epoxy materials. A cautionary note: These materials are not intended to make molds for pressure-forming techniques, as they do not offer the structural strength and stability to take the force usually induced by pressure-forming methods. Pressurized molds receive on the average 15 to 100 psi of pressure; this could turn these molds into potential bombs, causing them to rupture and even explode. No attempt should ever be made to use these weaker materials for pressure-forming methods.

Large molds for sheet-fed thermoforming operations can also be made out of fiberglass materials. This type of mold is usually made by the customary fiberglass lay-up technique. In this mold-making method, a pattern or sampling shape is used for the lay-up or spray-up of the fiberglass material. Naturally, the pattern or model surfaces are treated with release agents for ease of separation after the fiberglass sets. Fiberglass molds are generally made with thick gel coats to provide strong, solid surface coating; the fiberglass itself will provide the structural strength. Depending on the mold size, fiberglass molds can have from $1/4$- to $3/4$-in.-thick sidewall members to give sufficient mold integrity for the thermoforming process. The fiberglass material itself provides good heat-absorbing qualities when making surface contact with the heated thermoplastic sheet. Its cooling abilities can be improved further by fan-forced air cooling between the forming cycles. The use of fiberglass molds is employed mostly when large products are thermoformed, such as spas, boats, and large tubs or shower enclosures. The manufacturing speeds and production modes of these products are well suited to this type of mold structure. Fiberglass mold-making methods are also one of the most cost-effective ways to produce molds and to obtain duplicate molds from existing articles. For this particular mold-making procedure, no knowledge or equipment is needed other than the customary fiberglass technology, which is easily acquired. Numerous articles and books are available on fiberglass form-producing procedures.

Molds fabricated from any of the popular tooling metals can be produced by one of two basic fabricating methods. Metals can either be cast into shapes or machined out of a solid metal block. In both cases the metal material content will afford the most durability in strength and long-term service life. The high density of metals provides the best heat exchange rate and therefore the best efficiency in mold cooling. Of course, depending on specific thermoforming criteria, one metal could outperform other metals, which certain metals may not function well at all. Higher costs can also preclude the use of certain metals.

Actual metal molds consist of two basic body components: the mold body and its mounting plates. With metal molds it is most common to produce the mold body out of a softer metal such as aluminum while its mounting plates or surrounding frames are made from harder metals, such as steel. However, it is not inconceivable to choose aluminum for the entire mold construction. Also, there are molds for which most of the components are made out of steel, with portions containing hardened tool steel segments.

The actual body of the mold, which performs the thermoforming function, can consist of just female cavities or just male mold structures or both. The complete molding tool consists of the mold body, the mounting structures, and possibly, elevating "stand-off" legs. Each component of the mold serves a specific purpose. The mounting plates provide the mounting surfaces for the individual mold body or bodies when multi-up mold formats are concerned; they are also used for mounting the mold to the thermoforming machine platens. The stand-offs fill and compensate for distances that are left between the sheet line and the platen's daylight opening, so the mold can close right at the sheet line. The two halves of the mold body must meet when the platens close and must do so in close proximity to the sheet line for ideal forming conditions. If the mold halves mate out of the sheet's stretch range from the sheet line, they will cause undue strain on the outside edges of the sheet and may pull it out of the holding or even cause it to tear. Under- or oversized mold stand-offs always render incorrect mold matings and therefore prevent thermoforming. The correct mold surface elevation to the sheet line should be a part of the specifications for any thermoforming machine. This distance is established by subtracting the combined height of the complete mold body when closed, including the base and mounting plate thickness, from the total platen movement distance. For the purpose of mold weight reduction, the stand-off components can be made of aluminum instead of steel or can contain a number of large cutouts.

In making any metallic mold body, there are two basic manufacturing techniques. The first consists of making the mold configuration by metal casting. In the casting of any metal, a premanufactured pattern is used. The pattern for any metal-casting procedure can be made with any number of design intricacies together with the necessary shrinkage allowances but cannot contain undercut or reversed dart angles. As described in Section I.B, the pattern itself is usually made larger to compensate for both the natural metal cast shrinkage and the plastic material shrinkage, which occur after the forming cycle of the thermoforming.

The various types of casting procedures used for making thermoforming molds closely follow the customary metal-casting methods. The most popular casting method is the sand-casting procedure. In this type of metal casting, the pattern is inserted into fine-grained casting sand within a metal core box and then rammed down tightly to compact the sand around the pattern into an almost solid form. The procedure is performed in such a way that either the pattern or the core box is made with separating halves—when the pattern is detailed on both sides—or the pattern is placed adjacent to a core-box surface plate—when the pattern is flat on one side. In this way the compacted sand will duplicate only one side of the pattern. After the sand is rammed, the core box is opened up and the pattern is carefully removed, creating a female cavity impression of the pattern. Another core box is similarly rammed up to carry either a flat surface or the details of the remaining side of a two-sided pattern. The two core boxes are then assembled face to face. Pour-in funnels are created in the compacted sand for filling purposes. When molten metal is poured into this hollow mold through the filling funnels, the flowing metal will fill the cavity. After several hours of cooling, the poured metal will solidify in the cavity. After removal, the metal will retain the details of the sand cavity and duplicate the shape of the original pattern. The quality of the resulting sand cast is, of course, directly related to the ability of the maker. The use of fine-grained sand will make the surfaces and the details of the casting more refined. Skillful placement of the pour funnel and the risers used to eliminate "sinkholes" is very important for minimizing warping and distortion of the casting. Some castings, depending on their various configurations, will develop more shrinkage at their heavier, bulkier points, while thinner areas undergo substantially less shrinkage. Adding risers or body extensions to the heavier areas will draw the shrinkage from the riser itself instead of at critical body areas. Neutralization of shrinkage and warpage depends on the foundry operator's skill. Some foundries are capable of producing uniform and flawless castings, others produce a high rate of unacceptable castings. Quality castings free of distortions, casting flaws, and surface irregularities are vitally important to the suc-

cess of thermoforming production. Properly made castings will be consistent to one another in dimensions and will require minimal finishing work. There are improved metal-casting methods and casting variations that could further improve the basic sand-casting methods described here. Such casting methods use "plastic/rubber" casting techniques or "permanent mold casting" methods and in their own way, offer unusual design and quality benefits. However, with each improved casting method, the cost of casting will rise. The choice of casting method should therefore be predicated on the level of quality sought for the particular job in order to help keep costs down. Some thermoformed products, because of their material makeup, will display, from one casting to another, higher levels of postforming distortions within themselves than those of sand-casted metal molds. In such a case, the cast mold body tolerance variation would be far less than the postthermoforming results, which would indicate that the casting method to produce the molds is within a satisfactory tolerance level. More often than not, the casting method is well within the range of the thermoforming mold-making criteria and should be given full consideration.

Molds for thermoforming are generally cast of aluminum, but not exclusively so. Other high-quality metals, such as brass or beryllium alloys, are also used. Aluminum is the most popular metal because of its noncorrosive properties, light weight, ease of shaping, and excellent heat conductivity. It is also readily available at reasonable prices. The most popular aluminum alloy used in thermoforming mold making for casting purposes is No. 356 aircraft alloy. This alloy is easy to work with because it is neither soft nor overly rigid. This alloy does not require aging or stress relieving and does not readily gum up on or adhere to the metal-working tools' cutting edges. There are other metal alloys, with different characteristics, which can serve other purposes and may be more advantageous to use. The choice of alloys and base metals is entirely up to the thermoforming practitioner, who should maintain a close collaboration with the mold maker.

Any cast mold components, whether female or male, require some machining and finishing. The mold arrangement is produced in either a one-up or a multi-up configuration. Each set of mold units is cast as individual parts and then fitted and assembled into the common mold format. Each casting represents a single mold unit, and if in the matched mold arrangement, each will consist of a female casting and its separate corresponding male castings. To ensure proper seating on the mounting plates and proper side-to-side alignment between the individual castings, the mating surfaces must be prepared and machined flat and square—as well as parallel to the other mating surface—in order that they can be stacked and fitted together to create the combined mold setup. Since the castings do come out from the foundry with some degree of distortion and surface flaws, it is always good practice to order two or three more castings than the actual mold configuration requires. In this way, the toolmaker will have a number of castings from which to select. The castings will be checked for quality and those that contain higher levels of distortion or critical surface flaws will be rejected. It is also good practice to finish up one extra casting for the specific mold configuration and thus make the individual castings interchangeable. If any of the castings are damaged in the thermoforming process, the available ready-to-mount extra casting could become a real "lifesaver." If any of the nonessential or rejected castings later become obsolete, they can always be sold back to the foundry for their material value.

When any of the casting procedures are used to make molds for thermoforming, the cast surfaces require some preparation. Even with the finest-quality sand, the surface of the cast will retain some of the roughness of the sand grains. The surface of the casting must receive some smoothing and surface refinements. If the casting surface is left as is, the forming plastic will pick up the surface imperfections and display the same roughness. However, the casting surface, when refined, should not need a complete polishing job. Giving a high-quality polish to the surface could be just as harmful as leaving it untouched. A highly polished mold surface is not only time and labor intensive and adds unnecessarily to the tooling cost, but actually has an adverse effect on the mold release of the formed article. After the plastic article is formed, a highly polished mold surface will result in almost perfect surface contact between the mold and the article. Both surfaces are smooth enough to create a contact that will not permit air to penetrate between; this creates a vacuumlike adhesion that will interfere with release of the plastic article. This interference will remain in effect until the mold surface is altered by sandblasting or other means of surface roughening. It is always embarrassing when a highly polished and visually attractive mold has to undergo destructive surface coarsening to make it function better.

When casting surfaces are worked on for thermoforming molds, knowledge and understanding of the thermoforming process will make the mold maker's life easier compared to other mold-making techniques. In thermoforming, the heated plastic sheet forms only over male protrusions and into cavities that have air relief channeling. With any captive cavity that cannot rid itself of the encapsulated air that develops between a mold surface pit and the forming plastic sheet, the plastic tends to bridge over the cavity and not conform to the pits. Armed with this knowledge, the toolmaker should remove and smooth out the high points of the casting surface but need not work into the pitted surfaces to polish them out completely. If not overly heated or excessively forced by the forming forces, the thermoplastic sheet will bridge the pitted mold surfaces and come out of the mold with a smooth, glasslike finish. The same surface criteria will exist with any of the machined molds, and the highly polished finish is just as undesirable with them as with the casted molds.

Molds for thermoforming are often made by machine shops on milling machines, lathes, and other customary metalworking equipment. With this type of mold making, tool and die makers will be able to keep tighter dimensional control and far more uniformity among the cavities. In addition, an individual tool shop can have sole manufacturing and financial domination of the entire mold-making project. A mold maker using machining methods to make the mold does not have to get involved with pattern makers or a foundry, which would force a division of the mold-making profits among the shops. However, mold makers who attempt to make the molds by one method only cannot offer the various options of different methods, perform diversified tool-making, or provide a wide range of price options. In some instances the thermoforming plastic can have far more dimensional distortion by itself than a metal casting would develop. In such cases the casting method is the most satisfactory and economical mold-making method. On the other hand, where the thermoformed products demand outstanding dimensional control from one thermoformed product to another and such consistency is required for a large number of mold duplications, the mold making will rely on the machined quality mold-making technique.

The machined mold-making procedures basically follow all the customary metal machining techniques and are produced on the familiar milling, cutting, sawing, and lathe equipment. The molds are usually cut out of a block of metal which has been selected and purchased for that purpose. They are then subjected to machining, which removes all the excess material. The machine cutting and milling are usually done on a standard mill press using metal-cutting tools that in the milling process create metal shavings and chips. The block of metal is set up on the milling machine table, which moves in two or three axis directions. The depth of the cutting is usually guided by the rotating bit length and by the mill-head distances from the mill table. The movement of the milling table can be guided by an operator or through a computerized equipment setup. The actual milling of the metal, whether done manually or run by automated equipment, is always made gradually, in layers, with the cutaway ending up as an accumulation of metal shavings. The outside surfaces of the mold are milled with the usual multiple passes by the cutting tool until the desired shape and dimensions are reached. At the same time, the inner cavity milling is made by "hogging" out the entire inner cavity of the metal piece. To develop specific cavity contours or radii, special cutting tools are produced for the milling. When specified radii are required for the specific inner corners of a cavity, special cutting tools are ground for that radius so that the milling machine can cut to the exact corner form desired. For flat sidewalls, the milling machine makes a pass on the surface; for a ribbed pattern, the mill will cut indentations at preprogrammed intervals. Automated milling machines can be set up in a robotic way such that the cutting tool follows a preprogrammed path. With automatic milling machines, the entire mold configuration—even in a multi-up mold arrangement—can be cut out of a single piece of metal. Making thermoforming tools out of a single piece of metal is preferred by tool shops but is not in the best interests of the thermoformer. Certainly, it is easy to carve tooling out of a single block of metal using today's automated milling machines. However, if any damage should occur in the thermoforming process, tool repair can result in a lengthy production shutdown. For this reason, a mold assembled from several pieces is a better arrangement for the thermoformer, because the damaged segment can be replaced. The segments can be joined so well that no discernible mating lines will be visible. Although rare, there are mold configurations that cannot be machined out of a single block and are therefore either made with an insert or put together from several components. For that reason, when a single mold unit is pieced together, the adjoining surfaces must be made perfectly or else be made so that the design pattern hides the mating lines. Usually, the corner mating lines or incorporated rib patterns can provide good concealment for adjoining surfaces. Any misalignment between the components will show up on the formed article. This may not jeopardize its functioning but can render either a poor appearance or reveal its mold construction to a competing thermoforming processor.

The machine mold construction, whether in a single mold unit or multiup mold arrangements made from pieces or from a single block of metal, is made according to the customary metal machining procedure. Qualified metal machining knowledge is required for cutting and milling metals and mold production. Any mistakes made in the metal shaping will be transferred into the plastic through the thermoforming process. For most machined thermoforming molds, aluminum alloys are the most popular material. Again, aluminum is easy to work with, is lightweight, and has excellent heat transfer abilities. However, occasionally brass or brass and beryllium alloys are employed. These exotic and higher-grade metals are frequently justifiable because of their higher resistance to corrosion and surface damage. The machined thermoforming tools obtain a higher-quality surface finish than those made by casting.

There are no standards that could be given for suitable surface roughness. It is always dependent on the specific thermoforming application and the smoothness required in the resulting products. The surface finishes produced by the various machining methods vary over a wide range. Specifying certain metal-working equipment does not guarantee the same surface finish from shop to shop. Table 2 provides a comparison of surface roughness ranges obtainable by the various metalworking methods. Because of the various factors that can affect the surface finish produced by a given machining operation, each shop will provide slight differences in surface finish. It is better to specify surface roughness designation and measurement numbers than specific equipment. As the chart illustrates, an average mill finish will carry 125 rms, while

Table 2 Obtainable Surface Roughness (RMS)

Process	Micrometers (μm) [microinches (μin.)]							
	50 [2000]	25 [1000]	12.5 [500]	6.3 [250]	3.2 [125]	1.8 [63]	0.8 [32]	0.4 [16]
Sawing	··········▬▬▬▬··········							
Sand casting	··········▬▬▬▬▬▬▬··············							
Permanent mold casting			·········▬▬▬▬···········					
Milling	····················▬▬▬▬▬▬········							
Drilling		··········▬▬▬▬▬▬▬···········						
Laser		··············▬▬▬▬▬▬▬▬▬·········						
EDM		············▬▬▬▬▬▬▬···············						
Grinding			·············▬▬▬▬▬▬▬▬▬▬▬					
Polishing						··········▬▬▬▬		

smoothening can refine the surface to 30 rms or better. With a profilometer or tracer instrument, the surface differentiations can easily be established and the finished surface can be made to match most specifications. A surface finish between 30 and 50 rms is smooth enough to work with and permits passage of air between the mold surface and the formed plastic for easy removal. As for inner body imperfections, these will probably not be found in machined tooling. The minimum radius or taper angle of the cutting should not originate with the mold maker. Most specifications requested can be produced, and any limiting factors for those specifications should come from the thermoforming practitioner. Errors made due to overly specified criteria are discussed later in the chapter. The details of actual metalworking procedures are too involved to include here. For further information on these subjects, metalworking or machining publications should be consulted.

3. Mold Venting (Air Displacement)

Molds made of any of the materials or mold-making methods described previously must have sufficient venting holes to be functional in the thermoforming process. Through these holes, either vacuum or air pressure (or both) is introduced for the purpose of forcing the heated thermoplastic against the mold configuration. The vent hole placement in the molds is one of the most critical aspects of the molds; without them, the molds are not functional and the entire thermoforming procedure may be disabled. It is through these vent holes and their interconnected channels and plumbing system that the forming forces will be introduced and will force the heated thermoplastic to conform against the mold surfaces. The location and number of vent holes are as important as their size. In the case of a multi-up mold configuration, the interconnection between the individual molds is just as significant as the connection of the first mold to the forming source. The vent hole functions and the actions of forming were discussed in Section I of Chapter 3. The holes placed into the mold must be interconnected by channels in the back side of the mold that travel all the way back to a common plumbing system that connects to the source of the forming force. In the case of a vacuum, the source is a vacuum pump; with pressurized air, the source is an air compressor. The plumbing and all the channeling must be leakproof and be provided with flexible hoses that will not restrict the mold's reciprocating movement. At the same time, the total power of the forming force can be fully realized at the individual mold surfaces. There should be no flow reduction that would restrict the forming power between the mold and the source of the forming force. The errors usually made in this specific area of thermoforming are addressed in Section IV.

The placement of vent holes in a mold is done mostly by drilling. If the mold is cast out of epoxy materials, there is a good chance that instead of drilling, the holes can be cast right into the mold. The technique for making cast-in holes involves the use of small pieces of "piano wire" (hardened steel wire) cut into proper lengths, which are then partly hammered into the pattern surface, which will create an embodiment resembling a porcupine. The wires and pattern surface are well coated with a release agent (e.g., Vaseline) prior to epoxy application. After the epoxy has set, the protruding wires are pulled out of the epoxy cast and the pattern, leaving holes in the casting. The diameter of the wire determines the hole sizes. The determination of the number and location of the holes should follow the common practices used in typical thermoforming mold construction.

Vent holes are usually drilled with the help of power tools, which are either hand held or bench mounted. The actual drilling is done using common industrial drill bits. It is also possible that instead of drilled holes, the mold bottom is made of a separate piece which after installation leaves a narrow slot around its perimeter. Such a narrow gap can also be used for air movement.

a. Vent hole sizes: To begin with, the thermoforming practitioner must establish the maximum hole sizes that the thermoforming process can best handle. The larger the hole size, the greater the forming force that can be introduced through the holes. However, the larger holes can leave marks on the thermoformed article. In most instances, a buttonlike mark will be made by the oversized holes, giving an unacceptable finish to the plastic article. For this reason the mold maker must reduce the size of the holes drilled in the mold. The other factor in these buttonlike imperfections is the thermoplastic sheet thickness. The thinner materials can form into the smaller holes much more easily and therefore will be more visible than the heavier-gauge sheet materials. It has also been found through industry experience that different thermoplastic materials offer different levels of sensitivity, and some are more likely than others to pick up vent-hole patterns from a mold. Armed with this knowledge, only general guidelines can be given as to the correct choice of hole sizes. For particularly sensitive materials, nothing but the minimum drillable hole size should be used. Table 3 lists the selection of functional drill sizes for thermoforming. It is always easier to drill with a larger drill bit. However, there is less chance of obtaining unsightly button markings with the finer drill bits. Since the appearance of unwanted marking is closely related to the thermoplastic material thickness, it is acceptable to use a measure of half of the original material thickness as a guide for the maximum drill size selection. This is a good rule of thumb for sufficiently thick materials. When dealing with extremely thin sheet material or foamed materials, a No. 80 drill bit is the best size. The use of such a small bit can add greatly to the cost of vent-hole making because the bits break easily and their replacement cost is high. Unskilled personnel using these bits can easily run up costs above their actual wages.

Table 3   Drill Size Chart

Fractional size drills (in.)	Wire gauge drills	Decimal equivalent (in.)	Fractional size drills (in.)	Wire gauge drills	Decimal equivalent (in.)
	80	0.0135	1/32	66	0.0330
	79	0.0145		65	0.0350
				64	0.0360
1/64	—	0.0156		63	0.0370
	78	0.0160		62	0.0380
	77	0.0180		61	0.0390
	76	0.0200		60	0.0400
	75	0.0210		59	0.0410
	74	0.0225		58	0.0420
	73	0.0240		57	0.0430
	72	0.0250		56	0.0465
	71	0.0260			
	70	0.0280	3/64	—	0.0469
	69	0.0292		55	0.0520
	68	0.0310		54	0.0550
				53	0.0595
1/32	—	0.0312			
	67	0.0320	1/16	—	0.0625

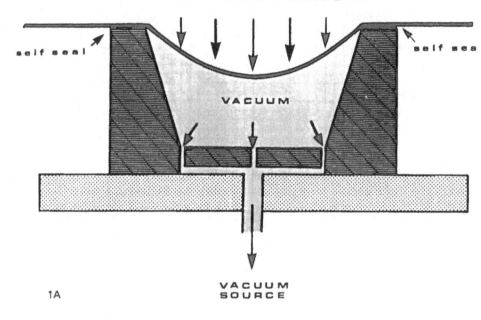

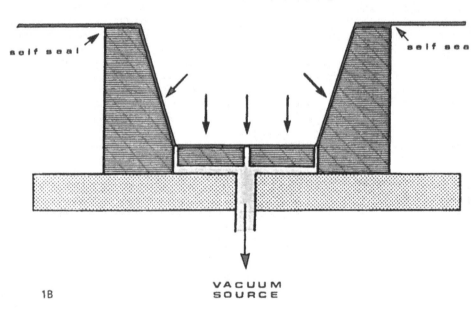

Figure 77 Mold with correct vacuum hole placement: (1A) forming begins satisfactorily; (1B) full detailed forming is made. Mold with incorrect vacuum hole placement: (2A) forming will begin equally well; (2B) centrally located holes covered by forming plastic with some trapped air behind.

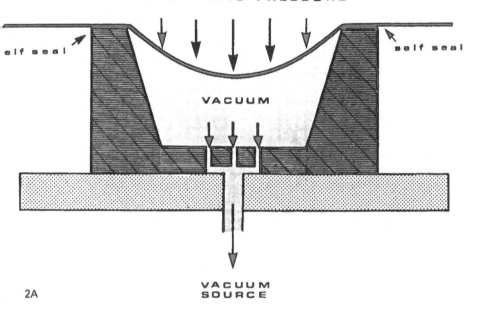

2A

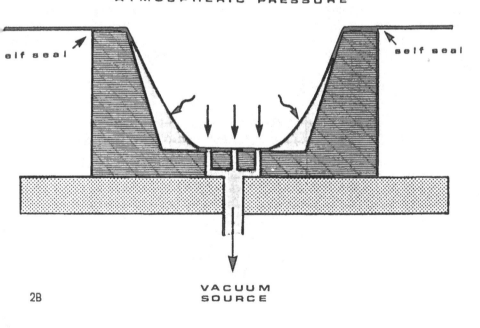

2B

b. Vent-hole locations: The second criterion in mold making is the choice of vent-hole locations. To achieve the utmost detail transfer from the mold to the formed plastic, all of the mold's inner air space must be evacuated by either vacuum or displacing pressure forces. Since there should be no trapped air left between the mold surface and the plastic, air must be removed from the vital cavity areas through the holes. Each hole is connected to the main evacuating chamber and consequently, either to a forming power source (vacuum or air pressure) or in some cases for proper venting, to the open atmosphere. For ideal air removal, venting holes must be placed along the corners as well as the bottom and sidewall intersections and also into all ribbing and stiffening detail structures. Holes must even be placed into the engraved patterns and lettering. To gain good detail transfer from the mold, every conceivable bit of trapped air must be removed from between the forming plastic and the mold surface. In this important effort, it is best to treat every mold configuration as a different and individual case where proper preplanning in the hole placement can create an enormous difference. Improperly placed holes (those placed only in a central location, for example) can be rendered useless by the partially forming plastic sheet plugging the holes and trapping the remaining air inside the mold. Figure 77 shows the importance of hole placement into cavity corners for full forming development in a thermoformed plastic part.

The same criteria exist with engraved lettering in a thermoformed product. For good legibility, each individual letter, depending on its configuration, must have one or more vent holes. Details of undercut or reversed-draft-angle forming are also very dependent on proper vent-hole placement and satisfactory air removal. Poor vent-hole placement or a weak airflow rate can reduce the effectiveness of undercuts.

c. The number of vent holes in a mold: The ideal location of vent holes to be placed in a mold involves a large number of holes within the mold structure. To further enhance the speed and quality of the particular thermoforming process, promptness in air evacuation is very important. With either vacuum or air pressure, such evacuation must rely on those holes, and their size or number can greatly influence air movement. The hole size is already curtailed by the thermoplastic sheet thickness to minimize the unwanted impression transfer from the holes. Therefore, the only remaining enhancement provided is an increase in the number of holes.

To determine the number of holes needed in a mold, the mold's volume capacity has to be established. This value actually represents the air displacement that will take place when thermoforming is performed. On simple geometric shapes, it is easy to calculate the mold's displacement value. However, with complicated or irregular shapes, the calculation is not as easy. The values

are estimated either by approximating the dimensions to the nearest simple
shape, or for closer determination, the mold cavity area can be filled with
free-flowing materials (i.e., bird seed, bean bag foam fillers, etc.). Such measurement techniques can help the practitioner arrive at the correct displacement values. In the case of multi-up molds, the value can be factored to calculate the entire mold area. The actual mold displacement value should first
be matched with the vacuum pump capacity. If the vacuum pump does not
have the necessary volume capacity in gallons per minute, it will either have
to be upgraded with a larger-capacity vacuum pump or the pump can be coupled with a second pump. When a very large amount of air is removed from
the mold, a "surge tank" can be coupled to the vacuum pump unit. The principle behind the surge tank is that a smaller pump will be able to evacuate the
surge tank when no vacuum is called for. Then, when the actual forming cycle
takes place, the emptied tank will provide the necessary evacuation capacity
to create the vacuum. Between forming cycles, the vacuum pump is given
enough time to reestablish full vacuum capacity in the surge tank.

The next factor to be established is the diameter of the vacuum outlet of
the pump. This pipe diameter must match the rest of the plumbing. Any restriction in the plumbing lines will cause a drop in the vacuum forces, creating
a bottleneck. Furthermore, the plumbing pipe cross section should first be
divided by the number of individual molds (ups) and then again by the chosen
drill-bit cross section. This calculation is expressed by the formula

$$\frac{P}{U} \div D = \text{number of vent holes required per mold unit}$$

in which P is the pipe cross section, U the number of ups in a mold, and D the
drill-bit cross section. For example, how many vent holes should be drilled in
a mold cavity with a six-up mold arrangement, a 1-in. plumbing line, and a
$^1/_{32}$-in. vent-hole size?

$$\frac{(1/2)^2 \times \pi}{6} \div \left(\frac{0.0312}{2}\right)^2 \times \pi = 171$$

This formula gives the minimum number of holes needed to match the pump's
vacuum forces within each cavity. It is always a good practice to add 25% more
holes to compensate for any holes that might plug up in the course of production. Depending on the hole size as well as prevailing plant conditions and the
length of production runs, it is not unusual to have a large number of holes
become inoperative through airborne dirt, lint, and moisture combined with
corrosion residue.

When the design features of the mold do not require maximum vacuum
forces, a simple gate valve installation in line between the pump and the mold

can reduce the vacuum force to satisfactory levels. Such valves not only provide the needed flow control, but also the necessary adjustments for fine tuning of the vacuum force.

The number of holes used becomes even more critical when pressure forces are used. Pressure forming is not only more forceful than vacuum forming but is also produced much more quickly. Straight vacuum forming (reaching a maximum 29.92 in. Hg) produces a forming force equivalent to 14.7 psi pressure forming. Most pressure forming, on the other hand, is performed between 50 and 100 psi. Table 4 compares vacuum and pressure force levels. When higher levels of incoming pressure force are applied against the thermoplastic sheet, the trapped air on the other side of the sheet must be ejected or evacuated equally fast. Any restriction in the trapped air outflow can just as easily interfere with the incoming pressure and can jeopardize the outcome of the thermoforming procedure. (The cause and effects of overpowered pressure forming were detailed in Section V.B of Chapter 2.) Overpowered pressure-forming results, coupled with restricted evacuation flow, can result in trapped air pockets within the mold cavity (Figure 78).

Table 4 Conversion Table: Vacuum to Pressure

Vacuum (in. Hg)	Pressure (psi)	Vacuum (in. Hg)	Pressure (psi)
5.00	2.45	18.00	8.66
6.00	2.94	18.36	9.00
6.12	3.00	19.00	9.32
7.00	3.43	20.00	9.81
8.00	3.92	20.40	10.00
8.16	4.00	21.00	10.30
9.00	4.42	22.00	10.79
10.00	4.91	22.44	11.00
10.20	5.00	23.00	11.28
11.00	5.40	24.00	11.77
12.00	5.89	24.48	12.00
12.24	6.00	25.00	12.27
13.00	6.38	26.00	12.76
14.00	6.87	26.52	13.00
14.28	7.00	27.00	13.25
15.00	7.36	28.00	13.74
16.00	7.85	28.56	14.00
16.32	8.00	29.00	14.23
17.00	8.34	29.92	14.70

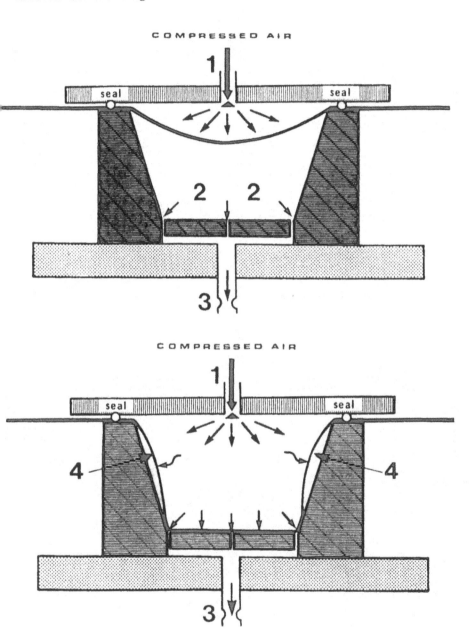

Figure 78 Overpressurized forming condition: (1) incoming pressure force; (2) undersized or small number of vent/vacuum holes; (3) restricted vent/vacuum channel; (4) trapped air pockets.

Table 5 Air Evacuation Flow Rates[a]

Gauge pressure (psi)	Cubic feet of air per minute for single line or combined orifice diameter (in.)										
	1/64	1/32	1/16	1/8	1/4	3/8	1.2	5/8	3/4	7/8	1
3	0.048	0.194	0.77	3.1	12.4	27.8	49.5	77.5	111	152	198
4	0.056	0.223	0.89	3.5	14.3	32.1	57.0	89.2	128	175	228
5	0.062	0.248	0.99	3.97	15.9	35.7	63.5	99.3	143	195	254
6	0.068	0.272	1.09	4.34	17.4	39.1	69.5	109	156	213	278
7	0.073	0.293	1.17	4.68	18.7	42.2	75.0	117	168	230	300
9	0.083	0.33	1.32	5.30	21.1	47.7	84.7	132	191	260	339
12	0.095	0.38	1.52	6.07	24.3	54.6	97.0	152	218	297	388
15	0.105	0.42	1.68	6.72	26.9	60.5	108	172	242	329	430
20	0.123	0.49	1.96	7.86	31.4	70.7	126	196	283	385	503
25	0.140	0.56	2.25	8.98	35.9	80.9	144	225	323	440	575
30	0.158	0.63	2.53	10.1	40.5	91.1	162	253	365	496	648
35	0.176	0.70	2.81	11.3	45.0	101	180	281	405	551	720
40	0.194	0.77	3.10	12.4	49.6	112	198	310	446	607	793
45	0.211	0.84	3.38	13.5	54.1	122	216	338	487	662	865
50	0.229	0.92	3.66	14.7	58.6	132	235	366	528	718	938
60	0.264	1.06	4.23	16.9	67.6	152	271	423	609	828	1082
70	0.300	1.20	4.79	19.2	76.7	173	307	479	690	939	1227
80	0.335	1.34	5.36	21.4	85.7	193	343	536	771	1050	1371
90	0.370	1.48	5.92	23.7	94.8	213	379	592	853	1161	1516
100	0.406	1.62	6.49	26.0	104	234	415	649	934	1272	1661
110	0.441	1.76	7.05	28.2	113	254	452	705	1016	1383	1806
120	0.476	1.91	7.62	30.5	122	274	488	762	1097	1494	1951

[a]This table gives only theoretical flow rates with 1 atm pressures at 70°F. Flow friction will cause a small reduction.

For determination of the ideal number of vent holes in a pressure-forming model, it is always a good idea to provide more holes than would normally be used in the vacuum-forming technique. If a mold does not have enough vent holes, the deficiency will show up in the forming results either in the part's appearance or in slower forming conditions. To estimate air evacuation from the mold and consequent flow rates at various pressure levels, Table 5 should be consulted. For determination of the individual mold evacuation rate, a consolidated hole measurement figure should be calculated.

Smaller holes do not necessarily need to be drilled through the entire bulky mold body. In fact, such drilling is not only time consuming and very difficult but the long, narrow opening can plug up more readily. In most instances the mold maker will mill out a large portion of the back side of the mold, creating an air chamber or interconnected air channels. A larger-diameter drilling can be made from the back side, to as close as $^1/_{16}$ to $^1/_8$ in. from the forming surface. The location of this larger hole is then mechanically transferred to the mold face to implement drilling by the smaller drill bit in the proper location. In this way, the tiny holes need only be drilled through a very thin mass, yet will still lead, by way of the larger holes, to the main air chamber. The larger air channels leading up to the fine holes can provide better flow rates, less chance of plugging, and therefore longer service life for the molds. However, any machining done on the back sides of the mold for the purpose of air chamber creation or air channeling should not substantially reduce the mold's surface contact with the cooling base, therefore reducing its heat exchange qualities.

II. MOLD TEMPERATURE CONTROLS

In the process of thermoforming, the mold has to perform double duty. The first function of the mold is to provide shaping for the plastic. The second, and equally important function is to absorb the heat of the formed plastic in order that the plastic can firm into its newly acquired shape. In any of the rapidly produced thermoforming cycles, the heated plastic will come in contact with the mold more frequently than otherwise. Throughout the vigorously repeated cycle, the mold by itself, or together with the thermoforming machine platens, cannot dissipate the mold's acquired heat. Without cooling, heat will accumulate in the molds and will eventually reach the same heat level as that of the forming thermoplastic sheet. Molds that get hot enough to match the temperature of the incoming heated thermoplastic sheet cannot provide cooling after the forming cycles. This will render the thermoforming process useless and ultimately force it to shut down. To regain the cooling effects of the mold, the mold (or at least the contacting mold base plate) must receive some type of effective cooling that promptly reduces the mold's temperature. An auxiliary cooling method in continuous, rapidly repeated production will maintain mold temperatures throughout the entire operation. Any auxiliary cooling method must be adjustable to increase or decrease its effectiveness. Adjustments in cooling must provide a wide range of temperatures to maintain any desired mold temperatures for the specific thermoforming process. The fewer the contacts made between the heated sheet and the mold, the less will be the cooling demand on the mold. A higher number of repeated contacts, on the other hand, will increase the importance of the cooling aspects of the mold.

In some instances the entire thermoforming operation can be managed without auxiliary mold cooling. The formed article's heat can be dissipated through the mold and into the thermoforming machine platens and frame structures. In addition, part of the heat is eliminated through the formed article surface into the ambient environment. This type of cooling can be enhanced further by fan-forced air cooling. Ventilator fans can be placed above the forming area and are usually operated only when cooling is needed, being turned off in subsequent phases of the cycle. The fans should not face the oven area, as this would create an artificial and undesirable draft. The fans should operate only within the cooling-time segment after the completion of forming. If the ambient air is too warm to cool the plastic effectively, low-volume-output fog-producing units can be positioned in front of the fans directly in the path of the air movement. The fog is created by a small orifice on the fog maker and ordinary city water pressure and is instantly vaporized upon air contact. The vaporization, with its normal heat-reducing characteristics, will reduce the air temperature, and the needed cooling effects are now provided.

When higher humidity conditions prevail, the method is obviously less effective than it would otherwise be. In fact, if not totally vaporized, the fog can produce moisture on the surrounding surfaces. The same adverse effects can be encountered with a higher output level from the foggers or sprinklers. For the best fog creation, greenhouse equipment suppliers should be contacted. These companies offer foggers with volume output rates as low as $1/2$ to $3/4$ gallon of water per hour. Anything more than this can create unwanted wetness, which can drench and damage the equipment and can cause heavy rusting of the thermoforming machinery.

There is one more option for cooling improvement, which can be implemented when fan or blower forced air cooling is employed. But before it is explained, the two most common errors must be discussed. One of the basic errors is found in the fan or blower location. The fans or blowers are often found right above the molding area, and usually located close to or above the oven's edge. Sometimes this location is chosen because of lack of a mounting place or just to try to avoid unwanted air movement into the oven. The excuse of the unwanted draft avoidance I can accept; however, when this spot is chosen, it is most likely the hottest area of the entire plant. We all know that hot air rises and will accumulate near the overhead area of the plant. Even if the plant has a high ceiling structure the heated air will rise and accumulate and very quickly will fill that upper space with hot air. Blowing this hot air onto the plastic is not an efficient way to cool a freshly produced part. Usually cooling will take longer and that extra time always increases the cycle time and adversely affects the economy of that particular thermoforming. In case the extra cooling demand is ignored, and formed articles are removed before full cooling has taken place, it will jeopardize the quality outcome. Most likely the inadequately cooled part will develop a warpage and postmold-distortions.

The second area of errors is made when the thermoformer is aware of the poor cooling results and concludes that more air movement (power) is needed to accomplish the cooling task. To remedy the presumed under cooling or the drawn-out cooling time, the decision is made to add more fans or blowers to the units already installed. They will attempt to increase the number of cooling motors or increase their air volume capabilities by horsepower. Sometimes you see not only doubling, but quadrupled units installed to provide the desired cooling. Just because there is a lot more noise of air rushing, it may not be necessary that better air movement is achieved and with it improved cooling is accomplished. First of all, the ambient air temperature has not been corrected and when it is hot it will not do a first rate job of cooling. With the presumed increased air movement, it is believed cooling can be accomplished with ease. But, what actually happens is the rushing air quickly fills the molded part's cavity and forms an impenetrable dome over the part, so that no additional air can enter onto the plastic surfaces. It is almost like an invisible dome cover over the formed part, that prevents any further heat exchange to take place. As the fans or blowers fill the cavity all cooling will stop and cannot force more air into the part. I have seen from time to time installation of as many as six to nine blowers with substantial horsepower, and still not doing an adequate cooling job.

It is mandatory to have sufficient airflow in to and out from the formed article to achieve the most efficient cooling. Fewer fans or blowers moving air in a well directed flow pattern is better than blowing air against each other that diminishes the development of an air breeze. Of course in the effort of achieving an ideal flow pattern the practitioner also must watch that there is no "shadow" effect left where air movement is blocked. This blockage or "shadow" is usually found in the cooling of large tublike product configurations, where the nearest wall, under the fan, just does not receive cooling air, due to the angle at which the fan blows the air, and misses that particular portion of the wall.

To create the most ideal solution for a rapid and thorough cooling of the formed article often times we should search for a colder source of air. Since the ambient temperature of the plant is already too hot to satisfactorily use for efficient cooling, it may be necessary to look for other sources of colder air. One way to reduce the temperature of the air is to use fog producers which spray a very fine fog like mist into the blowing air stream. The water as it is sprayed immediately vaporizes and in a low ambient humidity condition can have substantial cooling effect on the air surge. This may reduce the air current temperature by 10°–15°F or even more, depending on the prevalent ambient air's humidity. The negative effects of such a system is that it only works when and where low humidity conditions are present. The other bad side effect of this can be found when misters and not fog producers are used and its spray output rates are much higher. Fog producers emit $1/2$ gallon of water per hour versus misters that emit about three times as much, i.e., $1-1^1/2$ gallon of water per hour. Heavy spraying will make everything wet and very soon all iron parts will show signs of corrosion and rusting.

To obtain an even better solution for efficient cooling for the thermo-formed article one may not have to look any further than the source of outdoor air. Often the outside air temperature is significantly lower than the inside plant's air temperature. This may exist even at the height of summer, of course depending on the geographic location. To draw air from outside into the plant and channel it to the right place where the cooling is needed, will require some duct works. This can be accomplished easily by using standard ducting used by air condition and heating professionals. General purpose air blowers can be installed on the rooftop, right above the thermoforming machines and the air is ducted directly over the area where the part cooling is done. The duct should channel the outside cold air all the way down to the formed part and should exit the ducting in a central location of the mold, allowing it to flow into or over the formed part without any airflow restriction. If normal flow is main-tained a continuous cooling will be accomplished. In case a deep drawn female part is to be cooled, the duct system can be extended by a flexible accordion-like ducting, which retracts upward when cooling is no longer needed. Of course, power to the blower also has to be cut off at the "no cooling" times. By lowering the duct work, actually lower than the sides of the article, it will insure not only good cooling but permits excellent air movement and escape-ment from the cavity as well. The constant air movement and the cooler outside air should be sufficient for the most ideal cooling. With this forced outdoor air cooling it should reduce cooling time to the shortest time limits one can anti-cipate with any air cooling method.

The utmost preciseness and speed in cooling of a thermoformed article can best be accomplished through the mold surface. To provide both accuracy and rapid repetition in the cooling cycle, the natural dissipation of temper-atures must be made aggressively. This accelerated heat removal from the mold can be provided only by the circulation of coolant fluids within the mold body.

In the thermoforming process, a mold referred to as a "cold" mold is not necessarily chilling to the touch. The neutral temperature zone (neither cold nor hot) is always considered at the heat levels of the forming thermoplastic sheet. Mold temperatures lower than those of the softened plastic sheet are labeled "cold," and mold temperatures above the neutral zone are called "hot." For example, molds with an operating temperature range of 120 to 180°F or more can be referred to as cold molds as long as the forming temperatures of the plastic sheets are at higher levels. With mold temperatures that are closer to the forming sheet temperatures, the cooling cycle time must be extended further than with colder temperature molds. Any plastic article that is not sufficiently cooled due to improper mold temperatures or premature removal from the mold can develop deformations and distortions as it cools without proper physical support of the mold. Such thermoformed articles can develop unusable and unsightly distortions in time, well after their removal from the mold. On the other hand, for ideal thermoplastic sheet stretching, molds kept on the warm side would work more effectively, allowing better material distribution in the formed articles.

A. Mold Cooling

The accumulated heat in the mold's body must be removed for the mold to provide its full cooling functions for final setting of the plastic article's new shape. To achieve such a task, the mold temperature must be maintained below the softened sheet temperatures. The greater the temperature spread between the forming sheet and the mold, the shorter the cooling time that will be needed. However, exceptionally cold mold surfaces can produce premature setting, causing poor forming details or even undesirable and uneven stretching in the forming. When uneven stretching is encountered due to the overly chilling effects of the mold, the forming results will show considerably reduced stretching at the contact areas. In contrast to these prematurely set areas, the uncontacted areas of the sheet will undergo relatively greater stretching and with it, excessive thinning. Most "chill marks" can be caused by overly cooled molds causing a step-down (reduced gauge) at the border of the contact area. Chill marks or "chill lines" are measurable by a thickness gauge and can easily be seen as a line on the surface of a thermoformed article—an objectionable flaw. In an effort to eliminate unwanted chill marks, one of the steps that can be taken is not to let the mold become too cold. Another improvement to minimize such unwanted affects is discussed in Section II.B.

It is a well-accepted practice for the mold to receive some type of auxiliary cooling in higher-speed thermoforming productions. In almost all instances, the mold, or at least the mold base, is equipped with built-in cooling channels through which liquid coolant flows. The source of cooling can be provided by a closed-loop mold temperature control unit or from a centralized cooling-water source. In either case, cooling is accomplished through the circulation of coolant fluids. The general makeup of these fluids is mostly water with some chemical treatment, used mostly to inhibit corrosion. A glycol-base chemical can be added to the water as an antifreeze agent. It is rare but not at all inconceivable that a Freon formulation will be used in a closed-loop cooling system. The purpose behind any fluid cooling method is to absorb the heat inside the mold and with the flow of the coolant, carry the heat away from the mold structures.

The first requirement in utilizing this type of cooling is to provide sufficient flow for the coolant. The cooling fluids must be circulated with a pumping force that causes the liquid to move in and out of the mold body. The coupling of the plumbing system between the mold and the outside source of the cooling is usually made through flexible hoses. The plumbing lines are usually equipped with shutoff valves ahead of the juncture with the hoses in order to arrest the cooling fluid flow when the molds are removed and changed in the machine. It is also customary to use quick-disconnecting couplings for ease in fluid-line separation. The shutoff valves installed in both the in- and outflowing ports have another purpose in addition to cutting off fluid flow during mold changes. One of the valves can also be used as a flow regulator to reduce or increase the flow of the coolant fluids, rendering less or more cooling to the molds. Only the outgoing valve should be used as a control. The incoming fluid valve should remain fully open during the operation. In fact, it is highly advisable to paint the outgoing valve bright blue to eliminate confusion and the possible error of using the wrong valve for the flow control. Efficient and uniform cooling can be provided only with a fully filled mold cooling system. Air pockets and air bubble formation created inside a mold body prevent the system from providing sufficient cooling. For this reason, the incoming valve should not be used as a flow regulator. The fluid inside the mold must be circulated at a sufficient rate to prevent it from warming up and decreasing the efficiency of mold cooling.

The second requirement in this type of cooling method is closely related to the fluid flow rate. The flow pattern of the coolant placed into a mold must be designed with ingenious planning to maintain a uniform temperature throughout the mold body. If this requirement is not met, the mold will be cooler where the coolant enters and warmer where the used-up coolant exits. A cooling channel in a mold should not be made in such a way that it snakes through the mold from one end to the other or even so that it branches off at one end of the mold only to reunite at the other end. The ideal cooling conditions can be achieved only with an alternating flow manifolded from side to side or front to back. Specifying to the mold maker ideal coolant flow patterns in a mold can circumvent any suspected possible cooling problems that can be encountered with unknown molds (see Section IV.C and Figure 81). Measurements of mold surface temperatures cannot reveal most improper flow patterns. Even if the wrong coolant flow pattern is proven to be the source of difficulty, the remedy will usually demand major tooling changes. The measurement of incoming and outgoing coolant temperatures can be helpful; however, its interpretation can easily be misleading. The changes in temperature readings can be caused by many factors in addition to the change in thermoforming conditions. Examples are the flow rates of the coolant, plumbing and flow restrictions, coupling of other molds to the system, clogging of the cooling channels, flow-restricting cooling channel surfaces and flow direction changes, and stagnant pooling of the coolant within the mold.

In a high-speed operation where repeated contact with the heated plastic is made at least 35 times a minute, mold cooling must be made all the more efficiently. To achieve such rapid cooling, the majority of the mold body should contain as large a volume of coolant as possible. The mold can be made such that only thin wall structures will separate the coolant from the mold surfaces. In this case the cooling liquid must be pumped with high levels of flow speed and must also receive some turbulent flow characteristics to provide the high-speed cooling effects. Without this turbulence, which eliminates the stagnation in the flow, the cooling efficiencies will drop below satisfactory levels. If further cooling improvements are needed, mold cooling must be made with the help of refrigerants.

Negative aspects of mold cooling can be found whenever thermoforming operation is stopped and started up again. As with any process, the operation temperatures are no longer under control when the operation stops. In a continuous operation, as well as with a continuous cooled mold, the mold will experience more cooling during the stoppage than the process requires. It is not unusual after a shutdown for a mold to collect water condensation on its surfaces. Naturally, such cooled mold surfaces cannot initially be expected to function satisfactorily. However, through a few repeated cycles, the mold can again be raised to running temperatures. Further temperature elevations of the mold will be controlled by the mold cooling system. Production startups cannot generally be made instantaneously. Before obtaining ideal running conditions, numerous out-of-specifications articles will be produced unless both preconditioning and interim adjustments are provided. However, as soon as the running conditions are reached, all interim adjustments should be discontinued and the production line will settle into a smooth-running operation.

B. Mold Heating

There are occasions in the thermoforming process when the mold temperature has to be elevated. It is not unusual to elevate mold temperature to the prox-

imity of the heated plastic sheet temperature. In fact, such warmer mold structures can be beneficial in some specific thermoforming methods. First, in most
plug-assisted thermoforming, if the plug-assist temperature is cooler than that
of the heated thermoplastic sheet, unwanted cooling of the forming sheet can
occur. To avoid such chilling effects on the sheet, the plug-assist temperature
must be elevated to the same heat levels as the incoming softened thermoplastic. Second, there are particular plastics, such as the CPET materials (crystallizable polyethylene terephalate), which demand heatable thermoforming molds.
With these particular materials, the mold is heated after the normal thermoforming procedures to create crystallization of the plastic. Tray products made
of these materials on heated mold bodies, after the crystallization stage, are
capable of withstanding conventional oven temperatures (400°F) without destruction. Currently, such ovenable trays are the "hottest" product items for
the prepared frozen food and TV dinner industry.

     In the heating of a thermoforming mold, several options are open for
the processor. One of the most common initial approaches is to heat the mold
electrically. Usually, the back side of the mold is predrilled and fitted with a
cartridge-type heater. The heat levels are controlled with an embedded thermostat that may work directly on the heater element or, for better control, may
register its reading with an outside control panel where even visual temperature readings can be obtained. With most electrically powered internal heating
systems, the conveniently small equipment setup offers many possibilities.
However, as with most sheet-heating heater elements, the same temperature
oscillation will persist and its effects could easily interfere with good thermoforming practices. The oscillation levels can be checked with costly control
devices. To obtain mold heating without the slightest chance of temperature
fluctuation, either very sensitive electronic devices or closed-loop temperature
control units should be used. In this method of mold heating, the mold's elevated temperature is controlled using the same fluid circulating apparatus as
that used for mold cooling. Instead of cooling the circulated fluids, however,
the equipment will raise the temperatures of the fluids, and through their contact with the mold, the preset heat will be transferred to the mold body. For a
less costly mold heating method, the processor can use an ordinary household
water heater coupled with a closed-loop plumbing and pump system connected
to the mold's heating channels. The water heater's thermostat can be adjusted
to the correct heat levels, while the water volume and the action of the circulation pump can maintain the mold's heat at satisfactory levels.

     Both electrical and circulating fluid types of heating can produce ele-
vated temperatures in the mold body. However, with these heating methods,
the entire mold body will be heated. If the thermoforming operator only needs
to heat certain areas of the mold or does not wish to use any outside source of
heating, insulating mold materials should be used. Insulating mold materials

can consist of those materials which, through the act of thermoforming absorb the heat of the contacting thermoplastic sheet and retain that heat after contact has been broken. This insulation can be provided by wood, phenolic, urea, polyamide formulation, or most popularly, the "syntactic foam" materials. The principle behind these materials for the purpose of mold heating can be found in their ability to pick up readily the higher temperatures of the heated thermoplastic sheet. The material's heat gain is supplied through the forming procedures when contact is made with the heated thermoplastic sheet. Because of their heat-insulating characteristics, the insulating materials will also retain the heat longer in their bodies. The way in which those insulating materials absorb and hold the heat is closely related to their density. Denser materials reach their expected temperature levels more slowly than do the lighter-density materials and require a greater number of and possibly longer contacts in order to reach ideal running temperatures. For example, it is not inconceivable that the phenolic plug-assist material will require 10 to 20 forming cycles before their temperatures will match that of the sheet temperature. However, the syntactic foam materials can boast not only fast heating abilities but also outstanding temperature retention. Syntactic foams are applied to the thermoforming molds; their heat accumulation to reach full production running condition can be attained within a few cycles. Only three to a maximum of six forming cycles are needed before their temperatures will match that of the heated thermoplastic sheet. The real gain with syntactic foam insulating material use is most evident when particular metal mold surfaces are exchanged to this material. In most thermoforming processes, considerable effort is made to produce well-controlled thermoplastic sheet temperatures and to maintain such controls throughout the entire production. With the adaptation of syntactic foams in the mold, no additional controls will be required, because the foam temperature will exactly match the sheet temperatures. The use of insulating materials such as syntactic foam is adapted particularly well to the manufacture of plug assists. However, they are not the only single mold components in which such materials can be useful. To fully illustrate the usefulness of syntactic foam, in Figure 79 comparisons are made between them and standard metal plug-assist materials. Most thermoforming toolmakers shape the entire plug assist out of syntactic foam. This is convenient but not at all necessary. As Figure 79 shows, the plug assist makes contact with the heated thermoplastic sheet only by its leading face. Therefore, capping the plug assist with a $1/2$- to 1-in.-thick syntactic foam layer will work just as well as a solid plug assist. In either case, the syntactic foam must be made mountable to a metallic mold component. The foam can be drilled and the holes can be tapped and threaded with any ordinary thread cutter. The material can also be glued to the metal surfaces; however, with the adhesive mounting, the use of locating pins is always a good practice to avoid any misalignments. Syntactic foam materials can

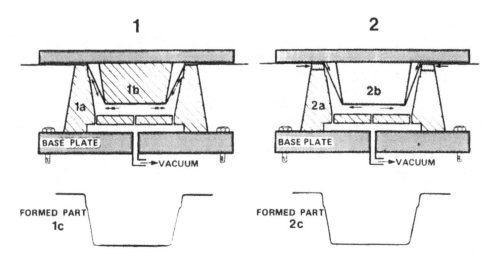

Figure 79 Benefits of using heat-retaining materials in a thermoforming mold. (1) All-metal mold: 1a, cooled all-metal mold cavity; 1b, cool all-metal plug assist; 1c, formed part with heavy flange and bottom, thin sidewalls. (2) Mold with heat-retaining material (syntactic foam): 2a, capped mold cavity lips (warm lip, cool mold body); 2b, syntactic foam plug assist (warm); 2c, formed part with even wall thickness.

be purchased to requested dimensions in premanufactured block form. Some knowledgeable tool shops can manufacture their own custom-shaped syntactic foams by casting them out of a mixture of a fine grade of hollow glass spheres and epoxy resin. The only criterion that need be considered is to obtain absolute uniformity in the syntactic foam density, or the material may separate in the casting. This can easily happen, since the hollow, lightweight, fine glass beads tend to separate out of the heavier, flowing epoxy materials. The two materials will separate as the epoxy is setting up. Any separation of this material results in density variations within the resulting syntactic foam casting. Variations in foam density within a thermoforming mold can interfere directly with the thermoforming results.

The use of syntactic foam materials for creating a warm mold surface is most appropriate with continuous web-fed thermoforming. In web-fed operations, syntactic foam can reach running conditions within three or more forming cycles. If no stoppage is experienced, temperature control will never be needed to maintain the ideal mold temperatures. The few flawed production cycles made at startup cannot be held against this mold-temperature-controlling method. The same temperature-controlling method with syntactic foam mold components can be made to work in a sheet-fed thermoforming operation, but only when the cycles are produced in rapid succession. In a sheet-fed

operation it is also necessary that the insulating mold material have an opportunity to receive heat and maintain it from cycle to cycle. If there are extremely
long intervals between cycles, the high cost of the thermoforming material
sacrificed in the beginning of the process will probably offset the advantages
of this mold-heating method in the sheet-fed operation.

C. Programmed Temperature Molds

There are basically two types of programmed temperature thermoforming
molds. The first changes temperatures along its entire body. For example,
within a single process cycle, a cool mold has to change from cold to hot and
back again. Necessary, radical temperature changes can be accomplished in
several ways. A mold can have the normal cooling channels with a coolant
flowing through them, and simultaneously have electrically powered embedded cartridge heater elements. The heater elements' heat will override the
cooling effects of the coolant. When the power is turned on, the mold gets hot;
when it is turned off, the flowing cooling fluids take over and cool the mold.
For additional efficiency, the coolant fluid can be shut off to gain faster heating
times. Either way, when a mold has to change its temperature radically, it will
take some time for the change to take effect. Mold temperature reactions will
be slower when the molds have higher body mass, since a lighter body mass
can accommodate temperature changes much faster. The only sensitivity to be
watched for with this changing mold temperature process is that the molds
should be constructed of materials that can take the rapid temperature changes
without falling apart. Most metal molds are not affected by the usual heat-re-
lated expansions and contractions. A minor problem encountered due to the
temperature changes is that of loosened screw fittings. However, molds with
a nonmetallic or multimaterial makeup can react adversely to rapidly alternating temperatures due to various rates of expansion among the materials. When
a mold is constructed from several different types of materials, the second type
of programmed temperature thermoforming mold comes into play. This type
of construction creates molds that develop different temperatures within the
same mold body as they are used. Instead of having a uniform temperature
throughout its body, the mold will have areas that are cooled and areas that
are either heated by an outside source or remain warm from the contacting
thermoplastic sheet. The principle behind these molds lies in the fact that the
thermoplastic sheet material will stretch more at the warmer contact points
(where no cooling is administered) and at the same time will resist stretching
where cooling is applied. This is the purpose of a programmed temperature
mold. The warm segments of the mold contacting the heated thermoplastic
sheet will not cool these areas and thus allow stretching to occur there. The
other sections of the mold will chill the thermoplastic material through the

contact; because of the cooling, the colder areas will not stretch as much in comparison and perhaps not at all.

The mold for this programmed temperature setup is constructed within the same configuration as any other mold. However, instead of single-material construction, a second insulating material that is like syntactic foam is added to the particular areas. The syntactic foam mold inserts are fitted into cutaway areas milled out of the original mold body, creating molds that have the intended mold configuration but with different materials incorporated into their body. For example, a male mold can have a "cold" aluminum body, while its four corner areas are made of syntactic foam, which becomes warm. In a reversed arrangement, the mold body is made of syntactic foam material and its corners are segmented out of aluminum. In this arrangement, the aluminum corner inserts are extended down the full height of the mold in order to attain cooling through contact with the mounting plate (Figure 80). Molds like this provide a combination of warm and cool areas, making up the mold structure in a preplanned fashion. Cleverly managed mold structures with preplanned warm and cool areas can prove just as effective a means as preprogrammed heating of the thermoplastic sheet. Between these two advanced heat management techniques—programmed sheet heating and intelligently designed mold temperature variations—the thermoformer can achieve outstanding wall thickness distribution in a thermoformed article. The management of temperature control within the thermoplastic sheet and on the mold surfaces can result in

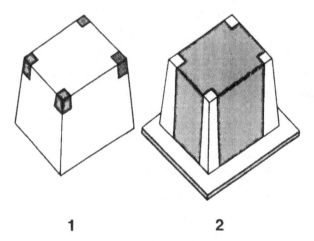

1   2

Figure 80   Programmed temperature molds: (1) cool aluminum mold body with warm syntactic foam inserts; (2) warm syntactic foam body with cool aluminum inserts extended all the way to a backup plate.

uncompromising quality and success for the thermoformer. Guidelines as to correct placement of the syntactic foam mold inserts are difficult to give because placement is highly dependent on mold configurations and design interactions. In many instances, a trial-and-error method can lead the practitioner to the most ideal mold segmentation. It is highly advisable that before metal cutting or milling is attempted in the mold structure, stick-on tape be placed on the selected cutaway areas for testing purposes. Although the tape's insulating abilities are not suitable for permanent use, the tape will give some indication as to whether the area's shape and size will permit proper functioning. If changes have to be made, it is far easier to try a different tape patch than to rework the mold. The use of tape can be considered a test measure only because the tape will leave its own impressions on the formed article, probably rendering it unusable.

III. SPECIAL MOLDS

Thermoforming often presents an out-of-the-ordinary challenge where routine thermoforming techniques with customary thermoforming molds cannot provide a specific need. Sometimes, a special mold construction or mold surface treatment is needed that can alter the thermoforming results in the favor of the processor. A special surface treatment or a different mold material can often establish major improvements within the process by making the product more eye appealing or possibly, easier to produce. Either of these benefits obtained through special mold applications will favor the specific or particular specialization. Most thermoforming procedures do not require specialized molds, and the higher costs of such molds can be prohibitive in the manufacture of the thermoformed article. Even though special molds are used only on rare occasions, they are worth mentioning. The knowledge and use of them can prove valuable when their special application is adapted to a mold. The thermoforming practitioner will have to make the final decision as to when and where to apply specialized mold features.

A. Mold Surfaces

When the thermoforming mold does not easily release the formed article, the slow release action may damage the product or hamper production speeds. Difficulty in the release of the part generally stems from deep-drawn, intricate design patterns or nominal sidewall taper angles. The surface binding that often occurs will happen sooner and create more problems with male molds; female mold configurations are less sensitive to release problems. Molds made out of the customary metals and carrying the usual mold surface finishes may not provide the necessary surface slipperiness that some design configurations require. In these instances the mold surfaces can be treated with a special surface coating that will make the surface slippery. There are instances when the mold surface must be made to resist frictional wear or even physical damage from abusive operating practices. Mold surfaces can encounter a harsh chemical environment. Certain chemical coatings or thermoplastic material compound combinations can induce corrosion of the mold surfaces. Some chemical attacks may not be readily noticeable on the mold, but can in a longer period, cause major destruction. Special mold materials or surface coatings can provide ample resistance to chemical damage. There are several sources of physical or chemical destruction that can be encountered in many of the thermoforming processes. When signs of damage to a mold are observed, it is most important that preventive steps be taken to solve the problem. Unfortunately, most problem solving addresses the damage, not the cause of the damage. The sensible way to resolve the problem is to pinpoint the true causes of the problem and then proceed to eliminate them permanently.

1. Release Coatings

Due to the effects of shrinkage, a thermoformed article often will simply not release readily from the mold. Such difficulties are encountered more often with male molds than with female molds. This problem can be further aggravated by a combination of mold configurations where male or female mold structures carry opposing mold configurations within themselves. Release is hampered even further when the mold designs are made with almost no sidewall taper angles. A thermoformer often realizes that all these negative thermoforming aspects will appear to some degree or other during most projects. Before using mechanical force to help remove the formed article during stripping, it is a good idea to apply a slip and release agent. Chemical slip and release agents can usually be applied directly to the plastic sheet. However, these chemical coatings may negatively affect the plastic and further use of the finished products. The chemicals applied to the plastic, for example, must have FDA approval if the thermoformed articles are to come in contact with food. Release agent chemicals could present a problem later in fabrication, gluing, and sealing of a thermoformed article or can even interfere with painting or labeling. Therefore, chemical coatings applied to the thermoplastic sheet or sprayed on the mold surfaces must be used with caution.

Many mold makers choose to use baked-on Teflon coating applied to the mold surface. This is a good method to use to provide slippage to nonslipping mold surfaces. The baked-on Teflon is durable (this is the composition and surface treatment used for "nonstick" cookware) and can provide a functional, long-lasting surface. The permanent Teflon coating, due to its baked-on finish, is applied only to metal molds, especially aluminum molds. In time the abrasiveness of some plastics can wear off the Teflon coating. Wear on the

Teflon surface is usually gradual and by no means uniform. In spots where more surface friction is present, more wear will be realized. If enough wear is encountered, the part release becomes a hindrance to production, and the mold must be pulled out of the thermoforming machine. The old, partially worn Teflon coating must be stripped from the mold's surface and a new Teflon coating baked on. The minute thickness of the Teflon coating is normally not a hindrance in the thermoforming procedure. Only in some critical high-speed operations will the Teflon coating interfere by acting as an insulating surface in the cooling phase of the thermoforming. Again, it is up to the thermoforming practitioner when and where to use Teflon coating on a mold surface, and a lack of full analysis of a particular thermoforming situation may produce an incorrect decision. The extra cost of Teflon coating becomes a luxury only when it is completely unnecessary.

2. Damage-Resistant Mold Surfaces

Thermoforming tooling placed in operation can often encounter a rough and potentially damaging environment. The thermoplastic sheet material used in a particular thermoforming process can have abrasive qualities that in the product removal cycles can cause substantial mold surface wear. Design configuration and plastic shrinkage contribute heavily to such mold surface destruction. The surface often experiences only minor wear, resulting simply in a polished surface area and is not significantly damaging. However, when critical portions experience wear, this can prove detrimental to the thermoformed product outcome. For example, the undercut and the stacking lugs can easily be worn away in constantly repeated production. Thermoformed articles containing undercut designs will usually resist removal during stripping from the mold. The thermoplastic material actually has to be forced out of the mold by a mechanical stripping action. Undercut formation stripping will itself increasingly force the plastic against the mold surface at these critical areas. The high pressures between the facing plastic and mold surface, combined with the sliding movement, will create unusually high levels of frictional forces. With softer, rubbery plastics, such stripping actions may not be noticeable at all; however, plastic materials having firmer consistencies or containing material fillers (pigments, stiffeners, or strengthening compounds such as chalk or glass fibers) can have highly abrasive tendencies. An abrasive material on the mold surfaces acts just like fine-grit sandpaper and will rapidly wear down most softer tooling materials.

There are also times when molds are damaged by people. When operating personnel experience difficulties with part stripping and the problem may not be a constant one, instead of resolving the predicament at its source, most operation personnel will resort to temporary measures to cure the problem. Without either consultation or authorization, they will attack the problem with a safety knife or a rigged-up "clawing" tool (a wooden handle with wire or a nail secured to its tip) to gouge out or disengage the stuck article from the mold. This practice is not only damaging to the mold surfaces but is dangerous to the operating personnel. Such practices should be treated as an unsafe act and immediately stopped for safety reasons unless acceptable "equipment lockout" is implemented during dislodging of the article. In most instances, mold surface damage will occur that indicates undesirable operating practices and should send a loud warning signal to supervisory personnel.

The mold damage caused by sharp, damaging tools can easily be resolved by using metals harder than aluminum or treating the aluminum with "hard-coating" anodization. This hard coating of aluminum is a chemical dip treatment that treats only the exposed surfaces of the aluminum mold body. The surface treatment is produced in a minute thickness (about 0.002 in.). Although its hardness is difficult to measure because of its thinness and the softer body backing, the treatment's resultant hardness is comparable to about 60 Rockwell hardness. In spite of its thinness, the hard coating will provide outstanding damage and wear protection for the aluminum mold body. Hard-coating anodization of aluminum mold surfaces should not be considered for all mold constructions; such an expenditure should be made only where a particular condition in thermoforming warrants such a surface.

One last type of mold surface damage worth mentioning is chemical attack. Airborne contaminants—thermoplastic-carried corrosive chemicals combined with the normal "sweat" (condensation) on the mold surface—can have corrosive results on most metal molds. Molds containing steel components will show rusting almost immediately, while those containing aluminum will show some resistance to normal corrosion unless a highly active corrosive residue is encountered. Mold sweat is almost pure distilled water and has the potential of reacting chemically with the aluminum. When antifog agents or other chemical treatments are used on the thermoplastic sheet surfaces, such materials can be deposited onto the mold surfaces to react later with the aluminum. The destruction of the aluminum molds may not be rapid but can be extreme if given enough time. When a corrosive environment is encountered, the thermoforming processor has two options: to flush or spray the mold periodically with a corrosion-neutralizing solution, or to build the mold out of a metal that is less vulnerable to the specific corrosive environment. Such metals would normally be a brass or beryllium alloy which will resist corrosion in any thermoforming situation. The use of these metals can increase basic mold costs at least two- to threefold.

B. High-Quality Mold Materials

Thermoforming molds that require extraordinary construction or materials for specific thermoforming uses, although rare, can have highly desirable benefits.

The use of exotic metal substances with highly specialized surfacing does have its place in thermoforming. These molds are not run-of-the-mill production tools and are produced only for specific needs, and their price tag will reflect this. A highly dimensional-toleranced thermoforming mold made of beryllium cavities and syntactic foam plug assists with trim-in-place hardened tool steel inserts together with all the accompanying manifolds for air, vacuum, and cooling can carry price tags comparable to most injection molds. Yet even with the equally substantial price tag, a thermoforming mold will probably have a greater number of ups within its mold configuration than will an injection mold. In addition, with the higher process cycles, thermoforming offers the advantage of higher output rates. High-quality thermoforming molds can boast accurate and refined details that closely duplicate injection mold tolerance levels. Molds made with this high degree of quality are specifically used when the customer is asking for high levels of detail and uniformity from one article to the next. The thermoforming equipment into which these molds are fit must have equally accurate and precise functions. The slightest deviation in platen movement and registration can cause misalignment between the two mating mold halves. Sloppy alignment can seriously damage the entire mold. When dealing with a precise mold arrangement, extra care must be taken when installing the molds into the machinery. Molds with nominal clearances between their female cavities and their male counterparts must have locating pins fitted into the mating bodies to assure proper mold alignment. The locating pins are removed after the mold is secured to the platens. Setting up a highly accurate mold in a critical thermoforming process should not depend on an "eyeball" method for alignment, as may be done in some general-purpose thermoforming mold alignment.

In an especially demanding thermoforming process when the basic sheet-forming technique incorporates borrowed techniques of compression molding, the common thermoforming limits can be further extended. With the adaptation of compression molding techniques, the two mating mold surfaces squeeze the thermoplastic sheet between them through the mechanical force of the platens. With this method, extra benefits can be gained. First, the thermoplastic sheet can be squeezed into much smaller-radiused corner forms than normal thermoforming methods will allow. Second, surface design patterns of a mold can be squeezed and transferred into the forming plastic, just as in a stamping operation. Third, a perfectly parallel and absolutely flat surface can be squeezed into a plastic sheet by the two opposing mold surfaces. This is most important when a close-to-perfect flange is to be produced on a thermoformed container, one that will offer an acceptable "double-seam" sealing capability with a crimped metal lid. This flat, parallel flange surface and its surface smoothness can be produced satisfactorily by this method only between the two opposing mold surfaces.

The same criteria will exist when the thermoforming goal is to develop an optically distortion-free bottom area on a clear thermoformed receptacle. In this case, the two facing mold portions, which create the specific bottom area, must mate with enough force to squeeze the plastic sheet and provide the distortion-free parallel surface. To provide a flawless impression on the plastic, the mold surfaces can be chrome or nickel plated. In this way, a perfectly smooth mirrorlike mold surface will be squeezed directly onto the plastic. With this compression-type molding technique, any excess material can also be squeezed out under the two mating mold surfaces and pushed into the sidewalls. When the stretching procedure is timed correctly, this squeezed-out excess material can easily be absorbed into the sidewall without any trace. If the stretching is insufficient, the material will show up as a circumferential step or bulge on the sidewalls near the squeezed surface area.

There is one more aspect of high-quality molds that should be discussed here. The best-quality molds and most delicate thermoforming machines are no guarantee of success if the thermoplastic sheet is not administered with equal preciseness throughout the thermoforming process. To satisfy such criteria, a mold-mounted clamping device must be applied. These specific clamping units are either built to surround the entire mold in a framelike fashion, or when individual sheet capture is needed for each mold, the clamping mechanism is built either in a grid pattern or with an independently operating clamp system (see Figures 13 and 14). All of these types of clamping mechanisms incorporated with a thermoforming mold require high levels of engineering, not only for the particular function but for compactness. Internal clamping units must be closely fitted around each mold body without interference from the mold actuations.

C. Special-Purpose Molds

Every so often, a special mold is called into service to be used for a specific product and its adaptation will fit only that situation or product line. Since they cannot be used for any other thermoforming method because of their specialization, some special-purpose molds are tied to specific industries or specific end-user categories. Each special-purpose mold has its own criteria, material makeup, and specialized thermoforming technique. Knowledge of these special-purpose molds and how to use them is most valuable when a thermoforming processor becomes involved with unusual tasks or project developments. Having this knowledge is just one aspect of maintaining a large base of helpful hints upon which to draw when working to solve a newly encountered task.

1. Porous Mold Materials

Thermoforming molds can be produced out of materials—metallic or non-metallic—that provide a high degree of porosity throughout the mold body.

The real advantage of using such porous materials is that the entire mold surface can become activated by the vacuum forces. In ordinary mold cavity air
evacuation, where holes are drilled for the introduction of a vacuum, no further
pulling of the material will be achieved on the areas between the hole patterns.
On the other hand, porous molds present an entire surface of holes for the
vacuum. As some pores on the surface become deactivated by the forming
plastic, the remaining pores will receive even more of the forming forces. When
using porous surface molds, the only criterion to be met is that all the nonessential surfaces, such as the mold's outer surface, be sealed off with a sealant
that plugs up the porosity in these nonessential mold surfaces. This will direct
all the vacuum forces to the actual forming surface. With this porous mold
surface, the vacuum forming force will apply all its power evenly along the
surface, creating surface detail transfer so good that even the finest surface
patterns of an animal hide can be duplicated perfectly. Since the surface of
porous molds contains numerous tiny holes, even the most delicate sewing-
thread details can be transferred to the forming plastic. The thermoplastic
sheet thickness remains a factor in detail transfer, of course; thinner materials
show detail better than do thicker ones.

     Porous molds can be produced of metallic materials such as aluminum
powder, seizing the fine particles together, or ceramic-type materials that are
baked. The main drawback to porous molds is their poor heat transfer ability.
Molds made of porous material cannot be used where rapid cycle repetition
is mandated. Due to their low cooling abilities, such molds are limited to use
in somewhat-low-volume productions.

2. Dual-Function Molds

Some molds serve a dual function. They are first used as thermoforming molds
to form a thin skin out of the thermoplastic material. After the forming is
completed, the formed sheet is retained in the mold and will index out of the
forming station together with the mold. The two will then register into a new
station, where the formed body is subjected to a second process. This second
process can apply various secondary applications, such as foam filling, coating,
adhesive spraying, and flock application. After the secondary application is
performed, the combined article is removed from the mold, which then returns
to the primary thermoforming stage. This type of mold can be placed into a
rotary thermoforming machine format or can be set into a single-station thermoforming setup. In both instances, after completion of the thermoforming
process, the mold will either index into a subsequent station or be flipped over
to an upward-facing position and subjected to the secondary process. Throughout these secondary operations, the thermoformed article remains in the mold
and is subjected to a subsequent process which finalizes the thermoformed

product into its finished form. Molds like this are most common in the automotive industry, one of the popular products being the automobile dashboard.
First, a thinly skinned surface structure is thermoformed; this is not removed
from the mold. The mold, with the formed skin, will index into a foam molding
operation. The original mold is closed on another mold half, creating a new
foam injection mold. Other product applications can be made just as easily
with this type of thermoforming technique.

3. Surrogate Molds

In this specialized thermoforming arrangement, instead of using a permanent
mold installation, a previously produced product body is substituted for the
mold. The actual forming of the thermoplastic sheet is made right on the surrogate mold form, which after thermoforming actually becomes part of the
manufactured article. For each thermoforming cycle a new surrogate mold
body is placed or indexed into the forming area. After thermoforming, the
combined structure is removed. During thermoforming, both the surrogate
mold structure and the thermoplastic sheet that has been formed on it will
undergo unification. Actually, due to the heat involved with the forming, a
strong bond can develop between the surrogate mold surface and the forming
plastic. The bond can be further improved by the use of heat-activated adhe-
sives. This feature of bonding can be judged as both an advantage and an
objective for using this type of thermoforming procedure. One of many products produced in this way is car door panels. In making this product, the surrogate mold consists of a composite "substrate" backing material. The material
is a wood-particle-based product which by some special process can be formed
into the desired configuration. The formed composite substraight panel becomes the mold. Through its porosity or any drilled holes, vacuum can be
applied for thermoforming. As the thermoplastic sheet is formed onto the
substrate, just like a skin, they bond strongly to each other. Because no mold
is involved and the composite material is not an ideal mold cooling medium,
only air cooling achieves proper cooling.

   There are other products that can be manufactured using the surrogate
mold thermoforming techniques, but they are more commonly classified under
packaging methods. One such method is "skin packaging," in which the actual
packaged goods become the surrogate mold. The plastic film is actually thermoformed right over the to-be-packaged item. The content becomes the mold
and the plastic film will enclose it, formed to its contours. The vacuum forces
are usually applied through the hole pattern, punched into cardboard stock
prior to the application of the film. Through these holes, vacuum is administered which pulls the film against the cardboard and the product to be packaged, creating a finished package.

4. Twin-Sheet-Forming Molds

For simultaneous thermoforming of twin thermoplastic sheet stocks into a single product, two opposing female molds are required. The two molds' cavity configurations do not need to be matching. The only criterion that must be met with these thermoforming molds is that the two molds must have matching cavity lip contours. These mating cavity lips are the key element in this thermoforming process since they form the sealing surface for unification of the two parts. The end results of this process closely duplicate those of a blow-molding process.

The concept of twin-sheet thermoforming is used mainly to produce hollow products from sheets instead of from a molten glob or parison. As in blow molding, these products are made to encapsulate air to provide flotation or for the storage of liquids. For example, small boat hulls, surfboards, motor fuel tanks, chemical shipping containers, and double-sided signs are good candidates for this type of thermoforming and mold construction.

Forming of the twin sheet is made similarly to customary vacuum or pressure forming. The only difference in twin-sheet forming is that airflow must be provided between the two sheets to accomplish the forming. If vacuum forces are used, air from the atmosphere must be allowed to enter between the two forming sheets. With pressure forming, the pressure force should be introduced between the twin sheets without air escape to accomplish full forming. Whether vacuum, air pressure, or a combination is used for the forming, the two molds must be squeezed together immediately to perform the sealing of the two thermoformed halves. There are several variations of this technique, but some are proprietary in nature.

IV. COMMON MOLD ERRORS

Most thermoformers who are closely involved with the production of thermoforming can find themselves in a competitive environment, concerned with the cost of molds and their trouble-free operation. Each concern has specific criteria that must be met to guarantee success. Errors made in thermoform tooling, especially in the mold, can halt most thermoforming efforts. They can not only cause problems for thermoforming producers, but could even affect subsequent product users. Difficulties in a thermoforming process can be caused by many things or by a combination of problems. It is also possible that a suspected fault is not the actual cause of the difficulties; hidden causes are also involved. Inaccurate, hastily made problem detection often misleads the thermoforming processor.

In the thermoforming process many forming difficulties can be traced back to the errors made in the original mold. This is especially true when new mold fixtures are first put into operation. In mold construction not all tool shops are equally familiar with the specific process requirement. Certain tool shops place more emphasis on mold appearance and overly specified dimensions than on proper function. On the other hand, if important mold features are not made up properly, it will cause faulty functions in the thermoforming. Tooling quotations received from various tool shops can often be very different, not only in pricing but in mold quality as well. Thermoforming practitioners must compare tooling costs and evaluate each quotation. Low-priced tooling which later needs rework or modification can in the long run cost more than higher-priced molds which will not need modification or rework.

Please remember that a good mold maker knows where to construct high degrees of tolerance in the mold and at the same time recognizes where less work effort and cost-saving procedures are acceptable. Most mold makers who are acquainted with this critical but generally not followed rule, can provide the best working-quality thermoforming molds without exorbitant cost. Unfortunately, not all tool shops can produce a wide range of mold construction at a competitive price.

The most commonly encountered tooling errors involve vent-hole placement, poor forming force connection, poor mold cooling practice, incorrect mold alignment, unfilled mold gaps, and overly specified molds.

A. Vent-Hole Placement

This is the most common mold-making error. Drilling holes into critical areas in the mold is not only time consuming, but not all areas are readily accessible. The mold's body often has to be tilted to clear the drill chuck in the drilling procedure. It is also possible that the body is not drilled all the way through to create an open channel for the forming force or that the hole size and number are insufficient for thermoforming.

Remedy: Additional holes or enlarged holes should be drilled into the mold body.

B. Poor Forming Force Connection

This could be blamed partially on the mold. The connecting hoses or plumbing lines between the mold and the forming source (vacuum pump or air compressor) are probably at fault. Insufficient line size or connections made with reduced size couplings can also cause poor forming forces. In both cases it will cause a reduction in the forming forces. The reduced airflow could cause the same results if flow reduction is incorporated somewhere inside the mold's body.

Remedy: Plumbing lines, fittings, and connecting couplers of sufficient size should be used to connect the forming source with the vent holes in the mold. The mold's body should be equipped with sufficient branching connections from the plumbing entrance to the venting holes. Adequate cross-section machined-in channeling (milled-in gating) should not cause restricted airflow; however, if any undersized gating is present, it should be enlarged.

It is also worth mentioning that in most thermoforming machine installations and operations the only vacuum gauge is available and installed right at the vacuum pump. Having a vacuum gauge and taking a reading near the pump may not provide an accurate indication as to what the vacuum level is at the mold. Often plumbing restrictions, longer distance between the vacuum pump and the mold, plus a multitude of plumbing turns with elbow installations can reduce the flow and deprive the force of vacuum at the mold. It would be advisable to have another vacuum gauge installed near the mold, that is visible to the operator throughout the entire forming cycle. The gauge next to the mold not only can indicate the maximum levels of vacuum available at the beginning of the forming cycle, but can give additional vital information as to how the vacuum and mold functions are behaving throughout the forming. It can provide knowledge about the mold: whether it has a leak or insufficient sealing is achieved between the thermoplastic sheet and the mold, if the vacuum forming is performed too slowly, and gives comparison to the previous forming speeds. It also can give a good indication to the vacuum system recovery rate for the next forming cycle.

This vacuum gauge when placed next to the mold must be hooked up to the main vacuum line. On a few occasions I have observed that this vacuum gauge next to the mold had a long separate line that is directly connected back to the vacuum pump. In such installations, the gauge is next to the mold, but it is reading the vacuum conditions at the pump. Such a reading cannot and will not provide the information the operator needs. The normal and customary reading of the vacuum gauge should indicate a substantial drop in vacuum levels as the forming is accomplished, but the vacuum level drop should stop at one point and then a recovery or increase in vacuum levels should follow. If the reading on the gauge drops down to zero, it is a good indication that there is a substantial leak in the system that needs to be repaired.

C. Poor Mold Cooling Practice

Most errors in mold cooling are caused by improper coolant flow controls. Time after time, the wrong valve is used to adjust fluid flow. For quick visual identification of the outflowing valve, the valve should be painted bright blue. The second error, an improperly designed cooling channel, is not as common, but its results are serious and far more difficult to resolve. When a mold has a wide range of temperature variations, especially from one area to the next, an

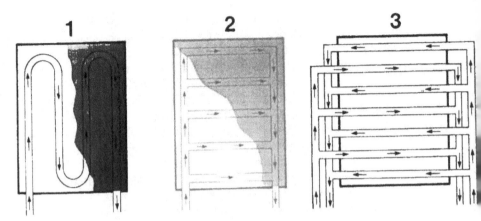

Figure 81 Cooling channel patterns: (1) undulatory cooling pattern; (2) branching cooling pattern; (3) alternating flow pattern with outside manifolds and two inlets and outlets.

incorrect cooling flow pattern is strongly indicated. It is also possible that the mold can absorb more heat at the side facing the oven.

Remedy: All mold cooling channel designs should be preplanned to use an alternating flow pattern with external manifolding (Figure 81). The mold must have sufficient flow rates. In certain molds where extreme heat removal is in demand, flow-turbulent-causing installation should be implemented. Turbulent fluid flow should be created within the mold body to eliminate the slightest coolant fluid stagnation.

D. Incorrect Mold Alignment

This is not a true mold-making error; however, the mold maker can ensure that such an error should not develop. The error is usually encountered when the mold is installed into a thermoforming machine and the two opposing mold halves are not aligned with each other. In an extremely severe case, the two misaligned mold halves, upon closing, can seriously damage each other. The problem is less noticeable if the misalignment is minute; however, it can still cause problems in thermoforming.

Remedy: Installation of locating pins or alignment clamps not only eliminates such misalignment problems but can reduce mold changeover time as well.

E. Unfilled Mold Gaps

This should not be judged as a mold-maker error; however, most of the time the outcome of the forming result will prompt the practitioner to check the mold or mold setup. This problem will exist only when using a thermoplastic foam material. The thermoplastic foam for thermoforming must be made of closed-cell foam structure. An open-cell foam tends to collapse in foam structure when subjected to heat. For this reason, closed cells or at least a high count of closed-cell-content forms, are adequate only for the thermoforming process. In thermoforming, only closed-cell foam sheet will expand in a thickness ratio of approximately 2:1 when exposed to heat. For example, a $1/4$-in.-thick heated foam sheet is originated from a $1/8$-in. thickness. If the foam material contains higher open-cell counts, or the sheet is subjected to some type of foam destruction prior to thermoforming, its expansion rate (blow-up ratio) will be severely reduced.

Expanded foam material in the thermoforming process is compressed between the two opposing mold halves. If the foam does not expand enough in thickness to fill the mold gap, the formed part wall measurements will be undersized. Undersized wall thickness in the forming is often blamed on mold-making or setup errors. However, the lack of material thickness is the cause of reduced thickness in the product, after forming, not a faulty mold or poor mold setup.

Remedy: First, the foam material thickness should be gauged. Second, the foam should be subjected to heat to gain its maximum expansion rate, then measured to see if there is any expansion. If the foam does not expand to fill the required mold gap, it should not be expected to thermoform in full detail on both sides of the article. The unfilled mold gap areas will remain unfilled even after the introduction of higher vacuum forces. Such foam material should be replaced with another sheet which has the proper amount of expansion rate or increased gauge thickness. An alternative solution is to reduce the mold gap. However, it will result in diminished strength of the finished article.

F. Overly Specified Molds

Thermoforming molds are often specified to extremely high quality standards. Corners and sidewall/bottom radius are often designed and specified in detail that the particular thermoforming method is unable to produce. Surface finishing could also be specified in a more refined texture than the particular thermoforming method would require. Short-run production molds or experimental molds made of high-grade materials or mold-making materials used in a "lavish" way are unnecessary. It may please the purchaser, but it will not enhance mold performance and can add to the mold's cost. As mentioned earlier, a good mold maker is one who knows how to provide top-quality, accurate molds while offering the best tooling pricing.

6
Economics of Thermoforming

I. INTRODUCTION

Like any other manufacturing method, the thermoforming process should be subjected to periodic economic evaluation and cost control analysis. Through these evaluations and cost studies, the process can not only be compared with other manufacturing methods aimed at the same market, but can have performance controls established. Having full economic control over the process can help to establish profitability and competitiveness. The goal of all manufacturers is to realize satisfactory profit levels as a justification for their efforts. The combined cost of material, energy, equipment, labor, and shipping is part of the cost of manufacturing, which could include the cost of merchandising. The many components of financial data can help in controlling the individual cost factors. If a certain area just cannot be improved upon, another controllable segment of the process will have to take over the cost burden. Overall, the average income must show satisfactory rewards on the bottom line or the project is not worth bothering with. There are as many outside economic factors that can influence the financial condition of a thermoforming project as there are conditions within the process itself.

In this chapter we do not intend to cover all aspects of a comprehensive financial study. Books on financial aspects of product manufacturing and merchandising are available from many qualified sources. In this chapter we concentrate on the special factors that contribute directly to the thermoforming process but do not give enough details for a general-purpose cost evaluation. Economic factors such as labor rates, energy costs, current material prices, geographic and competitive locations, and market conditions should be taken into account and where possible, revised with updated numbers. However, the analysis of financial data accumulated over a specific thermoforming project or product can reveal its profit potential and give sufficient clues as to whether or not a competitive situation has been realized. It is more important here to concentrate on economic factors within the control of the thermoforming processor that can be manipulated to influence directly the outcome of a thermoforming project.

II. THERMOPLASTIC SHEETS

One of the basic characteristics of the thermoforming process is that all thermoforming methods start with a thermoplastic sheet. The many types of thermoplastic resins used in the various sheet constructions can be produced in a multitude of sizes and more important, in as many thicknesses. The conversion of resin materials into sheet form can be accomplished in-house, within the thermoforming plant, or the sheet can be supplied by independent, outside sheet manufacturers. Some of the resin materials from which the thermoplastic sheet is made can be classified as general-purpose resins because their broad range of use and high volume of demand allow them to be produced in quantity. On the other side of the material use spectrum are specialty resin materials that qualify as high-grade engineering materials and are used only in limited applications. The second group of materials will not enjoy the same low pricing as the general-purpose materials. Since all the thermoplastic material used in thermoforming must be supplied in sheet form—whether in precut panels, roll form, or directly extruded web—thickness is one of its most important specifications. The sheet thickness or gauge is crucial not only to the thermoforming process and the outcome of the formed article but is also a key factor in the economic outcome. Less material is required to produce a sheet when using the lighter weight, thinner-gauge plastics. Conversely, the use of heavier-gauge materials consumes more material and tends to increase costs. In the successively produced cycles of thermoforming, a material gauge reduction or increase, implemented either purposely or accidentally, can strongly influence the product quality and/or economic aspects of a project. Gauge variation, reduction, or increase always has a major effect on the thermoforming process itself. However, the sheet material thickness will first affect the basic material supply costs. The material cost is one of the most influential economic factors in thermoforming. Whenever thermoforming is performed, whether with heavy- or thin-gauge material, the initial material cost can be a deciding factor in whether the particular product thermoformed ends up in the "red" or is profitable. Thermoplastic sheet purchased from sheet suppliers leaves very little opportunity for cost improvement. The supply source has to offer the sheet to the thermoformer at a current market price, which should, of course, include some profit for the sheet maker. Below-market-price sheet supplies should always be scrutinized for resin source or quality of makeup. There are three cost-reducing factors available to a thermoforming producer: lower quality resin supply, volume purchasing, or reduction in thickness. Individually or combined, any of these factors can have major implications not only on the economic outcome but on all aspects of the thermoforming process.

A. Material Requirements

There are two basic ways to acquire thermoplastic sheets for thermoforming. The first and simpler of the two, and usually costlier as well, is direct purchase from a sheet supplier. The supplier could be a distributor or the sheet manufacturer. In either case, some profit taking is to be expected, driving the sheet price somewhat higher than if the sheet were produced in-house. Buying sheet material from distributors in small quantities, especially if of different types and makeup, is highly desirable. The only drawback to such a supply source is that the sheet specifications cannot readily be altered and only stock sheet materials can be purchased. For larger-quantity purchasing, possibly some customization of specifications can be requested, especially when dealing directly with the sheet maker. With a larger quantity of orders, the opportunity to negotiate some sort of scrap recycling and returning arrangement can also be worked out. The opportunity for recycled scrap is a tremendous financial advantage. Of course, if the thermoforming operation has its own sheet-producing facilities, full control of specifications can be realized and all the financial advantages are also available to the thermoformer. In a true sense, with in-house sheet manufacturing, the inherited scrap can be of equal value to the virgin material purchased from a resin supplier. Only a nominal cost is involved in the recycling process.

To establish material requirements, it is a good idea to produce test samples or to use a reasonable estimate of the best material thickness for the specific article. The gauge of the sheet will strongly influence the finished article strength as well as the material requirement for that particular thermoforming program. Product dimensions can be worked into a mold layout arrangement that includes an allowance for the spacing between the parts and the needed edge area that together make up the entire forming area of one panel or a single shot size. Figure 82 illustrates the method for obtaining such measurements. Through measurement of the particular layout, the estimated product area, scrap area, and percentage ratios of the two can be established just as easily.

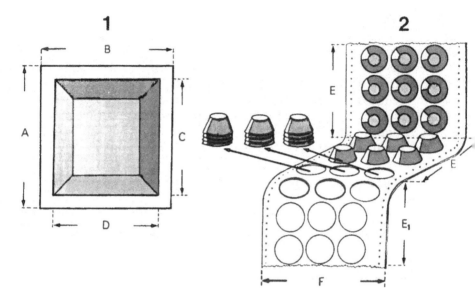

Figure 82   Thermoformed panel layout. 1. Formed panel from a sheet fed thermo-forming—A × B = panel size, D × C = formed part, [A × B] – [D × C] = scrap. 2. Formed web with scrap skeleton—F = web width, E × F = represents a single forming (shot) panel size, E_1 × F = scrap skeleton of a single shot.

As each individual thermoplastic is converted into sheet form, it will yield a different quantity of material. The area in square inches that can be obtained from a pound of plastic is directly influenced by the plastic's specific gravity and sheet thickness. It should be obvious that from a given weight of plastic, a larger sheet area can be produced at a lower gauge thickness than at a higher gauge. For example, 1 lb of polystyrene at 1 mil thickness will produce 26,300 in.^{2} of sheet, while at 10 mils it can make only 2630 in.^{2}, at 100 mils just 263 in.^{2} of sheet materials, and so on. Since the various types of thermoplastics have different specific gravities, their yields vary accordingly. Table 6 shows a comparison of some of the popular thermoplastics for 1 lb of material converted into square inches of yield measured at 1 mil thickness. Although these yield numbers are more meaningful to the sheet producers, simple calculations can produce useful data for the thermoformer as well.

The same numbers can be obtained from the formulas and the data contained in Table 7. The first formula will determine the weight of a measurable panel. For a panel weight in pounds, use the inch measurement and the factor number from the second column. If metric numbers are preferred, use metric

Table 6 Various Thermoplastic Yields (square inches per pound at 1 mil thickness)

Acetate	22,000
Polystyrene	26,300
Polyethylene	
Low-density	30,000
High-density	29,000
Polypropylene	30,300
Polyester	20,000–21,000
PVC	19,000–22,000

Table 7 Thermoplastic Data for Calculating Yield

Formula for computing sheet weight:
L × W × thickness × factor = sheet weight
For example, with ABS material:
42 in. × 48 in. × 0.125 in. × 0.038 in. = 9.576 lb/sheet

Formula for computing weight per square foot:
Specific gravity × 5.2 × thickness = pounds per square foot
For example, with ABS material:
1.05 in. × 5.2 in. × 0.125 in. = 0.6825 lb/ft^2

Thermoplastic material	Specific gravity (g/cm^3)	Factor (lb/in.^{3})
ABS	1.05	0.038
Acetate	1.30	0.047
Acrylic		
DR	1.15	0.047
MP.XT	1.12	0.041
Butyrate	1.20	0.043
PETG	1.27	0.0458
Polycarbonate	1.20	0.0433
Polyethylene		
Low-density	0.92	0.0332
High-density	0.96	0.0346
Polypropylene	0.90	0.0324
Polystyrene	1.05	0.038
Propionate	1.20	0.0433
Vinyl	1.41	0.051

324   Chapter 6

measurements, and the first column of specific gravity with the formula and
the gram weight of the panel will be obtained. Both columns contain average
numbers for the particular plastic types. For more accurate calculations, these
numbers must be readjusted to the exact specific gravity values. The exact
specific gravity of a specific plastic resin is supplied by the resin supplier, and
the thermoformer can use those numbers. The variation in the specific gravity
of any plastic is influenced by the resin molecular weight and molecular-weight
distribution ratios. It is an acceptable practice to verify the specific gravity
numbers of a newly supplied resin and then readjust the original calculation.
The second formula of Table 7 uses a single square root of area as the standard
of measure. With this formula, the weight of a plastic sheet made from a given
thermoplastic material and thickness can be calculated. This formula can provide fast comparisons among the different thermoplastic sheet materials. With
this formula, the area of any of the various types of thermoplastic material can
be calculated as long as the plastic's specific gravity is known.

For the purpose of comparison, Table 8 provides a list of popular gauge
thicknesses. To find a gauge not shown on the list, either add, average, or

Table 8 Pounds per Square Foot Yield Factors

Gauge (in.)	Acetate	Polystyrene	Polyethylene Low-density	High-density	Polypropylene	Polyester	Vinyl
0.005	0.033	0.027	0.024	0.025	0.024	0.034	0.035
0.0075	0.049	0.041	0.036	0.037	0.035	0.051	0.053
0.010	0.065	0.055	0.048	0.049	0.047	0.068	0.070
0.015	0.098	0.082	0.072	0.074	0.071	0.103	0.105
0.020	0.131	0.109	0.096	0.099	0.095	0.137	0.140
0.030	0.196	0.164	0.0144	0.149	0.142	0.206	0.211
0.040	0.262	0.219	0.192	0.198	0.190	0.274	0.281
0.050	0.327	0.274	0.240	0.248	0.237	0.343	0.351
0.060	0.393	0.333	0.288	0.298	0.285	0.411	0.421
0.070	0.458	0.383	0.336	0.347	0.332	0.480	0.492
0.080	0.523	0.411	0.384	0.397	0.380	0.548	0.562
0.090	0.589	0.493	0.432	0.447	0.427	0.617	0.632
0.100	0.654	0.555	0.480	0.496	0.475	0.686	0.702
0.110	0.720	0.602	0.528	0.546	0.523	0.754	0.773
0.125	0.818	0.684	0.600	0.620	0.594	0.857	0.878
0.150	0.982	0.821	0.720	0.745	0.713	1.208	1.053
0.180	1.178	0.985	0.864	0.894	0.855	1.234	1.264
0.200	1.394	1.095	0.960	0.993	0.950	1.371	1.405
0.250	1.636	1.368	1.200	1.241	1.188	1.714	1.756

interpolate between any pairs of numbers. It is important to remember that this table is based on the natural plastic resins without pigmentation or filler. The color pigmentation or filler materials can cause a 5 to 10% modification of these numbers.

With these formulas, a thermoforming processor can produce numbers to establish how much thermoplastic sheet material will be needed for each thermoforming cycle. Further calculation can also determine how much the product will weigh and how much scrap will be produced by each cycle. By using these numbers, the entire production can be calculated. The numbers gained by these calculations represent the exact quantity needed in the thermoforming process without any allowance for inefficiency. Extra material must be allocated for start-ups, normal production errors, and rejected articles of arguable quality. As a realistic expectation within the thermoforming process, some extra material must be allocated for such errors. It is customary to provide a 10% allowance to cover such discrepancies. However, some product criteria or manufacturing practices can easily raise the material's dependency on outside factors for providing a satisfactory outcome. Good manufacturing practice and the implementation of production controls should keep the thermoforming procedures within manageable levels.

B. Material Cost Management

Sheet material is one of the costliest factors in thermoforming. The sheet cost can be substantial when thermoformed products are either produced in small volume of heavy-gauge material, in tremendous quantities from thin-gauge material, or from initially expensive exotic materials. For example, a thermoformer who produces pickup truck bed liners could have materials requirement for each forming cycle costing well above $100 per panel. On the other hand, a large-volume producer using thin-gauge materials will realize high cost values not in each forming cycle, but in the repetition of the process, where large quantities of sheet material are used. Ultra-high-volume producers can have such high material consumption discrepancy levels that even a minute quantity of material per article (perhaps less than a gram each), can save or lose several rail cars of resin over a period of time. Maintaining desired quantities of material usage requires close control of the thermoforming procedure.

One of the basic rules of thermoforming is that the sheet supply is always produced by weight; however, thermoformed products are always sold by piece. With such a differentiation between buying and selling practices, a dangerous opportunity for error can develop under poor management. Thermoformers who purchase ready-made thermoplastic sheet from outside suppliers must watch closely their sheet costs and order quantities. Any overestimates of sheet purchases that may not be acceptable by the formed articles' customers can end up as leftover stock, thus tying up cash. This situation becomes most severe when no further product orders are received and a number of thermoforming projects have been oversupplied with material.

The actual pricing of sheet from various sheet suppliers is usually made in direct proportion to the cost of raw resin. Slight variations in pricing can be influenced by local market conditions, discounted volume buying, or by routine scrap recycling arrangements. In general, outside sheet suppliers are pretty much in line with each other on pricing. If one supplier consistently undercuts the rest of industry, it is highly recommended that the sheet from this supplier receive the utmost scrutiny before a purchasing commitment is made.

Thermoforming sheet material purchasers have limited authority to specify a resin supplier, especially if only small quantities of sheet materials are purchased. If sheet material supplies are ordered on a job-to-job basis in small quantities, neither the resin origin nor sheet thickness can be specified beforehand, and the purchaser must accept shipments from existing stock supplies or must await a combined order with other purchasers to justify a production run. Usually, a nominal waiting period is involved with a small or nonroutine order. An out-of-the-ordinary color change, for example, requires "purging" (cleaning by extra material feed) of the extruder, creating extra material waste and robbing production time from the line.

A purchase by a thermoformer of all virgin resin or a specified proportion of blends from an outside sheet supplier is one based on mutual trust. It is nearly impossible and economically unjustifiable to check or prove resin blend ratios after the sheet is made. Complete confidence between the sheet supplier and the thermoforming practitioner can eliminate questions and mistrust. It is true that an occasional unapproved blending of scrap or lesser-quality resin supply mixed together with choice resins to reduce costs can cause production problems with some products. On the other hand, it is also true that at the first signs of inability to thermoform, the thermoplastic sheet receives most of the blame. It must be remembered that maintaining the quality of sheet supply is in the best interests of sheet manufacturing firms. It is also important to remember that changing the sheet supplier, and possibly also the resin supplier, could cause changes in thermoforming performance and even in product outcome. A major change in sheet supply should be considered only after thorough testing of all phases of thermoforming.

Greater control of the sheet supply and further economic gains can be realized by having an in-house sheet-making facility. With a captive sheet manufacturing operation, the first benefit is full control of sheet making. With in-house sheet production, the resin purchase, virgin and scrap material blending, scrap recycling, and color choice become self-directed. The second benefit is the ability to create custom supplies of sheets that are unavailable from stock sources. The third benefit is that the profit structuring of the final product can also be realigned, providing a competitive edge over thermoformers who rely on outside sheet supplies. Unfortunately, in-house sheet manufacturing does not eliminate all antagonism between the sheet maker and the thermoformer; it only keeps it under one roof. The true benefits of in-house sheet manufacturing can be found in the three major areas mentioned: economics, ready availability, and most important, valuable scrap management opportunities.

C. Scrap Management

Although the thermoforming process begins with a sheet, not all of the sheet area is converted into the finished article. Some portions of the sheet are needed for clamping and holding the sheet, and some portions of the sheet are used as spacing between products. These areas make up the scrap skeleton and become "leftovers" in each thermoforming cycle. The production of scrap is an inadvertent result of each thermoforming cycle, and its reduction and minimization are ongoing goals of every thermoforming practitioner. There are very few instances in which these areas become part of the actual product and thus where no scrap is produced.

Scrapless thermoforming has intrigued thermoformers for many years, but to date, methods of achieving it have met with very little commercial success. However, as the industry moves more and more into coextruded layer sheet usage, especially in the fields of barrier materials for food, more emphasis will be placed on scrapless thermoforming methods. The demand for scrapless forming originates from the problem that some barrier-layer materials do not allow recycling of the scrap, resulting in scrap waste that has to be totally discarded. As yet, there are no commercially employed large-scale uses for scrapless forming, and interest in them remains only in developmental projects. However, it is quite possible that such thermoforming methods will become the front runner of the industry in the future.

1. Scrap Production

As the thermoforming practitioner recognizes, most thermoforming processes are burdened with scrap production, but it is also within a processor's control to minimize waste. The average thermoforming process produces between 10 and 15% of total weight as scrap. With some effort and proper implementation, this level can be reduced to less than 10%. However, it is not uncommon to realize much higher levels of scrap production, ranging to 20% or more. For example, in the case of a round product, nearly 40% scrap can be realized due to the configuration of the articles and the large areas left between them. Of course, there are no upper limits to scrap production. Odd configurations and various other production criteria can force situations on the thermoformer where the level of scrap will exceed that of the actual product. Abundant scrap production not only affects the pricing criteria of such articles but directly involves the scrap-handling limits of downstream equipment as well as influencing product quality.

With certain products, only a limited quantity of recycled material can be mixed with virgin material before product quality degradation results. Higher levels of scrap production can present problems in material recycling. Some products are too sensitive to accept recycled materials or can accept only limited quantities of them before adverse behavior causes their quality to deteriorate. For some products only strictly virgin material can be used. Products used for medical product packaging media, which must provide unquestionable integrity and sterility, cannot afford material failure due to recycled resin usage. On the other hand, there are thermoformed products that can accept very large quantities of recycled materials or even be made entirely of them. As noted earlier, high molecular weights are found with most virgin materials, and as they are subjected to processing and recycling, some degradation will take place.

The constant recycling of the scrap can theoretically render the quality unusable because of degradation. Of course, such extended recycling is rarely encountered, and most practitioners limit the proportions of their virgin and recycled blends to safe levels. In cases where more scrap is produced than normal production can absorb, either another product item is introduced to use up the scrap or the leftover scrap is sold to other material users. The option to sell scrap material has the widest range of opportunities, and scrap materials can always be sold in any of various forms. The least valuable stage, of course, is as produced, without further reduction processing.

2. Scrap Handling and Reprocessing

Scrap is produced in every thermoforming production cycle. In single-product manufacturing, the scrap will surround the product, while in the multi-up format, the scrap will produce a skeletonlike configuration where the products are trimmed and removed. The leftover and cutaway scrap is produced repeatedly as each cycle is duplicated. When thermoforming in a sheet-fed operation, the entire panel is transferred from the thermoforming machine and subjected to trimming. In this thermoforming operation, the scrap is produced in individual and framelike or skeleton configurations. In a continuous web-fed thermoforming operation, scrap is produced on a continuous basis and after product trimming, the scrap remains in skeletal web form. In either case, some type of procedure must be set up for handling the leftover material.

How involved one gets with scrap handling is directly related to the volume of products produced. With a limited quantity, the easiest and least costly method is to dispose of the scrap as trash. However, as production runs increase, so does the value of the scrap. In small- to medium-sized operations, it is not uncommon to collect the scrap in large Gaylord boxes or tie it into bundles. In this way, the scrap can be handled by hauling equipment that can carry the scrap to trash collection bins or set it aside for further scrap accumulation. When enough of the same type of scrap is accumulated, it becomes valuable. Orderly accumulated scrap can be reprocessed into new sheets or if that is not acceptable, into base material for other processes. These scrap materials can find a second life in many plastic processes other than thermoforming. The value of the scrap increases as opportunities for its use increase.

The value of the scrap increases as opportunities for its use increase. Because of its potential value, scrap recycling could be as important to the thermoformer as the making of the actual product. The first step is learning to see the scrap as valuable and not simply as a by-product of thermoforming. The scrap value increases with quantity levels and decreases with poor handling. As scrap is produced, it must be kept clean and segregated from similar-looking materials. For example, collection boxes must be covered in storage, as airborne and people-caused contaminants can fall into the boxes. Similar-looking plastics (clear OPS, polyesters, PVC) may look alike to untrained eyes but will not blend together due to differences in the characteristics of the plastic families. This can create havoc with recycling equipment and reprocessing.

The scrap produced by the thermoforming operation can be dumped or sold as is. However, it is clumsy to handle it this way since it needs lots of space and is vulnerable to contamination. This method requires the most effort to make the scrap reusable. The scrap produced out of individual scrap pieces or in a roll of scrap skeleton is best reduced to granulated chip form. In this process, any larger piece of scrap or scrap in a continuous web is reduced by granulating equipment. Such scrap granulators are usually in operation at scrap recycling facilities that buy up all the scrap from various small processors. The same type of granulator can also be set up by thermoformers, who can choose between small individual granulators for each thermoforming machine or can have one large, centrally located granulator for the plant. Each granulator setup is designed to cater to individual requirements and plant conditions.

As to which is better, the choice must be guided by several internal manufacturing, economic, and material usage factors. One method of granulation may be entirely suitable for one processor and yet may not fit into the scheme of another. A single centralized granulator system transfers the associated noise and equipment installation to a separate room. Scrap transfer to this room has to be implemented by underground scrap tunnels and moving belts or by motorized towing or carrying equipment. Individual granulating equipment installed at the end of each thermoforming line and placed into the path of advancing scrap skeletons must have pipes and blower equipment to transport the reduced scrap. The pipes will feed the reduced scrap either back in-line for sheet extrusion or into storage silos. The silos must have sufficient space and negative pressurization for air movement and a proper means of dust collecting for the exhaust. Each type of scrap-reducing equipment offers benefits over the other while carrying its own shortcomings.

From the thermoformer's point of view, any scrap-handling method is feasible as long as the scrap is processed without interfering with the thermoforming production. Scrap reduction and handling are closely related to thermoforming because the process creates higher levels of scrap than does any other process. Size reduction and scrap handling are highly specialized fields and supplementary readings in this area are recommended.

To thermoformers who do not pay enough attention to scrap evaluation and are selling all of the scrap produced, the following advice should be of value. A small granulator can be used in a smaller-scale operation to reduce the scrap skeleton into chips. This is a relatively inexpensive business practice, and the processed scrap will bring a high return on the small investment in a granulator. In very large thermoforming operations, on the other hand, all scrap materials are recycled back into products without loss of value. In cases where more scrap is produced than the normal reblending process can absorb, a new product is created just to use up such an overage of scrap. Such a product can often be made of nothing but recycled material. No thermoformer can afford to waste reusable material and should not wastefully discard it.

D. Reject Control

Although it is every thermoformer's wish to complete a job without once making a rejectable part, this is not possible. In fact, the thermoforming process has so many outside influences and interfering conditions that quite often it will turn out products of inferior or unacceptable quality. Even with the most careful planning, it is never known if the first attempt at thermoforming will produce a product that is up to par with regard to the planned product consistency. Furthermore, since the thermoforming process is tuned to correct errors while the machine is running (in most cases), the process will continue to turn out products of rejectable quality until proper specifications or conditions are met. Each time a thermoforming machine is shut down, for whatever reason, it will take time to restore the machine to its running condition. Even with the best preconditioning, which stipulates ideal conditions for the machine's running state, the start of the thermoforming operation will produce some rejectable parts. In addition, when equipment failure is encountered, further rejectable products will be produced before a shutdown can be implemented.

Switching from an old batch of sheet supply to a new batch from a different material source can have adverse effects on the thermoforming process. Most such effects can be tuned out, but in the interim, rejectable products will continue to be made. New molds, machine characteristics, and a whole array of known and unknown problems will all contribute to the unavoidable collection of rejects.

Of course, it is never expected that all cycles will produce an acceptable rendition of the intended process and product outcome. On the other hand, excessive reject making can destroy the advantages and profitability of any thermoforming endeavor. The goal of most thermoformers is to reduce rejectable part production to acceptable levels. Such levels will vary from one type of product to another.

First, acceptable levels and rejectable conditions must be established for a given product. Such criteria limits are usually set into a quality control specification and should be addressed by several circumferential factors, which are guided by the specific product criteria. Sensitive products may carry very tight specifications, while an array of general-purpose products are treated more leniently as to rejection rates. These limits and leniencies must be worked out before production standards or final products costs are established. Estimating these criteria is acceptable as long as the estimates can be readjusted. Experience gained in working with similar products will provide an advantage in the quality of the estimation. The process of estimation is nothing more than a great battle for perfection. No one should expect to achieve total perfection in product making, except over a short time span. However, most thermoformers wish to come as close as possible to the highest standard of efficiency. A 90% production efficiency can be considered good, and 95% is truly remarkable. In reality, 80 to 85% efficiency can normally be expected. It is within each thermoforming operation that these efficiency numbers become meaningful and will result in full gratification, simple satisfaction, or some disappointment, depending on the specific job. Within these statistics, production facilities and the individual thermoforming shifts or equipment lines are measured against one another. It is here that financial wizardry comes into play, especially when prior production estimates are compared.

Making an extensive number of rejects can be very costly and is often the cause of runaway costs and indicates poor managerial judgment. For example, a thermoformer who has individual panel costs in the hundreds of dollars cannot afford to run rejects cycle after cycle before deciding to shut down the operation, investigate, and correct the problem. In a web-fed operation, running five parts across the web and continuously discarding a single row of parts due to the same fault in the process is similarly unaffordable. That single row may have been made rejected on the basis of quality, but it represents 20% of production, which could be equivalent to that product's profit yield. If this practice is continued for any length of time, the product maker will find himself making that product for fun, not for profit. Tightly controlled production standards and comparisons will lead the practitioner to pinpoint the areas that cause the manufacturing and financial difficulties and provide for measurable and comparable production output data.

Just as with scrap, the rejected articles can be recycled and blended back into the thermoplastic supply. Of course, this opportunity to recycle is strongly diminished by the presence of contaminating material such as printing ink or nonblendable material composition. If these criteria do not interfere with the recyclability of the rejected materials, the materials can be collected, granulated, reblended, and used to produce newly made sheet. If the recycling is done in-house together with the sheet extrusion, the only loss associated with the rejected articles will be in their production and labor costs. However, even though losses can be minimized through recycling, a high number of rejects should be avoided because this could still take a substantial bite out of operating profits.

E. Plastic Recycling

The subject of plastic recycling is a much broader topic than the scope of this volume. Substantial portions of plastic products fall in the category of disposable single use items. Additionally, the thermoforming process itself produces a high percentage of scrap which comes from edge trim, scrap skeleton, and out-of-specs rejected products that are produced as part of the standard thermoforming procedure. When the operation has gone out-of-control or requires some tuning as improvements to the process are made, sometimes unusable products are made. Large manufactured items produce not a high number of rejected units but a substantial amount of plastic weight. On the other hand if the items are produced of lighter gauge material on a continuous high speed operation a large number of faulty units will be produced until the equipment is tuned to the correct running conditions. Since it is known that thermoforming always will produce a large number of products and this method of molding is expected to grow and penetrate more areas of product making it is most important to discuss and focus on the product's "after-life" and post-consumer aspects thereof.

There are many different types of thermoplastic materials used in thermoforming, but all have a common feature. All thermoplastic materials when exposed to elevated temperatures, have a tendency to soften and melt down. Due to this physical transition it can become processable or reprocessable again and again. Such characteristic features of this thermoplastic material are known and used by most thermoformers to handle their own production scrap, which is the result of edge trim, spacing trim, and the rejected parts, that did not meet the required specifications. Such wastes are normally recycled back in to the thermoplastic sheet stock.

In the early years of thermoforming most processors discarded the scrap and it was judged, as the unwanted, but necessary byproduct or waste of the process. This scrap was handled like any other waste—dumped into the waste collection system that most often has been sent to the landfill for disposal. In later years we learned that such scrap material, if handled properly, can be reprocessed and blended back into a newly extruded sheet material to make new thermoformed products. Thermoformers discovered that the scrap should not be treated as waste, but as valuable material source as the virgin plastic supply.

Scrap production in thermoforming can vary a great deal depending on the thermoformed products and the type of operation. For example, as it has been pointed out earlier, large and heavy gauge products produced mostly not volume but substantial scrap weight, while small high volume, light gauge products by high speed production methods, make a considerable volume of scrap production. Both processes can equally produce tangible amounts of recoverable waste. We must realize, that the configuration of the product also can contribute to the volume of scrap. For example when round configuration products are thermoformed in multi-up mold configurations, the minimum scrap skeleton will have a 40% scrap ratio to the 60% material needed to create this round shaped product. There are instances when the scrap ratio will be even higher. I myself at one time produced a particular product which had a scrap ratio of 60%. In today's economy we know that nobody can ignore the fact that scrap has to be recycled, except in a very few instances, when certain or specific materials can't be recycled at all. In those very limited cases, such as cross-linked polyethylene foams, or incompatible laminates and coextrusions. These scrap materials perhaps need research work to develop other methods of disposal or maybe we must accept that landfill disposal is the only way to discard such materials.

In the recent years there has been private and public pressure to solve the problems of environmental pollution and natural resource depletion. Studies and reports from small, private, environmental organizations to the largest governmental agencies, from classroom discussions to international forums show that such concerns are not fads, but genuine problems which will demand attention from now on. It is also should be recognized that most plastic products, whether produced by thermoforming or any other molding process, will have a specific life expectancy. Eventually such products are discarded and will end up in the waste stream. Almost all thermoformed products fall into this category.

As noted, the thermoforming molding process is one of the fastest growing segments of the molding industry. More and more products and product lines are producing products for consumers and there is no sign of slowing down. For that very reason, thermoformed products and product makers will be involved with increasing plastic recycling and even will see a demand to use postconsumer recycled plastic as well. Today's trend shows the producer will have the burden to identify makeup of the plastic material of their product and give disposal instructions or even will be held responsible to develop a recycling or disposal plan for the plastic products they manufactured in the past. For those very reasons I felt that these issues must be discussed in this book and at least give some guidance as to what to expect and what a thermoformer can do or cannot do. There are a lot of myths and misconceptions floating around, together with developable challenges and even good opportunities, to apply to plastic recycling.

Environmental pressures of recent years have produced much attempted legislation, from the Federal level to state and even county levels. Much effort has been put into this issue at the local level as well as in some very aggressive cities as they made their own laws on the issues of plastic recycling, banning certain plastic product usage, based on environmental concerns. However, most efforts have been postponed or the enforcement dates delayed and further studies requested. The lack of understanding of plastic recycling remained certain, especially by those individuals who had been the most eager to advocate such. There is nothing wrong with demanding the elimination of plastic waste and pollution from the environment, but only proper methods of reuse of recycling will work, achieving an attainable solution for plastic material recycling and reuse.

As the author of this book, I do not intend to take sides on this issue. I only intend to emphasize what can be done or what can not be expected of plastics when it involves the reuse and remolding of previously made plastic products.

The main objective of plastic recycling lies not in those product categories made in the factory and rejected because of out-of-specification status (we are already recycling that), but in those products already used by the consumer. These products are categorized as postconsumer status. Used plastic dishes, used containers and dispenser, or plastic packaging are product types classified as single-use items. And there are plastic products made for short life expectancy, used several times, but finally will be disposed of. At the other end of the product spectrum are those plastic products made to be "permanent"; however, they became obsolete, due to ever advancing technological developments. All these plastic products usually end up in the waste stream. This waste should concern us all.

Most people who are involved and working with thermoplastic materials know that plastic parts in a finished form just cannot be melted down and directly remolded into the same or a different new product. This product has to be reduced (ground) into workable form in order to put it back into the remolding process. But before we can reach this point there are many more things we should be concerned with. First is the collection of discarded plastic products. This is not an easy task; however, we have made many inroads to accomplish that task. Cities worldwide separate plastic discard from other refuse and collect postconsumer plastic products. Those collected plastic products are destined to be recycled and removed from the waste stream. The first commendable task is made.

The second task is to segregate the different types of plastic materials. Different types of plastics behave differently and have different melting points; therefore not all types of plastics can be successfully melted together. Different types of plastic should be segregated from each other; the mixture of them can and will affect most molding processes and the outcome of the product. So therefore, different plastic types have to be segregated into their own group or category. This is mostly resolved by placement of a common marking on the underside of containers that identifies the plastic type.

The third, additional task is that plastic products also should be segregated in to color groups. There are plastics molded into products in their natural color—either clear colorless or hazy-milky. And of course there are plastic products in all the rainbow's colors that merchandisers can think of. Today, we have made successful inroads in solving most of the problems of sorting by type. There are various types of equipment available to perform successfully this segregation of different colors.

The fourth area of concern and responsibility with the recycling of postconsumer plastic products is the removal of contaminations. It is an accepted fact that used plastic products will have some residue of foreign material that may be a label or printing ink or remaining residue of products that were packaged or served on it. There are even contaminants and chemicals, which have migrated into the plastic structure, and can't be seen, but are there: perfume, oils, pesticides, and other harmful materials, which need not be in high concentration, but can ruin a large quantity of recycled plastic material. Dealing with this kind of contamination is not an easy task, and its handling can offer success or failure for the entire plastic recycling effort. For example labels or loose and greasy food residues can be washed off, provided that the cost of washing and water justifies such an effort. Some food residues may be so difficult to wash off from postconsumer plastic containers that it would be unwise even to attempt, such as peanut butter or mayonnaise. In some instances solvent cleaning can be employed to remove hard-to-remove contamination.

Color printing can alter the original color of the plastic; therefore the printing ink color choice should be a determining factor for the future recycling. In the case of red printing ink the choice of plastic color should be pink; therefore, at recycling the plastic material will remain pink. If white plastic would be used with red printing ink the red print will contaminate the white plastic into pink recycled material, and never can be made white again. Instead using large and bold print format, with large outlined letter forms, will reduce the quantity of ink usage, resulting in a far lesser color contamination problem. Some residues left on the plastics or the chemicals migrated into it, should be attempted cautiously to recycle into products, where the contamination may be detrimental. It is important to know beforehand what contamination we dealing with. Then the proper decision can be made as to how to clean or discard that particular batch of material.

The biggest problem in plastic recycling is when you are dealing with unknown factors. Unknown dangerous chemicals can enter the recycling stream unexpectedly. For example, an ordinary soda bottle is used for a temporary receptacle to transfer some chemicals, gasoline, pesticides, etc. It is a totally irresponsible act but it does happen, sometimes by accident or even purposefully, and discarded together with the rest of the recyclable materials.

The source of contamination in postconsumer plastic material is not totally under the control of the recycling processor and therefore must receive extra careful scrutiny. The resulting problems for the thermoformer or any of the other molders as well, can play havoc, not to mention the legal implications that can come into the picture.

Another important fact and perplexity of plastic recycling must be understood, involving recycling. This particular fact, that governs all thermoplastic materials, is totally ignored and brushed aside by those who propose, demand, and write legislation involving postconsumer plastic recycling laws. Even when the above mentioned contamination problems are solved one more major obstacle could be encountered: plastics molecular degradation, an obvious problem when plastics are repeatedly reprocessed. It is accepted that such molecular destruction and degradation is unnoticeable and should be a "no concern" after a few recycling procedures. What concerns me is deteriorating product quality as a result of the ongoing repeated process.

The industry has often been forced by law to recycle its own product line. This is called "closed-loop recycling" when the same products are made of the previously manufactured same postconsumer product. This is a common and well established practice, performed successfully all the time with glass containers and metal cans. The glass and metal recycling and reprocessing is done at such elevated temperatures that most of the contaminations are destroyed and burned off. Additionally the reprocessing does not cause any destruction in the material's physical makeup.

On the other hand, plastic material reprocessing is made at substantially lower temperatures that do not remove or destroy contaminants. So in order to remove any existing contamination the plastic parts or the plastic granules must be subjected to wash or other means of cleaning prior to the melting procedure. As pointed out earlier the plastic articles must be reduced into reusable material and must go through a shredding or granulation process, just as the scrap produced by the molding process. This segment of recycling and processing is a must in order to mix and blend the plastic. In many instances chips or a granulated form of plastic do not blend uniformly with virgin plastic material in pellet form. The different physical shapes tend to segregate out one from the other, no matter how well they have been blended together originally. To overcome this problem it is often necessary to repelletize the grounded plastic chips. In the recycling effort, the more times the plastic is subjected to the different process segments the higher number of molecules will experience degradation. Basically the original molecules mechanically will be torn, resulting in more smaller molecular weight material. To this molecular weight destruction, add the number of repeated recyclings, and in a very short time span some of the plastic molecules can and will go through the reprocessing an undeterminable times. Theoretically and in real practice the molecular destruction can be so devastating that the plastic product's intended function and behavior or life expectancy will be altered. Deterioration effects can be lessened up to a point by additives or blending the recycled material with virgin material, but eventually the molecular destruction will be evident again. For this very reason, some plastic products, e.g. medical purpose products, should not be made out of any recycled content materials at all.

It is also important to know that there are substantial differences in the various types of molding methods, which also require different types of plastic resin with different molecular weight, and preferred molecular-weight distribution. That is why resin manufacturers classify their resin as injection molding grade, blow-molding grade or thermoforming grade, that makes the particular plastic molecular makeup mix more suitable for one molding technique than for other molding techniques. Using the proper molecular weight and molecular weight mix ratio blends will provide an ideal molding condition and a more functional finished product.

Thermoforming is almost always made out of a preheated plastic sheet, held by its edges and then stretched over or into a mold. The success of the process is totally dependent on the ability of the resin molecule to withstand the stretching without rupture. Shorter molecules (smaller molecular weight) resins and resin blends tend to tear and rupture before full forming can be achieved. Also the finished product that has the longer (higher molecular weight) plastic resin makeup will not rupture or break as easily, as the opposite type (lower molecular weight material) and will have higher levels of structural strength. The high molecular weight material has the longest molecular chains and therefore it has the best integration and molecular interlocking features. In case the plastic material will undergo many, and sometimes an unknown number of repeated recycling procedures, such material cannot and will not behave properly in the thermoforming molding and will jeopardize the finished product's functionality and quality. This will be more apparent and noticed when the products have the deeper drawn configuration. Knowing this, the thermoformer should be more confident to form articles of a shallow configuration made out of a higher recycle content or from more frequently recycled material. For that very reason, selecting and choosing products made of recycled material or material blends should not be allowed by those who know nothing about the particular molding procedure or have no direct responsibilities and may not face significant consequences in the final product outcome. Please remember that by the time postconsumer plastic waste is treated as trash, it will perform most likely in the recycling effort as trash!

Manufacturers, I think, are responsible enough and if there is a pricing advantage to use recycled plastic material they will do so. It is much easier for thermoformers who have a multiproduct line to select those products less sensitive to recycle contents. It is also possible that one may develop a new product, manufacture and market it, just for the purpose of using up excess capacity of recycled plastic, and may even make a profit of it.

I hope with this short and somewhat simplistic write-up on plastic recycling I am able to shed some light on what can and what cannot be expected in the use of recycled plastic material, especially in regards to thermoforming. As for cross-the-boards close-loop-recycling today and in the near future, I find it somewhat unrealistic, and only a very few and limited products qualify for closed loop recycling. However, there is some hope, with some ingenious chemical manipulations, it may be possible to depolymerize and then repolymerize, where plastic can be remade to its original molecular structure and reused again as virgin material. Of course all has to be economically feasible. Who knows what we will be capable of doing in the near future?

III. QUALITY CONTROL

As the thermoforming process comes to dominate more and more industries and penetrate more and more markets due to its sophistication and demanding quality levels, thermoformed product quality has undergone an equal degree of scrutiny. With today's higher speeds of production output, the chances for error have also been increased. Refined details and quality tolerances have made the industry more aggressive in entering into product making that requires discriminating design patterns and meticulous configurations. Such critical product complexities have stimulated more sophisticated manufacturing methods and equally precise quality control practices. Setting up quality control standards for particular articles and product lines is up to the thermoforming producer and the product purchaser or user. Any tightening or relaxing of those quality standards should be made with the agreement of all parties involved. The thermoforming process should be allowed to produce within the limits of realistic expectations, and some tolerance levels should be included when setting the standards. Quality control operators should actively enforce such standards as originally set and should be discouraged from increasing their stringency to unrealistic levels. After all, unreasonably high quality standards could diminish profitability considerably.

On the other hand, it is also the responsibility of quality control management and their staff not to give in to pressures from manufacturing personnel to increase production volume at the expense of quality. The primary goal should always be first: customer satisfaction followed by proof of product function. If these goals are met, the rest of the quality control details should be implemented easily. Since the thermoforming process is involved with such a wide range of products, with an equally wide range of performance criteria, it is difficult to set specific quality control rules. The procedures for checking and statistically tabulating quality control data are no different from those of any other manufacturing industry. For setting up and practicing proper quality control procedures, the many publications available should be consulted.

Applying quality control standards to specific thermoformed products may require some modification of testing procedures or slightly different instrumentation. Testing for quality need not be limited to finished products but could be performed at the beginning and midpoints of the thermoforming process. For example, checking the thermoplastic sheet quality before attempted thermoforming or checking the quality of formed goods before the trimming stages of production can save much effort and production expense. Specified test methods that follow specific steps and formulas have been developed and listed under ASTM (American Society of Testing and Materials) test procedures. Beside such sophisticated and universally accepted testing methods, a thermoformer can implement practical quality control methods. Such simple methods may not have across-the-board usefulness, but in a particular situation can offer "quick and dirty" check results that can yield more meaning to some line operators than more elaborate testing methods would. As thermoformers become more and more acquainted with the aspects of their own product and its particular criteria spectrum, quick and feasible test concepts can just as easily be developed for their testing. One of the easiest and quickest ways to test for variations in the thermoplastic sheet from supply to supply is to cut a strip of given width and subject it to repeated flexing at the same point. The number of flexes is counted until breaking occurs. Using these numbers, a comparative test can be set up between different sheet materials. A test strip from a future plastic sheet order that survives fewer flexes before breakage could behave differently under the same thermoforming conditions or render an adverse result. If this is the case, a "quick and dirty" quality control test has been established. Obviously, such a crude test can be meaningful only when larger numbers of flex differences occur among test samples in conjunction with thermoforming or product failure.

There is another simplified test that is just as effective as the previous test when applied to thermoplastic sheets for a check on orientation. This test consists of making a representative cut from the stock sheet supply. The sheet is first marked with an arrow representing the sheet's machine direction (MD). Next, a simple but geometrically accurate design (a circle or a square) is drawn on the material's surface to provide a visual indicator of any resulting distortion. Then the sheet is exposed to heat without clamping its edges so that it can freely react to the heat. If the sheet does not shrink in reaction to the heat, it has no orientation at all. If it shrinks equally on all sides, it has an even orientation. Any shrinkage difference between sides will indicate an uneven orientation. Observing the effect of shrinkage on the geometric design drawing will also give valuable clues to the material's orientation. The distortion is correlated with the indicated machine direction of the sheet to establish the direction of orientation. Uneven orientation in a sheet may not present a problem in thermoforming. However, if it does cause difficulties at some point, the thermoformer will have a means of checking the sheet supply for uneven orientation prior to thermoforming. Such simple tests may not provide all the answers to quality control or make predictions accurately, but definitely can be a useful indicator for the thermoformer. It is best to give some leeway to the operating personnel and clearly define what is acceptable and what is not. Well-trained operators can make the proper decisions for themselves. It is a serious mistake not to involve the actual thermoforming process personnel in the quality control tasks. No one knows the product and equipment better than the operator, and no one can better resolve a problem or adjust the running conditions of the thermoforming equipment. Unfortunately, it is too often the case that production quotas and production running speeds and their joint influence on bonus rewards have caused an imbalance between quality control and production demands, in favor of production.

It should be the goal of the thermoforming practitioner to produce the best quality product and to do that in a repeated fashion. They cannot stay in business very long with a high number of rejected parts or selling inferior quality parts to their customers. Even if those rejected parts can be subjected to a complete recovery by recycling, the wasted time and energy is not recoverable. The inferior parts also can jeopardize customer relations and could lead to permanent loss of business. Critical variables must be known ahead of time and the customer must be made aware, instead of letting the customer discover the problem later. Quality validation is a two-way street and it is just as important to all parties involved.

For these reasons, preproduction testing of molding must be performed after the design work is completed. Samples must be produced on a sample making set-up until all problems are resolved. Based on this effort, even product and production standards can be established, possibly in a preliminary form of product standard leading to the final Quality Control specifications. These preestablished material standards should be used when comparisons are made on the lot-to-lot resin variations, or in case the thermoplastic sheets are purchased. This may be a critical factor in vendor selection, and such a decision should not be made by pricing alone.

After the preexamination of the sheet material the practitioner must turn to inspect the thermoforming results. The thermoformed articles must be closely examined for part-to-part variations, quality of detail transfer from the mold, and scrutiny of corner radius development in the forming. Finished products are measured to gauge thickness and ideal material distribution, i.e., no thin spots are developed. Product weight must be in line with the specified weight of the specification. Good detail development usually indicates top quality and well-performing process controls.

How tight of a window of process control the practitioner has depends on the particular product. In some instances the process control window is so narrow that a minute change in the process can have detrimental results. On the other hand this process window can be so wide that the thermoformer almost has to make a major error to produce an out-of-specification product.

When problems arise, the thermoformer must make a determination as to in his or her opinion, what went wrong in the process. Some of the problems encountered may have a single cause or could have a multitude of causes. Sometimes a hindrance can mislead the practitioner, because it shows results that mimic other problem-causing factors. Only a thorough investigation and quality checks can guide the practitioner to the right direction to solve the problem. And of course nothing is better than having years of experience to pinpoint a mystifying cause of a problem. Please refer to Section VII, Troubleshooting, later in this chapter.

Last, the thermoformed product must be subjected to product functionality testing. At this point the thermoformer practically does what his customer will do with the product. This way there is no distress that could come about from the customer rejecting the product because it does not perform the way it was intended. Depending on the product type, thermoformers and their quality control staff should check out the products completely, from the shipping and warehousing to the end use application and even consumer satisfaction point of view as well. For example, if the product is an egg carton, eggs must be packed into the egg carton and various tests performed to satisfy the intended use criteria. If the product is a sky window the thermoformer must perform some type of an impact strength test to find out the performance limits of that particular product. This type of a testing may be done in collaboration with the customer or performed together with competitive products to establish whether the products are comparable or not.

The aim for a high quality standard and with it a high quality product should start right from the conception of the project. In fact the product should have designed-in quality. An ideal design should have offer more than basic functionality, and a good appearance, but shape and form, that lend itself to a repetition production with minimum difficulty. Product design must contain reasonable radius corners and satisfactory transition angles between the different mating surfaces, ideal mold configuration, good material flow in the forming and stretching of the thermoplastic sheet, and proper taper angles for easy part removal. The product designer must know the limitation of the thermoforming and choose to cater to all the other process functions to make the thermoforming reasonably easy and to have an attainable quality.

If quality is put forth, quality is what you should get out! Reasons for the failure must be checked out and eliminated. If necessary, learn to work around the problem and respect the true limitations of the thermoforming process.

IV. EQUIPMENT PURCHASES

To enter into a thermoforming manufacturing business, the prospective practitioner must have some knowledge as to what type of products or markets are involved. The preliminary information in regard to product must include knowledge of size, gauge thickness, and color. In regard to the market, information must define expected volumes as well as potential growth rates. If the calculations indicate that the project has merit and a reasonable prospect of success, investigations into available equipment must be made next. The first decision to make is what type of equipment will best suit the project. Once the equipment has been chosen, the next decisions are whether to produce the product one at a time or in a multi-up format and whether to run the process on a single-cycle basis or extend the manufacturing capabilities into a continuous type of operation. When these questions have been answered, the next decision is whether to set up production on new or used equipment. All decisions will be subject to financial requirements and cost evaluation factors. Many options are available to the practitioner, of course, but pursuit of the best possible deal can make the difference between profit and disappointment. Productivity and the economic requirement for a return on investment must be satisfied to retain a flourishing business position.

Having the wrong type of equipment can create a devastating situation, about which many horror stories have been told. It should be accepted, for instance, that a shuttle former is not intended to operate or even replace a rotary former or take the place of continuous web-fed machinery, or vice versa. Substituting one type of machine for another always causes shortcomings and unacceptable manufacturing conditions and thermoforming results, together with exorbitant expenses. There are specific reasons why the machines have been built the way they have, and any alteration to them to change their intended purposes can, at the very least, be disappointing.

A. Life Expectancy of Thermoforming Machines

The decision to purchase any thermoforming equipment, new or used, should be guided by careful consideration. The main factor here is the duration of the particular project. Another important question that needs to be raised is whether the type of business will be constant and repeating or only be a short-lived venture. Constructive planning can be implemented only when armed with proper knowledge of business terms and project prospects.

With a thorough understanding of today's equipment, an extensive evaluation and instantaneous appraisal may not be necessary in order to estimate wisely the purchase of thermoforming equipment. Any thermoforming machine will have four distinctive "life expectancies." Each one can be measured by different criteria and as a result of those criteria, can be just as easily interpreted from diversified points of views. The first of these is the "technological" life expectancy, which only lasts between 5 and 8 years and as the technology advances, can rapidly decrease to 3 to 5 years. The second is the "useful" life expectancy of a machine, which is the longest of the four: 15 to 30 years. Abusive operating conditions and environments or major equipment alterations can dramatically reduce the useful life. But with normal maintenance procedures and available replacement (stock) parts, most equipment can exceed its normal life span. The third is the "financial" life expectancy, which can be estimated as 1 to 3 and possibly 7 years. Such financial aspects of a machine are directly dependent on how the cost write-off of the specific equipment is handled. Such factors as depreciation, interest rates, and investment tax credits can have major implications as to how long a thermoforming machine can enjoy a favorable financial status. When a thermoforming machine has been written off, it obviously does not become worthless. The machine still remains productive for awhile or can be sold as used equipment. The fourth life expectancy is the machine's "practical" life expectancy. The practicality of a machine is the most difficult factor to determine, but can be estimated at an average of 5 to 10 years. The reason for the difficulty in establishing the practicality of a thermoforming machine lies in the many influencing factors, such as market conditions, competitive circumstances, and the manufacturing plant's regional or geographic location.

A thermoforming equipment having the utmost state-of-the-art features can still prove impractical for a particular thermoforming job or manufacturing setup. The competitive condition can force running equipment in a nonsuitable way, which is just as harmful to the equipment as to the business itself. Such an impractical situation is most often observed when a small-output machine is forced into an overextended production run to produce larger volumes than those for which the machine is intended. Such overburdening is usually encountered when runaway sales outstrip manufacturing capabilities and when restrictive management behavior will not permit timely equipment upgrading. The decision to purchase new equipment or renew old equipment should be made when the present thermoforming machine does not have the potential to function properly to produce in the newly demanded way. If the thermoforming equipment has proved impractical or can no longer be upgraded, it will usually end up in the used equipment market and probably will be moved to a foreign country, where it will still be found useful.

B. Thermoforming Machine Upgrading

When deciding on rebuilding or replacing thermoforming equipment, the same life expectancy factors should be kept in mind as those discussed previously for new and used equipment. The only thing that favors equipment already owned is that besides its used machine value, its performance and reliability are already known. A good service record or reliable operating history is a good indicator to justify upgrading older equipment. Upgrading means not just replacement of old, worn components, but the addition of state-of-the-art instrumentation and accessories. The decision to upgrade existing thermoforming equipment or invest in new machinery is based primarily on economic conditions. The current state of the economy, interest rates, and even the company's capital expenditure allowances can all be influential. The ever-changing technology can also be considered as a factor in this decision. Older thermoforming equipment usually undergoes periodic evaluation, judging for technological impairment, energy efficiency, and processing ability, which indicates cost-efficiency. On the other hand, thermoformers who buy new equipment will probably obtain better productivity through faster cycles, better instrumentation and controls, and longer trouble-free running. All these improvements also come with a higher price tag.

To rebuild and upgrade thermoforming equipment is not an easy task because many internal factors must be considered in the decision. Each piece of equipment must be observed on an individual basis, and a detailed cost study should indicate if it is worthwhile to rejuvenate or retrofit thermoforming equipment. A thermoforming machine that has been abused or has undergone major modification (i.e., when the frame structure has been altered so that it is no longer recognizable and cannot be traced back to its original maker) is difficult to restore to its original condition. Such machines, as well as custom-built machines, often cause problems for their new owners. On the other hand, thermoforming machine suppliers will often offer upgrading kits or retrofit components as new technology is developed. Thermoforming equipment less than five years old stands a good chance of being upgraded, and new machines come equipped with add-on provisions for future computerized connections. Updated information on the latest trends and innovations can be acquired through advertisements, brochures, frequent visits to trade shows, and attendance at meetings and seminars.

In upgrading thermoforming machines, heater units, controls, and indexing mechanisms are the most common recipients of improvements. Replacement of old heater elements with better heater units is very popular. Calrod-type heater elements can be upgraded with ceramic heater units, which can offer mosaiclike programmed heating. In a shuttle thermoforming machine, replacing tubular heaters with quartz heater elements can significantly improve the heater's cost-efficiency. Clamp-frame structures are also a good candidate for replacement by improved designs and operating clamp frames. The most innovation can be achieved in the thermoforming controls. Controls that operate the heaters and also operate the thermoforming machines can be a most important aspect of the upgrading program. These instruments and equipment can offer the most value to the thermoforming process and are also the easiest to retrofit. The new controlling systems are capable of handling many different tasks within the same time span, taking over most operating duties and self-diagnostic procedures together with providing excellent reliability.

C. Electronic Process Controls

The industrial revolution has progressed to the miraculous age of electronics. With the development of microchips and microprocessors, "up-to-date" mechanical controls and timers have been superseded by more reliable devices. In fact, microprocessor controls can take over the entire process control from a human operator. The system can be implemented to control, oversee, and even perform diagnostic functions. The new electronics will eventually push mechanical relay controls out of the operation entirely, leaving them to be categorized as outdated and outmoded. This is the age of solid-state circuitry, microchips, and microprocessors, and the current trend is to use these sophisticated control systems as much as possible. No current or future economic conditions can affect this progress despite higher costs. The justification for using the new technologies is based on direct labor savings and the more precise control results that are made possible. Even the current depressed thermoformed product market acts to stimulate the use of more precise electronic devices to help cut costs. Such electronic control devices are available not only with new equipment but also as retrofitting kits for older machinery. Such kits are available from several manufacturers and are designed to be applicable to many different makes of machines. Furthermore, some of these sophisticated systems are not only capable of providing controlling functions for a single machine but can be coupled into a centralized computer bank to control a multitude of machines simultaneously. The sophistication of computer controls will someday allow them to handle the entire thermoforming process with fewer human attendants, particularly during the unfavorable second and third operating shifts. However, such futuristic concepts remain to be seen, but present thermoforming, to some degree, still depends on active and well-trained thermoforming operators.

A microprocessor programmable controller (PC) does not actually need to handle any more aspects of a thermoforming process than ordinary mechanical controllers do. It just does it more precisely. PCs can offer a range of control possibilities, from simply relay replacements all the way to top-of-the-line full command of decision-making programs. Their efficiency comes from their fast operation and unerring reliability.

The largest single group of control functions in thermoforming is involved with the heater elements and ovens, and it is here that PCs have the greatest potential for improvement. But they are equally adaptable for use as cycle-initiating and measuring timers. Integration of these functions with one another and with other outside controller units makes PCs even more functional. Further integration of controllers with digital or analog indicators and recorders makes the system even more valuable. Full computer control functioning is gained when a complete interplay of equipment is integrated into a single system that couples machinery with a main controller (PC), encoders, proximity devices, potentiometric position sensors, temperature and color sensors, and so on. Once all the sensory devices are feeding combined inputs into the main controller, which provides a continuous and discrete regulatory control as well as some computational capabilities, a true computer function will be evident. Now further automation and some type of robotic manufacturing can evolve.

With PCs, the control and application parameters can easily be preset and even stored into a memory bank for future setups. The system also allows communications between other operating systems so that their data can be transferred between machines through distant plant locations. With the self-diagnostic features, hidden problems can quickly be brought to an operator's attention. Routine preventive maintenance schedules can also be programmed into the system. These technically advanced systems can save time and money through a high degree of sensitivity, responsiveness, and their interactive logic functions. Microprocessor-based programmable controllers are the most energy efficient systems, and this benefit carries over to the entire thermoforming process to which it is linked. The savings are even greater in view of the reliability and minimum breakdowns of PCs.

It is also fair to give some caveats with regard to these instrument systems. Although the virtues of PCs make them a desirable tool in thermoforming, it is important that their capabilities not be oversold. Even though microprocessors represent an exciting new tool for the thermoforming industry, they are not for everyone, nor does every thermoforming process need them. There are definite cost considerations that must be settled first. Also, due to their especially sensitive responsiveness, PCs can be driven to provide unwanted chaotic reactions. For example, when unpredictable drafts exist in a plant, the controller can react almost frantically to the resulting temperature fluctuations; a supersensitive piece of equipment, when forced into a less-than-ideal environment, can react adversely. Simpler, less sensitive thermoforming processes and short production runs may not justify such elaborate equipment controls. Furthermore, there are still many thermoforming plants and machines that operate without the latest innovations and will continue to do so for many years to come. At the same time, a complete in-line thermoforming system presents such complexity (the operation starts with the pelletized raw material, which must be accurately blended and fed into an extruder that produces a sheet for the thermoforming, and then enters the trim press, after which the thermoformed articles are trimmed, collated, counted, and automatically packaged) that it cannot be accomplished without sophisticated interacting programmable controllers.

D. Mold Mounting and Installation Devices

It is a well-known fact that the molds built for thermoforming must be placed and secured into the thermoforming machinery. As a customary and standard installation procedure, screws and clamps are used to secure the molds to the platens of the equipment. The platens either consist of a solid plate with a number of plane holes drilled or having threaded holes, or the entire platen is made of I-beam structures. The holes or the leading edges of the I-beams are used to secure the clams or clampdown mechanisms between the mold and platens. Whatever method is used to secure the molds to the platens should not allow the mold to shift in its fixed location. It would be detrimental if the slightest shift in location occurred, because it can not only alter the forming but can render a complete mold "wipeout," especially when matching molds are used. The screws, clamps, or clamp-bars are together and their tightness makes a secure alignment in the molding after the molding procedure. On the other hand those screws, clamps, and clamp-bars must be made and used such a way that the mold installation and mold removal can be accomplished rather fast, without cumbersome time delays. Mold changes are a part of the business in most operations that may be done as often as a daily basis because of short run projects. Time spent on the changing of molds are nonproductive time.

Charges for that nonproductive time must be included in the price quotation, to cover the incurring expenses, whether it is charged to the customer or absorbed into the operating expense. In some plants, to make the mold change-overs as efficient as possible, there are specific personnel who assigned to the task of performing nothing else but making the mold changes.

In case there are upper and lower mold halves, each side of the mold must be installed onto the respective platens. There are certain rules and sequential procedures used when these particular molds are installed into a thermoforming machine. The implementation of "mold bases" or "mold stand-off" are utilized in order to achieve the correct mold mating and matching the sheet-line with the mold's parting line. Molds sometimes can mate a small distance above or under the sheet line, but eventually if extended any farther away from the sheet-line, that procedure can not stretch the material without causing problems. Having mold mating a somewhat greater distance away from the sheet-line than some material allows, can jeopardize the stretching and even can cause sheet-pull-out from the sheet clamping fixtures. The mold base or mold stand-off is the interposition connecting part between the mold itself and the platens of the thermoforming machine. Since the thermoforming machines are made to accommodate the maximum depth-of-draw part specification, the platens are made to open as far as the machine's "daylight" opening allows. When the thermoforming is made on a shallow part, the mold's stand-off (legs) is used to compensate for the height difference. This stand-off is cheaper to build than making the entire mold's body extend all the way down to the platens. It often becomes an integral part of each mold, remaining attached to the mold; together they are often installed to and removed from the thermoforming machine. It even can have the built-in clamping attachment in order to speed the times of mold changes.

In earlier times mold installation was very simple, and such simple practices still remain in some shops today. The mold is lifted onto the platen of the thermoforming machine, then "eye-balled" to the center-line to line it up. Then the clamps are secured and the thermoforming is run. Any slight misalignment usually is corrected with hammer-taps until satisfaction in alignment is achieved. By no means can this be called a fine tuning operation. In shop after shop no one admits to this practice today; however, the remaining tap marks on the mold's sides are good proof of such practice. This mold installation technique is further aggravated when matched molds are a concern. The slightest misalignment between the two molds can cause all kind of problems to the finished product. One of the most severe problems, or maybe accidental benefits, as a result of poor mold alignment is discussed under the heading Section IV.C.2: Partially Matching Contoured Molds.

Any of the old fashioned methods of mold alignment described earlier are outdated today and not only results in poor quality of mold alignment, plus time taking, but never produces repeatable results. To produce a repeatable and accurate mold alignment and to achieve that practice in a fast-paced operation the thermoformer must have equipment that can do that. There must be devices that are parts of the machine platens and have mating counter-fitting parts on the molds. For this fast mold mounting technique several "quick-mounting" devices have been invented and implemented. Most of these quick-mounting devices are based on a tapered form of locating pins or channels. These tapered pins or channels are mounted on the machine's platens or platform, and matching counter fitting parts placed on the molds. As the molds are being lifted near the platens or platform these tapered locators will accept the slight misalignment variation. As the mold and platen distances are reduced because of the matching tapered locators the misalignment discrepancies are corrected automatically. There will be perfect alignment each time the mold is placed over the tapered locating fixtures. This method will eliminate any need for "eyeball" readjustment and it is fast too. The goal should be to make a side-to-side alignment as well as a front-to-back alignment at the same time, quickly. Therefore, more than one tapered locator is used or in case tapered channels are the choice they must be placed crisscrossed in two directions.

To enhance the benefits of such a device even further, the manual securing and clamping operation can be also automated. The locating pin's cone shape can be interrupted with a machined-in groove, using this groove as a lock-on ridge. The female counterparts of the locating pins have a matching taper with a built-in locking device that mates with the groove. Usually this locking feature is pressurized air operated, making it a one-button operation to engage or disengage. Four or more of these locating devices are employed to locate and hold the mold to the platens of the thermoforming machine. In order to make a quick mold change, all the operator has to do is to locate the mold above and near the locating pins, then lower it onto the pins. This will locate itself, then activate the air operated locking device and the mold is secured.

When the thermoforming involves matching mold procedures, the placement of the mold is slightly more complicated. First of all, a perfect mold alignment between the female side to the male side must be made perfectly and that alignment must be maintained at all times. The platens must be moving without any side-to-side discrepancy against each other. It is easy to assume that most machines function that way, but any worn out bushing and bearing surface will cause enough slop to develop discrepancies. How big of a misalignment discrepancy a thermoformer can live with its entirely up to the specific product he or she produces. But any misalignment present never gets better with time; it just gets worse and soon will be a problem.

To insure accurate alignment between the male and the female mold halves the molds must be secured together with locating pins. These pins, a minimum of two placed into two diagonally opposing corners of the mold and installed by the mold builder's facility, are cylindrical shaped hardened tool steel pegs, fitting into hardened steel inserts placed into each mold half. Being made of hardened tool steel, any wear factor is eliminated. Of course the locating pins' purpose is to locate and hold the two molds together without shifting out of the proper alignment. The procedure to install that mold into a thermoforming machine is as follows: The two molds are secured together with the locating pins lifted onto the lower platen, then first clamped on to the lower platen. Then, the top platen is closed on to the mold and at that time the upper platen is clamped on securely to the upper mold. Now the platens can be opened and the locating pins removed. With this procedure a perfect alignment is made between the two mold halves as long as the machine platen move without misalignment. When the mold is to be removed from the thermoforming machine, first the locating pins should be refitted into the locators, then the mold should be closed, and after that the clamps to the platens can be removed. This way the mold's alignment is preserved until the next mold change.

There are other methods that can be used to achieve perfect mold alignment and make the mold changes rapidly, but none as simple as this. However, as long as the mold changes are made fast enough for the individual business and do not cause extra long "down-time" from production, it can be maintained and not changed. The purpose of using such devices is to minimize the nonproductive times that mold changes are causing.

Mold alignment between the male and the female halves gets more and more critical as the thermoformed parts get equally more complicated and this alignment criteria can be even further scrutinized and emphasized for importance, when downstream trimming is involved. There are thermoforming operations in which in-line trimming is produced, right in the same equipment. Mold alignment and the trim tool alignment must be made so that they will match each other's "foot print" in order to make the thermoformed part according to the specification set forth. In this case, both the mold and the trim tool must have gone through the same installation and locating procedure to insure a perfect alignment. The shrinkage of the plastic, after molding and in the subsequent indexing, must be accommodated for. Often thermoformers tend to ignore the shrinkage taking place at the areas in-between the multi-up parts. The spacing area must be cooled by the mold as well as the formed parts, otherwise there will be different shrinkage rates in the parts than at the spacing, and this will cause distortions and can pull the trim indexing out of the intended alignment. This criteria is only important to those thermoformers who trim multi-up configuration parts out of a common sheet and make the trim registration once on the entire panel "shot." Those who register each individual part by itself only have to worry about distortion wrongfully developed in the forming itself.

V. COST AND FINANCIAL FACTORS

The most important aspects of any process is to have financial control over all production functions. The thermoforming process is no different, and its manufacturing conditions should always undergo some financial analysis. After all, the bottom line is that the project prove profitable. To achieve satisfactory bottom-line figures, specific individual cost limits must be maintained. If any expenditure shows disappointing numbers and gets out of hand, the discrepancy must be absorbed into the remaining cost areas. However, excessive cost overruns cannot be distributed easily or be allowed to burden other cost components unreasonably. It is up to the individual practitioner to determine which cost components are unreasonable. This decision depends on many factors, including labor rates, energy costs, material prices, equipment cost, and even the cost of packaging and the plant's geographic location. All of these factors are directly related to actual manufacturing costs. There are, in addition, merchandising costs that involve sales, warehousing, advertising, and shipping costs as well as promotions and discounting when such sales tools are used. The direct cost figures are strictly related to the specific item being manufactured. In the context of this book, the discussion will focus on direct manufacturing costs.

A. Direct Manufacturing Cost

Cost estimates must be made for any product when approaching a brand new market or entering into an existing market. With knowledge of the costs of a manufactured item in hand, further calculations can be developed to arrive at appropriate pricing. But before serious pricing attempts can be made, all foreseeable manufacturing costs should be considered. To develop a successful and profitable business that will be capable of meeting future competition head on, detailed cost studies must be made. It is not the intention of this book to provide a marketing strategy or pricing guidance nor to outline methods of calculating profit structures and gross margins. Neither price competitiveness, revenue income, nor the return on investment will be discussed here. These are strictly marketing functions and the reader is referred to numerous books on these subjects.

Rather, the goal here is to establish uniform criteria for cost estimating and to provide guideline figures for the estimator. Each figure is contributed from several primary supply sources involved with the thermoforming industry. The basic thermoforming manufacturing cost comprises the following five direct operating costs found in all thermoforming projects: material, labor, energy, and production equipment, and packaging costs. Each of these five direct cost components makes its contribution to the overall cost figure; however, each has its own level of contribution. Depending on the economic situation

Table 9 Material Cost Estimation

95% HIPS @ $0.65 (Natural Color) per pound	
5% Color Concentrate @ $0.95 (Some up to $10.00) per pound	
Virgin resin supply	$0.6175
Color concentrate	$0.0475
Estimated scrap recycling cost	$0.0350
Total Material Cost	$0.700 per pound ($1.54/kg)

and regional conditions, one cost component may affect the others. The following calculations and tables provide guidelines for most cost-estimating practices. The numbers that are applied to the formulas should be adjusted and corrected to actual and more current cost figures in order to gain valid, up-to-date cost estimates. Each of the five cost components has, in turn, its own contributing factors, which come from various sources.

1. Material Costs

For the thermoformer who must purchase sheet supplies from converters or sheet distributors, a published price list is offered. However, suppliers generally give discounts to some customers, for various reasons. The cost of freight and the recycling of the scrap can also be negotiated. All in all, sheet pricing must be favorable to the thermoformer and stay within comparative quality classifications to remain competitive.

When the sheet is produced by thermoformers for their own captive use, cost estimation will begin with the raw resin supply. For example, when HIPS

Table 10 Labor Cost Estimation

Job function	Hourly wage	Hr/yr 8 × 5 × 52	Shifts	Direct labor cost/yr
Operator	$14.00	2080	3	$87,360
Packer/helper	$9.50	2080	3	$59,280
Setup technician	$16.00	2080	1	$33.280
Process engineer	$19.50	2080	1	$40,560
		Total Direct Labor Cost:		$220,480

In case there are several machine lines operated the Setup Technician and the Process Engineer's cost will be spread over the number of lines operated.

Table 11 Production Hours Estimation

Hours per day: 24
Days per week: 5
Weeks per year: 50
Tooling changes: 16 hr each — estimating 10 per year
Efficiency: 85%
Total hours per year: $24 \times 5 \times 50 = 6000$ hr
Actual production hour: $6000 - 160 = 5840 \times 0.85 = 4964$ hr

(high-impact polystyrene) sheet is produced in color, the calculation will follow the format given in Table 9.

2. Labor Cost

Table 10 provides an example of possible labor costs for a single line of operation. Actual labor rates can vary somewhat from one geographic location to another, and these numbers are subject to change over a period of time (Table 10).

3. Energy and Production Cost

Samples of production time and energy cost estimation are given in Tables 11 and 12. Both tables are self-explanatory.

4. Equipment Cost

This group of numbers should be used only for example purposes. Actual equipment pricing varies a great deal from one manufacturer to another, since basic equipment and options are widely diverse. Comparing equipment is not an easy task until the comparisons can be made under the same specifications. Standardization and comparative evaluation are difficult since what one manufacturer offers as a standard feature may be an option or even unavailable

Table 12 Power Consumption Estimation

Equipment	Production (hr)	kW	kWh usage	kWh cost	Total cost
Thermoformer	4964	100	496,400	$0.08	$39,712
Support equipment	4964	45	223,380	$0.08	$17.870
			Total energy cost per line:		$57,582

Table 13   Equipment Cost Range

Thermoformer machine:	$25,000–1,250,000
Trim press:	9,500–160,000
Tooling:	8,000–180,000
Mold temperature control unit:	13,000–65,000
Air compressor:	32,000–125,000
Scrap granulator:	20,000–95,000
Installation (approx. 10%):	11,000–180,000
Total equipment cost:	$118,500–2,055,000

The lower range represents mostly used equipment pricing, while the higher range is new equipment prices.

from another equipment supplier. If equal price comparisons are to be made, all the different equipment features must be similarly designated for the various makes. The price ranges given in Table 13 represent a wide spread of equipment costs that reflect variances in equipment size, optional features, operating differences, and even supplier-to-supplier pricing inconsistency.

After compilation of all cost figures, the numbers must be combined into a common denominator. The cost of a thermoforming project must be presented either on the basis of elapsed production time or in terms of the number of units produced. Whether time or output is used as the measure of production, all numbers must be represented in comparable figures. When elapsed time is chosen for the productivity cost calculation, a 1-hour period is generally used as the standard of measure. This calculates how much product is made at a given cost within an hour of time. When article output is used as the basis of cost calculation, 1000 parts is generally used as the standard of measure. The latter method of calculation is the most popular because unit cost figures are easily tied to selling prices, which are also counted on a 1000-unit basis. The calculation and conversion of cost figures into product units makes cost control simpler because almost all thermoformed products are bought and sold by this common 1000-unit denominator. This figure is adopted from the packaging industry, which is the highest-quantity user of thermoformed articles. Of course, this is not a mandate, and many thermoformed articles that come in a larger shape and smaller volumes are priced out, sold, or bought in units of 100 or even less.

VI. PERIPHERAL FACTORS IN THE THERMOFORMING PROCESS

There are a few factors that play only a fringe role in a successfully managed thermoforming operation. They may not be critical to the process itself, but through their joint contribution, by eliminating most time-consuming process setbacks and the risk of accident, can make the entire manufacturing process trouble-free. In today's complicated manufacturing and competitive environments, these peripheral factors have actually become accepted as routine components of product-making practices. Thermoforming practitioners cannot ignore the safety aspects of an operation. Safety measures are also enforced by state and federal government agencies to protect the work force.

In keeping up with the latest technical developments, process innovations, and product designs, it is beneficial for the thermoformer to use outside consultants and expert advisors, as well as employee training. Using and relying on all available help can keep the thermoformer abreast of the most recent knowledge in thermoforming and ensure their success.

A. Safety

The importance of safety awareness and safe working conditions cannot be stressed enough, especially in thermoforming. The process largely involves reciprocating mechanical motions both in the forming process and in the trimming of formed articles using sharp instruments, which provide many opportunities for accidents. It is of utmost importance that the company provide employee training in the handling of thermoforming equipment. Further instructions, warning signs, and built-in safety measures can help minimize misuse and abuse. Recurring hazard evaluation could result in the elimination of foreseeable dangers, and planning for safety is an ongoing concern.

The use of safety guards on equipment is imperative, and their design and implementation require just as much effort and skill as does building the equipment itself. With each new innovation added to existing equipment, additional implementations to provide for worker protection are required. Making a thermoforming operation safe is the duty of all conscientious operations managers. Placing protective guards over all moving components is costly and difficult but necessary and should be designed so as to provide full protection throughout the entire operation. At the same time, during equipment repair or tooling change, guards should be capable of being dismantled without causing extensive interruption to production. The installation of satisfactory safety guards often falls short of expectations. Designing the ideal safety guard equipment is still an open challenge for the industry. Some safety guards are coupled to the machine's electronics and the entire system will shut down if they are removed or opened. There is sophisticated equipment which accepts commands only through computerized controls from authorized code numbers. Without the proper codes, the equipment shuts down or alert signals are generated. However, precautionary measures for safety must be extended beyond the customary electrical shutdown. Even with a total electrical shutoff, certain equipment components can still move, posing a danger to unsuspecting personnel. For example, heavy platens or mold- and die-equipped platens can be dropped downward against the lower platens by an interruption in the air supply, completely independent of an electrical disconnection. Safety blocks and platen lockout devices must be implemented and added as safety precautions. One of the most ideal safety measures that can be offered concerning moving platens is to design the protective guards so that they always open with a hinged motion, thus causing not only electrical disconnection but also, through their motion, automatic rotation of safety stop blocks between platens. As the guards are closing, the safety stop blocks are moved out of the way through the same rotating motion. When the guards are closed, electric power is restored.

A protective guard must be placed to cover all moving components, not only those acknowledged to be dangerous. Even moving parts not in proximity to the normal operating area could prove to be hazardous to other plant personnel working around the equipment. The guards should be designed to provide protection against carelessness as well as for personnel who are in training. Many accidents are caused by a momentary lapse of safety consciousness when a worker makes a sudden automatic but unthinking, impulsive move. Most basic guarding should reduce the chance of injury while the machinery is being used and is the best accident prevention device.

On the other hand, while thermoforming machinery guarding must be designed with worker protection in mind, it should not hinder production or equipment servicing. Protective guarding on a thermoforming machine is provided in varying degrees from different equipment suppliers. Some machines provide barely any protection, while others are more than adequately shielded with safety guards. Safety measures should not be taken lightly and should be a joint responsibility among equipment makers, thermoforming plant operators, and their employees. Accident prevention is an all-level everyday activity that should involve everybody. This is the only way to create a sound foundation for a safety program. The active elimination of hazards is best accomplished when all working personnel become safety conscious and involve themselves in an accident prevention program.

B. Help from Outside Experts

Thermoforming practitioners locked into a manufacturing routine who work with a specific type of thermoforming equipment occasionally run into situations that demand a different thermoforming approach than the one to which they are accustomed. For those thermoformers who are willing to accept the challenge but have limited knowledge, an experimental attempt could be difficult as well as time consuming. Although they will eventually solve the problem, experimentation often wastes precious time and valuable sheet material.

For this reason, a thermoforming practitioner should seek help from an industry expert. Most suppliers of plastic raw material and thermoplastic sheets, as well as thermoforming equipment manufacturers, will be more than willing to give advice and help to their customers. On some occasions, the supplier's or manufacturer's expertise may be limited to their own specific area and may not be broad enough to cover more complex problems. In such an instance, seeking out professional consultants is the best remedy. Although the number of available full-time consultants is small, there is sufficient help available to handle all levels of problems. Professional publications and associations as well as trade directories can provide complete information on such professionals. The annual *Modern Plastics Encyclopedia* is a useful source of listings and information on practicing consultants in the United States.

There are two common reasons why many thermoformers are apprehensive about bringing in an outside consultant, neither of which is justifiable. The first concerns the process origination and is based on the secretive philosophies and practices of the packaging industry. These have insinuated themselves upon the thermoforming industry due to the close relationship between the two industries. As the thermoforming industry develops and matures, such practices have begun to fade and soon will be left behind. As the various professional trade organizations work to bring common interests and discussions to open forums, the information barriers will be eliminated. The second objection to using outside consultants and advisers is caused primarily by the extreme pride of most thermoforming practitioners. Most operating personnel are too embarrassed to admit to even the slightest lack of knowledge. In their view, calling on an outsider for help is precisely such a declaration and they choose instead to go through a program of experimentation. It is true that given time, the correct solution to the problem will be found. But meanwhile, valuable time and material have been used up. The very purpose of using outside advisers is to tap into someone with an exceptionally wide range of expertise and knowledge, who can offer immediate problem solving. At the same time, they can provide the latest manufacturing innovations, giving the client a competitive edge.

At this time I would also like to issue a warning to those thermoformers thinking of hiring outside consultants to help them out in a process predicament. First of all, choosing a qualified consultant is not an easy task. The available consultants are few and their geographical locality, coupled with their specific expertise, may be a limiting factor as to whether the consultant is capable of fitting the needed qualification criteria. The expert's background and years of close involvement with a specific thermoforming practice is what makes a person qualified and his or her advice dependable. For example, there are experts, whose background is only in the heavy gauge area of thermoforming. They by no means would qualify to give advice in the high speed, high volume (thin gauge) web fed thermoforming process. Equally their experience and knowledge is most likely inadequate when it involves thermoforming foam material in the same respect as well, unless they had the necessary practical involvement in the past. Of course this lack of expertise may not stop someone from trying to get involved or bluff through a lesser known project assignment. Secondly, it is also important to know the prospective consultant's service record. This will indicate that the thermoformer is dealing with an established consultant and not with one of those "wannabe" persons who may be in between jobs and in the mean time wants to try out to become a consultant, until a suitable position is found. Today's industry standards are a lot more complex. The thermoforming process has outgrown its "art form" stage and has become sophisticated, more scientific. Today's practicing consultants have invested a lot of years and effort to gain the knowledge of this industry. Faced with great challenges, they can access important data that cannot be obtained overnight, but only with experience.

The lack of expertise behind poor quality advice can be harmful. Dissatisfying results can be costly for the thermoformer, who may not only lose material and effort, but especially time—possibly the most critical factor in keeping the customer happy. Unfortunately the very same poor results will hurt bona fide consultants as well, because after the sour situation most thermoformers permanently lose their desire for using an outside expert.

C. Training

To maintain any successful thermoforming operation, operating personnel must receive sufficient training to handle the process and the equipment. All too often, such training is limited to the specific area or fields in which the personnel will be involved and may not cover the entire thermoforming process. Without thorough training, workers cannot be expected to handle the job satisfactorily. The process itself may pose inherent dangers to the workers. Lack of proper training will only serve to endanger workers as well as jeopardize the equipment on which they work.

The first objective of training should be to teach personnel to obey all safety rules. Workers and operators must learn how to develop safe working habits for themselves and the people around them. They should respect the thermoforming and supportive equipment, heed warning signs and barricades, and maintain all protective guards and other safety measures. These are the most important aspects of safety, together with good housekeeping practices. When workers are well trained and become safety conscious, most typical everyday accidents will be avoided. Attitude and behavior can be trained.

The second objective of training is to make sure that workers have the necessary skills to perform their task. Thermoforming operating personnel must receive adequate training not just as to how and where to turn a machine on and off but also to know how to operate it, to understand what it does, and to learn what types of adjustments can be applied and why. Unfortunately, due to the shortage of a trained work force, unprepared and undertrained workers are often placed onto production lines prematurely. Training on the job often falls short of adequate instruction, and underpreparedness usually results in costly mishaps. The cost of serious mold damage or operator-inflicted damage is often far greater than the cost of training. For this reason, the management of a thermoforming operation should consider providing the necessary and most important training for their operating personnel. Equipment-operating training programs are usually offered by equipment makers when a new piece of equipment is purchased and are generally sufficient as a base. Further education is available through seminars sponsored by professional magazines, societies, and universities, and is offered periodically in most regional areas. Such educational programs are also available on an invitational basis to factories and plants that wish to offer them to employees as in-house programs. Training and its benefits can strongly influence the outcome of the thermoforming process. In addition, meetings can provide the best opportunity to hear about and review the most recent thermoforming trends as they occur day-to-day.

The relentless economic pressures and the constant battle for funding employee training overwhelms all manufacturing enterprises. This problem is not at all new and it is not going to be any easier in the future. The shortage of available trained personnel is going to be more of the problem in the future and will force the thermoformer to make the best of the situation. They have two options to fill positions in their factories. One choice is to hire anybody who is willing to be trained, but the employer has to bear the cost of the training or try to hire someone else's already trained personnel. Both methods take an established route, but with different consequences. Unfortunately, today's companies are guilty for not spending a decent amount of their profits and reinvesting it back into the training of their own employees. Even those firms, which boast that they have offered sufficient training programs for their employees, are usually under budget crises in their ratio of profit margins versus spending on actual training. What percentage of the profit should be going back into the training budget is also not easy to pinpoint. This criteria should be judged individually by every firm, with some periodical self-review together with hard and honest self-criticism. It is not unusual to expect to have training costs for a group of employees to run as much as $10,000.00 for a comprehensive classroom style training package. Training cost, however, on the average may cost somewhat less than that, depending on your locality, when inviting an outside instructor. Recent surveys indicate that close to half of the existing firms conduct and contract training out "occasionally," one third "seldom," only less than the fourth offered training "often," and the last remaining very few companies fall into the hardened group who do it "always," or absolutely "never" do training.

Seeking knowledgeable and already trained personnel is a popular method to fill positions. This is also a well established path for many firms. After all, a lot of Personnel Recruiters and "Head-Hunters" make a good living out of that particular effort. There are just as popular as advertising for needed employees in trade journals or local or nationwide newspapers as well. This is a simple and cost effective way to lure someone else's trained personnel away from them, especially when snaring with higher salaries. This is even more effective, when the advertising thermoforming factory is located in or near a large metropolitan area where competing firms are established, but could be less successful, when the thermoforming plant is located in a smaller rural community. It is not a forbidden practice to advertise to fill an open position, not in the local paper but instead in the local paper where the targeted similar firm is located. This way the aim is focused on obtaining the specific job-skilled person, and is made directly toward the specific people. In recent years lots of factories moved and located near small towns for better and advantageous start-up costs, lower labor costs, and for all the other enticements offered to the company by the local establishment. Surely they may have found a good and steady labor force, who are fully committed to staying with the firm (often the only employer in town). But, this situation sooner or later produces a technologically stagnating environment, even if the initial training and knowledge has been sufficient for the start-up of the specific thermoforming practices. As they tend to remain with their original know-how, without exposure to ever-evolving new developments, it will come evident that something is missing. To remedy the situation, self- and internal training make sense initially, but eventually will lead to the same stagnated conditions. These types of thermoforming operations are the best candidates to have training exported to them, or their employees should be sent to public seminars that will update their knowledge to equal levels of the other competing thermoformers. The company will be able to keep up with the latest developments and techniques, while obtaining the necessary knowledge to remain competitive in this demanding business environment.

Knowledge of the process is the fuel of any manufacturing endeavor, which is very much true in thermoforming as well! Anybody can buy and set up a factory, purchase the latest machinery, and install a power supply, but knowledge is the most precious resource, capable of providing the motivation and result for a successful and profitable business!

VII. TROUBLESHOOTING

In the thermoforming process, just as with any other manufacturing method, a well-producing line will go out of control every so often and experience difficulty or produce poor results. The trouble is most obvious when equipment breaks down, particularly when it is observed visibly. However, much controlling apparatus contains sealed mechanical or electronic components and malfunctions cannot be observed from outside; breakdowns can be observed only through the thermoforming results. Difficulties in the process can be so severe that they cause or demand instant shutdown. Conversely, they could be so minute that they do not force a stoppage but simply create out-of-spec articles. In many instances the entire process can be so sensitive that the slightest change in heating levels or just a small alteration in the forming force power can have major implications for the thermoforming. Ideally, good production consists of shot-to-shot uniformity and a consistent quality product outcome. Problems encountered in the process can result in faulty production either intermittently, frequently, or continuously. If the problem occurs frequently, the thermoformer must make some adjustment against the unwanted results. The problem encountered may have a single cause or it may have multiple causes, with confusing and not easily recognizable sources. When a problem occurs or unwanted changes take place in the thermoforming process, a decision must be made as to what type of adjustment will remedy the situation. It is the best policy to make only one adjustment at a time. If simultaneous adjustments are made, without acknowledging the single adjustment's result, nothing but confusion results, not elimination of the problem. With all types of readjustment possibilities, it is easy to lose totally the setting in which the problem began and also the opportunity to start at the beginning to make and determine the proper adjustment. This situation can be worsened by too many people providing assistance and making their own adjustments. A good rule of thumb should be to stop the machine and retrieve for evaluation the last product made. A plan of action should be devised, and adjustments should be made one setting at a time.

A different view of expectation should be given for thermoformed products that have been produced once satisfactorily, as opposed to those that have never been produced before. It may take some doing to duplicate exactly the precise conditions that resulted in that one satisfactory forming, but once these have been found, the problem is solved. However, with a newly acquired thermoforming product or project, the outcome could be uncertain. A multitude of trial approaches has to be made before satisfaction can be claimed. In the event that a thermoforming problem cannot be resolved or minimized by adjustment, the project should be realistically evaluated as to whether it is a design or a manufacturing defect. Each problem can be resolved only within its specific area of difficulty. A design-caused problem is usually resolved when design changes are implemented. For example, a change in a sidewall angle or enlargement of a corner radius can render satisfactory results to an earlier thermoforming failure if the problem was a result of the earlier design.

On the other hand, manufacturing problems may require complete re-tooling, a new thermoforming method, or even a change of equipment to achieve satisfactory results. For example, a single-platen thermoforming machine has no means of offering plug-assisted thermoforming techniques, and improved deep-drawn wall thickness distributions cannot be made. A change of machinery is therefore needed to achieve the desired results. As in most other manufacturing methods, certain material, equipment, and design limitations exist in thermoforming, and as a good manufacturing practice only realistic expectations should be made.

With the recent growth and increased popularity of personal computers, it is obvious that some companies or individuals will develop and make available specific computer software programs to help thermoformers in various areas of thermoforming manufacture. Like any other manufacturing methods the more information that is known and tabulated, the easier it is to make sensible decisions, modify plans, and even predict the final outcome. Therefore it is natural to look to the abilities of the computer to help, where in the past only unpredictable results were obtainable.

Recently I learned that there are extensive efforts being made to develop software programs specifically for thermoforming. Up to now, when thermoforming produced unexpected results, the cause has been mysterious. For example, most thermoformers wish to know the finished product's wall thickness before forming and wall thickness variations of the sheet material distribution over the entire product's surface. A better material distribution can greatly improve product quality or it may allow using a lesser thickness sheet that can reduce material cost. All of this together could lead to improved thermoforming practices. Naturally all this improvement is for the betterment of the thermoforming business. It is a highly desirable task and it is hoped that with the aid of computers the unknown mystery is eliminated from the thermoforming process. At this writing I do not know what the intended software programs can and will encompass or how many aspects of the thermoforming criteria will be incorporated within them. But I am not sold on a specific function of a computer program that can exactly predict actual wall thickness results. It may be capable of giving a theoretical result of a given size of sheet material stretched into or over a known mold's surface. Such calculations would be reasonable to expect from a computer and from specific data inputs, when the sheet dimensions are known and the mold's surface is calculated. The surface area enlargement could then be determined by the computer. From these inputs the computer will provide an excellent estimation of material thickness results. Until it is proven reliable, I am skeptical that such results can match "real-life" results. Complicated mold configurations, variations in the extruded sheet material thickness, in heating and drafty plant conditions, mold temperature, and even some thermoplastic material behavior can and will change the material's stretching ability. The actual stretching of thermoplastic sheet into or over the mold never results in an identical stretching from one forming to the next forming. Even if the above mentioned factors can be put into the computer as data the complexity of the actual stretch can not be identically replicated. Keeping the input and comparison factors too narrow the program most likely will miss the target. As time goes on, we will find ways to refine our controls over the thermoforming process and may predict the outcome results much better and closer to "real life." But for now and the near future I recommend the use of any of the upcoming computer programs as no more than an estimating device or tool. I know all of us would like to open the computer program or data base and pull out exactly what is needed, without dealing with compromises and "fudge factors." Of course that is why the computer experts invented the "fuzzy logic" computer concept.

The following troubleshooting list is produced to guide the thermoforming practitioner through the most common and simplistic problems. It lists possible causes and solutions to remedy them. The list utilizes the following abbreviations: P = problem, C = causes, and R = remedy. Where there is more than one cause for a given problem, each cause must be investigated and eliminated. This list represents only the most common difficulties that occur in the thermoforming process. Many more, sometimes puzzling difficulties are possible. To remedy out-of-the-ordinary problems, good, persistent investigative skills, and clever, unique solutions must be found.

P: Loss of Detail in Forming
   C: 1. Underheated thermoplastic sheet condition.
      2. Drop in the forming force level.
      3. Plugged up vent holes or channels.
   R: 1. Increase heater temperature or extend resident time in the oven.
      2. Check the reasons for the loss of vacuum or pressure forces.
      3. Clean and reactivate the plugged-up vent holes or channels.

P: Hole, Slit, or Rupture in the Sheet
   C: 1. Faulty sheet supply (pinholes made into the sheet).
      2. Underheated sheet rupturing under the forming force.
   R: 1. Check sheet supply; if pinholes appear repeatedly, reject.
      2. Increase the heating of the sheet and decrease the level of forming force.

P: Overly Detailed Forming, such as Mold Roughness and Vent-Hole Markings Transferred from the Mold to the Formed Articles
   C: Overheated thermoplastic sheet introduced into forming together with excess pressure force.
   R: Reduce the heating levels of the sheet as well as the pressure of the forming force.

P: Excessive Thinning of the Plastic after Forming
   C: 1. Overheated thermoplastic sheet.
      2. Using the simplest thermoforming technique without plug assist.
      3. Attempting deeper forming ratios than the process technique will allow.
      4. Insufficient sheet thickness.
   R: 1. Check and adjust thermoplastic sheet temperature.
      2. Upgrade the thermoforming method to use plug-assisted forming.
      3. Use a warm plug assist and mold lip.

P: Thin-Spot Development on the Thermoformed Articles
   C: 1. Uneven heating of the thermoplastic sheet.
      2. Hot spots on the sheet created by faulty heating elements.
      3. The thermoplastic sheet has irregular orientation or a patchy mo-
         lecular-stress-relieving condition, which activated under heat.
   R: 1. Check oven and heating elements for uniform heat distribution.
      2. Check sheet behavior under heating. Draw parallel markings onto
         the sheet in both MD and TD directions prior to heating, then send
         the marked sheet through the heating cycle. Any distorted markings
         will reveal material shifts or irregularities, and it is difficult, if not im-
         possible, to make corrections.
      3. Check another batch of thermoplastic sheet and compare.

P: Formed Plastic Article Sticking to the Mold
   C: The mold temperature is too hot.
   R: Apply means of cooling to the mold (metal molds should be substituted
      for wood or synthetic mold materials) by installing a cooling jacket to
      the back side of the mold; increase the flow of coolant if more cooling
      is required.

P: Stripping Difficulties
   C: 1. Using a male, instead of a female, configuration.
      2. Not enough sidewall taper angle.
      3. Use of undercuts or reversed draft angle.
   R: 1. Convert male mold to female mold configuration.
      2. Make an oversized drawing with shrinkage measurement incorpor-
         ated. Determine from the drawing if shrinkage can neutralize the
         size of the undercut or reversed draft angle.
      3. Apply mechanical strippers to the mold, such as a spring, pneumati-
         cally or hydraulically activated stripper bars, or complete plates.

P: Obtaining Various Stack Heights
   C: 1. A constant side-to-side sheet thickness variation in the thermoplas-
         tic sheet supply.

2. A constant but rather widespread temperature variation in the oven, which develops routinely, causing uneven sheet temperature. Thinner articles will be produced out of the hotter areas, while thicker products from cooler areas. The thin articles will nest in a shorter stack than the thicker products.

R: 1. Check the thermoplastic sheet for side-to-side gauge variation. (A tapered sheet or gauge band condition will usually be present.)
   2. Control oven temperature to create uniform levels from side to side.

P: Poor Definition in Engraved Patterns
C: 1. Underheated thermoplastic sheet.
   2. Poor forming force level.
   3. Lack of sufficient holes or channeling for air evacuation in the engraved areas.
R: 1. Increase the heat of the thermoplastic sheet.
   2. Increase the power of the forming force.
   3. Add additional holes into the engraving areas; open or increase the sizes of channels between vent holes and the forming force ports.
   4. Check the size of plumbing lines and restricting connections.

P: Hot Spots on the Sheet
C: 1. Heating element or control problem.
   2. Possible drafty condition in the plant.
R: 1. Replace damaged or poor-quality heater elements and controllers.
   2. Apply screen shielding where hot spots occur.
   3. Provide absolute draft protection.

P: "Fisheyes" on the Formed Article
C: Spotting or "fisheyes" will develop on a formed article where the heating of the sheet is interrupted either by poor surface contact with the heater (if a contact heater is being used) or foreign material is present on the surface of the sheet which absorbs the heat (e.g., water droplets on the sheet surface).
R: Check the heating of the sheet and remove any obstacles (wipe or dry the surface).

P: Excessive Sheet Sagging
C: 1. Using unoriented thermoplastic sheet made by a method other than extrusion.
   2. Using large thermoplastic sheets that cannot support their own weight under heating.
   3. Any overly heated sheet will develop a sag.

R: 1. Controlling the heating and the sheet's resident time in the ovens are the most common methods used to control sagging.

2. Occasionally, use of an expanding clamp frame or pinchain rails can tighten the developing sag.

3. The elimination of sheet sag is constant work, requiring heat adjustment and readjustment throughout the process. (Sag in a thermoplastic sheet can rapidly develop to a heating problem due to changes in distance between the heater and the sheet. A slight sag may be acceptable as a sheet prestretching method or as a readiness signal for thermoforming. However, development of a large sag is very rarely acceptable and should be avoided.)

P: Loss of Fracture and Impact Resistance in a Formed Article

C: 1. Using lower-quality thermoplastic material (change of resin supplier, change in scrap recycle ratio).

2. Loss of orientation levels in the plastic material.

R: 1. Check material supply source.

2. Check orientation levels before and after thermoforming.

P: Uneven Wall Thickness in Formed Articles

C: 1. Improper heating of the thermoplastic sheet.

2. Poor prestretching of the sheet.

3. Ineffective forming method.

R: 1. Check heating conditions.

2. Reevaluate forming method.

P: Irregular Sheet Thickness Entering into the Forming

C: 1. Poor, uneven thermoplastic sheet thickness.

2. Through the heating cycle, the sheet is experiencing material shift from one area to the next, usually caused by molecular stress or orientation in the sheet.

R: This problem could be tolerated or may require rejection of the material, because improvement in sheet quality is not within the control of a thermoformer.

P: Small Holes in the Thermoplastic Sheet

C: Pinholes or small voids are usually caused by lack of material flow in the calendering or extrusion process. (This type of a sheet is faulty.)

R: There is no remedy for this problem; however, the processor can live with an occasional hole and void, but if repeated too often, the sheet should be rejected.

P: Poor Definition of Detail on Formed Articles

C: 1. Underheating of the thermoplastic sheet.

2. Insufficient forming force levels.

3. Lack of air volume capacity (need of a surge tank).
4. Poor seal result, causing a loss in forming force.
R: 1. Check the forming force efficiency for force and volume.
 2. Check for air leaks or premature forming force actuation.

P: Weight Differences Among Formed Parts
 C: 1. Incorrect sheet gauge or variation in gauge.
  2. Poor material distribution between cavities (robbing material from each other).
 R: 1. Check the gauge thickness.
  2. Add preclamping mechanisms to capture material, either around the entire mold or at the individual cavities (see Figures 13 and 14).

P: Heavier Bottom Wall Thickness on a Formed Female Article
 C: In forming, the heated thermoplastic sheet is cooled by a cold plug assist (see Figure 79).
 R: Apply heat-retaining material (e.g., syntactic foam) to the plug assist.

P: Above-Normal Flange Thickness
 C: Using a straight vacuum-forming technique with cold mold lips, causing zero material stretching at the flange area.
 R: Use a plug-assist forming method or any of the prestretching techniques described earlier (see Figures 41 to 47).

P: Extra-Thin Sidewalls
 C: Using cold female mold and male plug assist [see Figure 79(1)].
 R: Consider the use of heated or warmed plug assist and mold lip [see Figure 79(2)].

P: Postmolding Distortion
 C: 1. Undercooled article or article prematurely removed. (Any retained heat within the formed article can cause warpage as the article is prematurely removed from the mold and cooled without the mold's support.)
  2. An unusual shape can cause deformation and warpage, especially when male and female configurations are combined in one design.
 R: The best advice is to cool the formed article thoroughly within the mold while maintaining a full application of forming force. All the retained heat must be removed, not just from the surface but also from the core of the plastic.

P: Postmolding Part Expansion
 C: This is also the result of premature part removal from the mold, which affects only the thermoplastic foam articles. The residual heat in the

foam core is causing the foam to expand after the removal from the mold. The expansion can be so severe that the formed articles no longer retain their original dimension.

R: Extend cooling time until all heat is extracted from the foam article.

P: Strong and Weak Areas Found in the Same Thermoformed Article
  C: 1. Material thickness differences.
     2. Changes in the molecular orientation of the plastic.
  R: 1. Check for material thickness variations.
     2. Alter the forming method for better material distribution.

P: Spots (Fisheyes) on the Formed Article's Surface
  C: Foreign material, most commonly water, on or within the plastic sheet.
  R: Predry the sheet stock prior to thermoforming if the cause is water. At the presence of other contamination, the sheet stock should be discarded.

P: Blemished Surface on the Finished Article
  C: 1. Mold surface blemishes imprint itself onto the formed surface.
     2. Plastic chips pushed out by the pin-chain penetration, which carried into the forming area and formed into the plastic.
  R: 1. Clean and refinish mold surface.
     2. Apply heat to the plastic sheet's edges to ease pin penetration.

P: Drag Marks (Scratched Surface)
  C: The mold is damaging the plastic surface if the product is removed forcefully. This condition is more frequent with male molds or undercut configuration.
  R: 1. Refinish the mold surface.
     2. Apply a Teflon coating to the mold surface.
     3. Use a release agent.

P: Sheet Pulled from Clamp Frame or Pin Chain
  C: 1. Overheated clamp frame or pin chain.
     2. Mold edges placed too close to the clamp frame or pin chain rails.
     3. Extremely deep-drawn forming, causing excessive pull on the sheet.
  R: 1. Apply additional heat to the thermoplastic sheet to enhance stretching.
     2. Implement heat shielding or cool the clamp frame or pin chain rails to prevent overheating.
     3. Increase the sheet size or web width to gain an additional buffer zone at the clamp areas.
     4. Make sure that the mold's parting line matches the sheet line.

P: Poor Trim Registration
   C: Registration error at the indexing.
   R: 1. Acknowledge the equipment's trim limitation and compare with the actual trim variation. Define a realistic expectation which can relate to the specific equipment's and article's trim configuration. Certain equipment may not be capable of providing the expected trim quality.
      2. Check for the cause of misregistration, shrinkage variation, feed acceleration or deceleration, feed finger stroke, web bounce, pull-back errors, and so on.

P: Wrinkling or Webbing in a Formed Article
   C: 1. Access sheet material (larger than the mold surface) due to overstretched sheet material prior to forming.
      2. Webbing can develop between male mold configurations if they are placed too close together.
   R: 1. Limit the prestretching of the sheet. Usually, reduced air pressure or prestretching time can eliminate the wrinkles.
      2. Allow more space between male configurations (more than a 1:1 draw ratio).
      3. Use matched mold or mechanical pushers to force the material to form.

P: Inconsistent Sheet Advancement
   C: Equipment error.
   R: Define realistic stroke specifications for the equipment. Any deviation from the expected consistency requires correction.

P: Undersized Articles
   C: 1. No shrinkage allowance was built into the mold.
      2. No material thickness allowances were built into the mold, usually when a male mold is converted to a female mold without compensating for material thickness.
      3. The article is produced on a mold that has been copied directly off a previously thermoformed article.
   R: Building a new mold.

P: Forced Production Slowdown
   C: 1. Cold temperature in the plant.
      2. Cooler thermoplastic sheet temperatures (brought into the plant).
      3. Overheating the plastic sheet will extend the heating cycle as well as the cooling cycle.
      4. Inefficient forming force (vacuum or air pressure).
      5. Leaking mold.

R: 1. Check for temperature changes in the plant or in the sheet and oven.
   2. Compensate in temperature levels but not in cycle time.
   3. Check the forming force levels (not at the source but at the mold).
   4. Inspect all seals in the mold.

P: Angel-Hair Development
  C: The development of angel hair, like plastic slivers, appearing at the edges of the trimmed articles, usually is the result of cutting, and it is more prevalent with one type of plastic than with others. Dull and worn cutting edges is one of the reasons for the angel-hair development, although extremely sharp knives can cause the same problem.
  R: Cleaning the removing the angel hair is not only difficult, but the attempt will increase the static, which will redeposit angel hair somewhere else.
   1. Minimize angel-hair development by minimizing cutting-plane variations (no more than an 0.002-in. variation in knife height).
   2. Limit any overtravel of cutting knives.
   3. Dry cycling of newly re-knifed tooling is highly recommended.

P: Grease or Other External Contamination
  C: 1. Excessive greasing of the equipment.
   2. Chemical, antifog, or slip-agent application on the sheet prior to thermoforming.
  R: 1. Do not overlubricate equipment.
   2. Control the application of chemicals (limit spray volumes, nip or squeegee the sheet surface prior to heating).

7
The Human Element in Thermoforming

The thermoforming process, just like any other, involves supplies, manufacturing and supportive equipment, and, most importantly, people. Personnel is the most important element in the manufacturing endeavor, from the actual labor force all the way to company management, including the owner, president, or CEO.

I feel that this book covers all the technical elements of thermoforming. In this second edition, I would like to address the human element as well. Throughout my career as a consultant I have been involved with many firms and their people, not just as a problem-solver but as a constant observer. This chapter is intended merely to convey my observations and perhaps focus awareness to these subject areas with the hope that some redirection may be implemented away from today's increasingly impersonal industrial environment.

None of the following observation or concerns are new practices or qualify as business-threatening crucial mistakes. We have seen them in our manufacturing environment for some time now, but their increased prevalence and frequency are cause for concern. If these business practices continue unconstrained, the seriousness of economic insecurity will soon catch up with us and we will all miss the boat for a better and more prosperous social and working environment. It isn't easy to keep hopeful and optimistic about the future and not cave in to today's more materialistic social influences, which tend to focus on business and business alone, forgetting about the human interest within the work force.

Like most of us, I often wish for a dependable crystal ball that could predict the future and give precise guidance as to which direction to proceed. It is a well-known fact that trends and fads come and go, yet as soon as there is the slightest indication of a breakthrough, all others jump on the bandwagon trying to duplicate the same unproved procedure without spending any time to study it. The thermoforming industry is no different from other industries often found heading in the wrong direction, following a poor leader. Too many factors guide our decision-making, and sometimes outside influences such as economical or political conditions will cause major directional changes for our industry.

Economists and statisticians are great believers of charting accumulated data and providing predictions based on the results of their charts. Following the developing line of the chart and estimating the established trend, they anticipate that the next generation of data will follow the very same trend. On the other hand, most pessimists, doomsayers, and believers of the so-called Murphy's law might predict the complete opposite, again due only to the chart indication based on previous data. It is time to reverse this trend. No matter which side we find more acceptable, we must recognize that until the collected data are in it is just a prediction, only a guess.

The different fields of thermoforming are governed by different specific rules and subjected to different competitive and economic influences. It can very well be that while one segment of the thermoforming industry is basking in good economic fortune others may be under severe economic pressure. This pressure, whether it is just economical or politically inspired or both, can play havoc and be so intense that it is capable of paralyzing a previously striving company.

The bases for the economical or political pressures are mostly manmade and mainly result from human influence and involvement. As for a thermoformer it is not enough that he or she has to worry about the basic demand of the process and its success, like the cost of material, cost of energy and labor, as well as the pressure of the competitive forces, now the industry has to face many outside influences which can derail the entire thermoforming endeavor. These areas concern not only the thermoforming industry, but are equally true in other molding process industries as well. Several specific issues exist, which I encountered throughout my years as a consultant. It is my humble opinion that the following trends will have great implication on the future of most plastics industries, including thermoforming.

In many companies, top management is comprised mostly of business people who are rarely technically inclined. Most management personnel are well trained in many aspects of management techniques and practices. They usually have a respectable MBA degree or years of management experience, which is practically a must these days. But when it comes to technical knowledge, they rely mostly on the advice of subordinates. I feel it is an imperative for the president or vice president of a company to know at least a little of the technical background as well. I am astonished when I discover that a company president's background is other than thermoforming (I have seen them come from the paper, soft drink, ice cream, and snack food industry, and even from candle manufacturing). I can accept that there are similarities between these industries, and all therefore demand just about the same management skills. However, specific technical knowledge is vital for recognizing the limitations and capabilities of thermoforming and absolutely necessary to make decisions at the helm of such a company.

Over the many years of seminar instruction I have conducted, there have been very few occasions when management people are in attendance. What could be the reason for this? Do they think that the "technical stuff" is not that important for them? Maybe they think attending such seminar programs may indicate to others how little they know of the thermoforming process. Or perhaps their pride does not allow them to attend. I find it almost humorous that after a brief conversation with most high-ranking officials, I discover how badly they are in need of some basic education in thermoforming. It is also my finding that many of the practicing individuals involved on a daily basis with thermoforming know their daily process routines perfectly, but never had the time and training on the basics, and therefore their troubleshooting usually depends on myths or guessing, rather than on facts. Almost all thermoforming company's human resource managers are constantly complaining how hard it is to find trained personnel to fill needed positions in their organization. Few companies have found an answer to this problem by initiating in-house or out-sourcing training programs for their employees. Even fewer are the humanistic, liberal companies that specifically organize central training opportunities sponsored by several local firms, combining their training resources and even arranging for some governmental subsidy to underwrite this effort. I would like to re-emphasize my earlier statement that nothing is more important than having well-trained knowledgeable people in the thermoforming operation, individuals who are also familiar with modern management practices such as TQM (total quality management), SPC (statistical process control), JTM (just-in-time manufacturing), ISO 9000 guidelines, etc. How can any company equip itself with the proper tools of the trade if their people do not know the basic thermoforming principles?

In today's corporate management it is very rare to find individuals who grow up and mature with the same industry and gain all the first-hand practical knowledge of the thermoforming process. The trend indicates that it may not be necessary to have some, or even any, technical knowledge to manage the company's day-to-day operations. This thinking, in my opinion, is disastrous for the company's long-term survival, and is often a cause of friction between top and mid-management, which can even trickle down to the actual operating personnel. It all comes to a head when the *profit now* desire collides with the *how* mentality. Squeezing profits out of a business is acceptable as long as some of it is reinvested back into the future of the company.

I have seen too often severely curtailed spending trends; underbudgeting for important expenditures; and cutbacks on training, research, and new product development. Businesses get by year after year out of an existing antiquated and somewhat declining product line. There are no plans in the works for replacement product and no effort is made to come up with new products to take the place of the aging product line. Even the larger firms that used to have substantially sized product development departments have let them dwindle to just a few people dedicated to new product development. When innovations and new product ideas are generated by chance or luck alone, we are in trouble for the future. Too often today's policy of budgeting for new product development is: if it takes longer than one budgetary year, it is pushed back onto the "back burner." Projects must produce immediate results within a short budgetary confinement. The larger firms are too sluggish and expect an enormous return for their new product development efforts, a return way too large to encompass many developing product ideas. On the other hand, small companies that may be more inclined to take risks and would be more interested in smaller volume product lines have limited budgets for experimentation. This financial constraint shuts out many companies, leaving only a few brave or desperate companies. The current patent laws and practices, together with most companys' reluctance to deal with an outside inventor's submissions, also constitute a major hurdle to overcome this blockage in inventiveness. The proof of this is the dramatic decline of patent applications currently generated from the United States.

Another area of concern is the future of educational opportunities in thermoforming. This subject was addressed briefly in Section VI.C in Chapter 6. Presently very few educational opportunities exist in the field of thermoforming. Few of us who dedicate our time for instruction have actually worked extensively in this industry. We have worked very hard to lift thermoforming manufacturing methodology out of the job-shop environment into a precise engineering discipline. Our accumulated knowledge is based largely on practical experience gained on the production floor operating equipment. Experience and knowledge acquired over the years, however, is not sufficient; qualified instructors must also have the ability to convey and present that accumulated knowledge. It is for these very reasons that a very few higher educational institutions offer short courses on thermoforming. They are usually offered only as a continuing educational course open to all persons who may register. The granting of higher level degrees in plastic engineering covers other subject areas, like extrusions, injection molding, and even blow molding, but thermoforming is often only briefly mentioned among the other subject areas. There are few professors and other instructors well-acquainted with the thermoforming process, and these are often limited to laboratory-scale operations, not full-fledged production systems.

Other learning opportunities exist, specific society-sponsored programs which conduct two- or three-day-long multidisciplinary training seminars. Such programs may go slightly further than an introductory/refresher program for individuals already employed in thermoforming as well as those with no previous training in thermoforming. In order to maintain the thermoforming industry's growth rate, the educational opportunities must be increased and the operating companies must bear the cost. Otherwise, we will have to cope with untrained and undertrained personnel. Imagine the costs a lack of knowledge can incur when operating a $100,000 piece of equipment with an $80,000 mold or a high-speed operation running reject-quality product for an hour or more, not to mention the safety aspects.

My other major concern can be found in today's headlines: companies are down-sizing, announcing major layoffs. Dwindling sales and a drop in profit for various reasons result in cutbacks on expenditures, which obviously includes cutting the labor cost and letting people go. The reduction of the work force immediately improves the company's profit margin, and most of the time automatically increases the company's stock value, which elevates the profit margins overall as well. But there is another side to these practices. Any major down-sizing indicates a previous uncontrolled growth practice and perhaps other ignored problems. It is most likely the result of uncontrolled overstaffing, lack of new product development, or even overestimation of business. Then the company goes into consolidation mode. Whatever the reasons for the consolidation and cutback, it most likely should have been dealt with and averted before the problem got out of proportion.

In my opinion, there is nothing more important than to have ongoing new product development. This is the only answer to continually climbing out of the impending rut created by the drop in sales. Even in poor economic times, those companies that had products that customers needed and accepted had good fortune. Those ideal products do not require costly advertising-based revenues to force onto buyers. It may be a total departure from old and outdated convictions as to how the product should look or function in order to make it so attractive to the buyer.

The research and new product development efforts must be made independent from the sales department's influence. Cooperation is fine, but the new product development cannot and should not be under the control of the sales department. Salespeople tend to put emphasis on easily salable products and are often unwilling to accept unfamiliar territories or different product areas. New product development criteria and specified target parameters on research must come from top management, redirecting the parameters as necessary according to the current business environment. Making extremely long-term commitments over several years is also unwise due to the rapidly changing social and economic conditions of today.

The potential growth opportunities will come from our ability to innovate in thermoforming. We can change sheet materials without a new mold, we can change and even rework molds without purchasing new ones, we can adopt different forming techniques, and more just by rechanneling the forming forces.

The other long-term damaging effect of major cutbacks in personnel and downsizing is an obvious chain reaction. As one company goes through its downsizing, they are actually getting rid of other companys' customers. When the lack of sales hits the subsequent companies, they will be forced to do the same thing, eliminating yet more people from the workforce. Downsizing and layoff practices are only momentary relief options and should be practiced with caution.

The next discussion involves tooling and mold purchasing for thermoforming. There are thermoformers who have in-house mold building facilities. These are the lucky ones. They have full control over the entire mold making, not only the physical expect, but over the actual cost as well. Making the tooling for their specific equipment, scheduling its timing and controlling the quality and outcome, whether the specific tooling will be used for a short run or endure long-term production, is up to the company's tooling department.

On the other hand, there are many firms who depend on outside sources to build their tooling and molds. Purchasing a mold from an independent source can be risky, since there is less control over the tool and mold-making outcome as well as delivery timeframe or even the proper function of the mold. It is best to work with the same tool shop all the time, one that is familiar with the thermoformer's specific needs and equipment, fitting installation, and product line criteria. In this case tooling rework is nearly nonexistent. However, the majority of thermoformers tend to farm out their tooling to different mold-making facilities, shopping for best prices as needed. Although this practice is more common, it is not without risks. Mold-making shops are numerous and some have a great deal of experience. However, just because a shop is capable of making tools for injection molding, it does not qualify them to build molds for thermoforming. And just because a mold maker is familiar with vacuum forming, it does not automatically qualify its capability to build molds for pressure forming. Although the shop practices and metal working equipment are the same, the molding criteria between injection molding and thermoforming are not. Knowledge of the specific molding criteria is imperative to produce well-functioning molds and tooling.

As an experiment, I requested quotations from ten different tool shops on the same thermoforming mold. To my surprise, the mold cost deviated greatly. Between the lowest and highest figure was tenfold difference. After careful analysis, I came to the following conclusion. Those giving a lower price may not know what criteria lie ahead and may run short of money soon after the first supply purchase. Whereas the higher bidders know how to quote and then some to make the job extra worthwhile to come back even from a tropical semi retirement.

Those whose quotations are in between are the ideal shops to deal with. Their discrepancies may be explained by the differences in shop organization. Some shops, for example, are equipped with more up-to-date equipment or CNC (computer numerically controlled machining) that requires higher skilled labor but often result in much more accurate and precise mold tolerance levels.

Some mold makers may tend to "pad" their estimates because they are unfamiliar with the specific thermoforming criteria that lies ahead to include allowance for rework. Shipping costs are also a factor when the mold maker is at a great distance from the thermoformers. I have seen molds shipped from Europe and Israel to U.S. installations for thermoforming.

One of the major expenditures in mold purchasing is the cost of tooling (mold plus trimming die). The outstanding tool makers are the ones who know the process well and build a functional mold with high tolerance and quality where it is needed without wasting time or resources on unnecessary extra work. When the thermoforming mold does not require extremely high-polished surface finishing, for example, why do it! It just needlessly increases the cost of the mold.

Last but not least, I should mention herein regulatory and legislative aspects. The industry is governed by legislative and mandated regulations put in place to protect people's health and welfare as well as the environment. Whether the industry is over- or underregulated is the subject of an ongoing debate. In some instances it is claimed that some areas are so overly restrictive that it is nearly impossible for certain businesses to compete. On the other hand there are some areas in need of governmental control or regulatory adjustments. Whether an area is in need of regulation or removal of previously enacted regulations is constantly argued by politicians and lobbyists. Many times dissatisfaction with an existing controlling measure can cause individual states, counties, or even cities to enact their own regulatory laws, which may go further than the federal regulations. It is the responsibility of each thermoformer to untangle the multilayered regulatory maze and find the best path to comply with all relevant legislation. This is not an easy task, and it often requires specialized personnel and lawyers to interpret the written rules and regulations (sometimes it may even be necessary to argue one's case in court).

If they are not met satisfactorily, the threat of hefty fines lingers over the heads of noncompliers. This issue is further complicated by competing pressures from those who do not operate under the same jurisdictional regulations. Thermoformers that are not under as restrictive rules are able to offer identical items at a lesser price. This dilemma applies not only to localized inconsistencies, but also to global international competition. It has been observed among European countries as well as in Central and South American and Asian thermoformed product imports. Truly the lower costs are beneficial only to the current buying public. Over the long term, however, this situation could induce a total phase-out of any locally manufactured supply, and as soon as this happens the initial low price might be increased to even higher to what it was originally.

The concerns I have voiced here in regards to thermoforming really affect all of us. We must not forget that we all in this together. The issues of better work relations, education, and employment benefit everyone. Product makers need customers and customers need good functional products, including the many thermoformed products.

Glossary

Air-Assist Forming Method of thermoforming in which pressurized air is used to partially preform the plastic sheet prior to the final forming; also called pressure bubble forming.

Air Pocket Air trapped between the mold and the formed part.

Blister Pack Thermoformed product usually made of clear plastic, used in packaging.

Blowing Agent Foaming agent used in the extrusion of foam sheets.

Blow-Up Ratio Foam sheet expansion ratio from cold dimensions to the preforming hot dimension.

Breakaway Mold Insert Mold segments within the mold, which can be pulled away from the mold with the formed article, then refurbished back into the mold for the next cycle.

Breakdown Machine inoperable due to breakage or wear.

Calender Machine that has a series of counterrotating rollers to prepare sheets of material (e.g., vinyl sheets).

Cavity Depression in a mold made by casting, machining, hobbing, or a combination of these methods; depending on number of such depressions, molds are called single cavity or multicavity.

Chill Marks Marks caused by uneven temperature of the heated plastic sheet caused by improper heating or draft of air.

Cycle Complete repeating sequence of thermoforming process; consists of four stages: heating, forming, cooling, and stripping.

Daylight Opening Clearance between two platens of a press in the open position.

Density Weight per unit volume used in foamed sheets: pounds per cubic foot, grams per cubic centimeter, etc.

Depth-of-Draw Ratio Ratio of mold opening to depth of mold.

Die Cutting Blanking or cutting shapes from sheet stock using a "steel-rule die."

Die Set (1) Matching punch and die; (2) device that holds the punch and die in perfect registration through the use of guide pins and die plates.

Dimensional Stability Ability of a plastic part to retain the precise shape in which it was molded, fabricated, or cast.

Downtime Nonoperation of a thermoforming machine due to normal shutdown or breakdown.

Draft Degree of taper of a sidewall necessary for the removal of parts from a mold.

Drape Forming Method of thermoforming whereby a thermoplastic sheet is clamped to a moving frame, heated, and draped over a male mold; vacuum is applied to complete the forming cycle.

Draw Down Ratio Ratio of cavity opening to the depth of draw, normal limit 1:1.

Drawing Process of stretching or pulling a heated thermoplastic sheet into a mold cavity.

Edge Trim Area of the sheet used for clamping or pin-chain holding.

Extrusion Process by which a thermoplastic sheet is produced for thermoforming.

Extrusion Coating Process of extrusion when a substrate sheet material is covered with another sheet material and done in a secondary extrusion process.

Fabricate To work a material into finished form by machining, forming, gluing, or by means of methods other than thermoforming.

Family Mold Multicavity mold with similar component parts.

Feed Cage Device that guides and feeds the formed web into the trim station.

Feed Fingers Steel fingerlike extensions that push the web into registration and to its location between the punch and die set.

Female Mold Cavity or cavity side of the mold.

Film Usually an extruded thermoplastic film from $1/2$ to 10 mil thickness.

Flammability Measure of the extent to which a material will support combustion.

Gauge (1) Measurement of material or sheet thickness; (2) an instrument to measure material thickness.

Infrared Part of the electromagnetic spectrum between the visible light range and the radar range. Radiant heat is in this range, and infrared heaters are much used in sheet thermoforming. Wavelength, 2 to $10\,\mu$.

Laminated Plastic Plastic sheet material usually consisting of two or three layers of thermoplastic materials which have been bonded together by means of heat and pressure. Most often the lamination is put onto a foamed material to improve its surface characteristics.

Locating Pins Hardened steel pin used with matching hole locations drilled into the opposing mold halves, forcing them to align with each other.

Male Mold Plug or protruding side of the mold.

Matched Mold Close-fitting metal molds with predetermined mold gap. Male and female molds approximately follow each other's contour. This mold is used mostly in the thermoforming of foam sheets.

Mold Gap Space remaining between the male and female mold, after closing; this space is taken up by the thermoformed plastic sheet.

Mold Height Overall dimension of a closed mold.

Molding Cycle Period of time occupied by the complete sequence of operations on a thermoformer for the production of one set of product.

Mold Release Parting agent, usually a silicone compound.

Monomer Relatively simple compound that can be made to react with itself and couple to form polymers.

Movable Platen Vertical or horizontal platen or platens to which molds have been installed. Usually moved by pneumatic cylinder, hydraulic ram, or toggle mechanism.

Multicavity Mold Mold with two or more mold impressions; produces more than one molding per mold cycle.

Nip Rollers Counterrotating rollers that pinch and move the plastic sheet.

One-Up Mold Mold that produces one article at a time.

Orientation Alignment of the crystalline structures in polymeric materials; biaxially oriented sheet; stretching and alignment of polymer molecules.

Oven Chamber used for heating thermoplastic sheet; the sheet is either shuttled or moves through.

Parting Line Where mold halves meet in mold closing or where molds separate when opened.

Pinhole Very small hole in the plastic sheet.

Plastic "Fluff" Ground-up plastic foam materials.

Platens Mounting plates of the thermoforming press to which the entire mold assembly is bolted.

Plug-Assist Forming Same as plug forming, using the assistance of a plug.

Plug Forming Thermoforming process in which a plug or male mold is used to preform a part before forming is completed.

Polymer High-molecular-weight organic compound made by the polymerization process of joining monomer molecules together.

Postconsumer Plastic Plastic material recycled from plastic products that has been used by the consumer, collected, and re-processed again.

Postmold Expansion   Thermoformed foam parts expand after molding when removed from the mold; increase in dimensions.

Preheating   Heating the sheet prior to entering into a thermoformer to facilitate reduction of the mold cycle, or heating the mold to eliminate startup scraps.

Preprinting   In sheet thermoforming, random or distorted printing of sheets before they are formed. During the forming operation the print assumes its location or proper proportion.

Pressure Forming   Thermoforming process wherein pressurized air is used to push the heated sheet into the cavity.

Rail Jam   Thermoplastic material jamming the chain-rail system.

Recycled Contents   Indication of plastic material blend consisting of a given portion of virgin and recycled material mixed together (usually the same family of plastic materials).

Recycled Material   Plastic material which has been reprocessed from previously made plastic products.

Rejects   Products that do not meet the product standards set.

Reprocess   Reusing material previously subjected to thermoforming.

Resins   Another name for polymer or plastic materials.

Rib   Design pattern used to reinforce or strengthen sidewalls in the thermoformed product.

Sag   Drooping of thermoplastic material when exposed to heat.

Sandwich Heating   Method of heating a thermoplastic sheet prior to forming at both sides.

Scrap   Any leftover material from a thermoforming operation which is not part of the finished product; trim-off, which can be reground and reused.

Scrap Factor   The comparative percentage of product versus the scrap it produced by weight or by square area.

Setting Temperature   Temperature at which point a formed part retains its shape.

Setting Time   Period of time during which a thermoformed part is cooled by the mold to retain its shape.

Sheet   Uniform thickness of thermoplastic resin in various sizes of panels or on continuous rolls (approximately 10 to 250 mils thick).

Sheet Line   Level at which the thermoplastic sheet is traveling in a thermoforming machine.

Shot   Yield from one complete thermoforming cycle, including scrap.

Single-Cavity Mold   Thermoforming mold producing one part at a time.

Slack Loop   Material loop created in a web when feeding and advancing.

Snapback Forming   Thermoforming technique in which a heated plastic sheet is allowed to stretch in or over a mold and then snapped back to form against the opposite mold surface.

Spacing Area of sheet which is not part of the primary product and is between each thermoformed part.

Stacking Lugs Added configuration formed into a thermoformed article in alternating locations to prevent them from nesting too close together.

Stand-Off Mold legs that elevate the mold to meet at the sheet line.

Stationary Platen Platen that does not move during thermoformer operation.

Stretch Marks Surface or skin breaks on thermoformed foam parts.

Syntactic Foam Manufactured material made of hollow glass spheres embedded in epoxy resin.

Synthetic Materials Manufactured materials.

Thermoforming Any process of forming thermoplastic sheet in which the sheet is heated and forced to contact a mold surface, then cooled until it retains its shape.

Thermoplastic High-polymeric materials made by polymerization; capable of softening or fusing when heated and of hardening again when cooled.

Toggle Action Mechanism that exerts pressure developed by the application of force on a knee joint; used to close or open the platens of a press.

Twin-Sheet Forming Thermoforming a product by using two independent sheet stock and forming it either simultaneously or one after the other and uniting the two sheet forms into a single but dual walled product.

Undercut Negative draft or taper created by having greater dimensions than those of the mold cavity opening; therefore interferes with the product stripping.

Vacuum Forming Method of thermoforming whereby the heated sheet is clamped and drawn into or over a mold using vacuum.

Virgin Polymer Plastic material used or worked on for the first time.

Warpage Dimensional distortion after molding.

Bibliography

Modern Plastic Encyclopedia, McGraw-Hill, New York (1979–1995).
Modern Plastic Magazine, McGraw-Hill, New York (monthly).
Plastic Design & Processing, Lake Publishing, Libertyville, IL (bimonthly).
Product Design & Development, Chilton Book Co., Radnor, PA (monthly).
Safety & Hygiene News, Chilton Book Co., Radnor, PA (monthly).
Plastics Design Forum, HBJ Publications, Cleveland, OH (bimonthly).
The Industrial and Process Control Magazine, Chilton Book Co., Radnor, PA (monthly).
Compressed Air and Gas Hand Book, Compressed Air Gas Institute, New York (3rd Edition).
Hydraulic Handbook, Colt Industries, Kansas City, KA (9th Edition).
Plastic Engineering Handbook, Society of the Plastic Industry, Inc. New York (4th Edition).
Machinery's Handbook, 22nd Edition (E. Oberg, ed.), Industrial Press, Inc., New York.
Plastic Products Design Handbook, Part A (E. Miller, ed.), Marcel Dekker, Inc., New York.

Index